Classic Set Theory

FOR GUIDED INDEPENDENT STUDY

Classic Set Theory

FOR GUIDED INDEPENDENT STUDY

Derek Goldrei
Open University
UK

CHAPMAN & HALL/CRC

A CRC Press Company
Boca Raton London New York Washington, D.C.

Library of Congress Cataloging-in-Publication Data

Catalog record is available from the Library of Congress

Direct all inquiries to CRC Press LLC, 2000 N.W. Corporate Blvd., Boca Raton, Florida 33431.

Visit the CRC Press Web site at www.crcpress.com

First edition 1996
First CRC Press reprint 1998
Originally published by Chapman & Hall

International Standard Book Number 0-412-60610-0
Library of Congress Card Number 96-85139

CONTENTS

PREFACE

How to use this book

This book is intended to be used by you for independent study, with no other reading or lectures etc., much along the lines of standard Open University materials. There are plenty of exercises within the text which we would recommend you to attempt at that stage of your work. Almost all are intended to be reasonably straightforward on the basis of what's come before and many are accompanied by solutions – it's worth reading these solutions as they often contain further teaching, but do try the exercises first without peeking, to help you to engage with the material. Those exercises without solutions might well be very suitable for any tutor to whom you have access to use as the basis for any continuous assessment of this material, to help you check that you are making reasonable progress. But beware! Some of the exercises pose questions for which there is not always a clear-cut answer: these are intended to provoke debate! In addition there are further exercises located at the end of most sections. These vary from further routine practice to rather hard problems: it's well worth reading through these exercises, even if you don't attempt them, as they often give an idea of some important ideas or results not in the earlier text. Again your tutor, if you have one, can guide you through these.

The book is also peppered with notes in the margins, like this! They consist of comments meant to be on the fringe of the main text, rather than the core of the teaching, for instance reminders about ideas from earlier in the book or particularly subjective opinions of the author.

If you would like any further reading in textbooks of set theory, there are plenty of good books available which use essentially the same Zermelo–Fraenkel axiom system, for instance those by Enderton [1], Hamilton [2] and Suppes [3], all covering roughly the same material as this book, while the books by Devlin [4] and Moschovakis [5] go some way further, looking at some of the modern set theory built on the foundations in this book. For a short outline of many of the key ideas, Halmos [6] is invaluable.

Acknowledgments

I would like to thank all those who have in some way helped me to write this book. First there are those from whom I initially learnt about set theory when studying at the University of Oxford: Robin Gandy and Paul Bacsich. Then there are my colleagues at the Open University who have both taught me so much about maths and the teaching of maths, and encouraged me in the writing of this book, especially Bob Coates. Next there are those who have given me practical help in its production, not to mention a contract: Nicki Dennis, Achi Dosanjh and Stephanie Harding at Chapman & Hall, several anonymous (but not thereby any less deserving of my thanks!) reviewers, and Alison Cadle and Chris Rowley at the Open University, for refining my LaTeXstyle files. Richard Leigh copy-edited the book and made many very helpful comments. Sharon Powell at the Open University helped me greatly with printing the final version of the book. Last, but foremost in my mind, there are all my old students at the Open University and at the University of Oxford, especially the women (and now men!) of Somerville and St. Hugh's Colleges. In

As is customary, I would like to absolve everyone else from any blame attaching to anything in this book or any of its inadequacies. But whether I mean this is entirely another matter!

particular I'd like to thank the following for their comments on parts of the book: Dorothy Barton, Kirsten Boyd, Cecily Crampin, Adam Durran, Aneirin Glyn, Alexandra Goulding, Nick Granger, Michael Healey, Sandra Lewis, J-B Louveaux, David Manlove, Chris Marshall, Nathan Phillips, Alexandra Ralph and Axel Schairer.

This book is dedicated to all students at both OUs, but especially to M2090737.

1 INTRODUCTION

1.1 Outline of the book

The language of sets is part of the vocabulary of any student or user of mathematics, acquired very early in one's schooling. Words like 'intersection' and 'union' have become part of everyday mathematics. But the 'set theory' in this book is about much deeper and more complicated ideas about sets, at the heart of any discussion of the foundations of mathematics, especially about the place of 'infinity' in the subject. Many of these deep ideas have their origin in the work of Georg Cantor, a German mathematician at his most active in the latter half of the 19th-century. Cantor had many remarkable (and controversial) insights to the nature of infinity, producing a theory involving two sorts of 'infinite number', each equipped with an arithmetic incorporating and extending the familiar arithmetic of the natural numbers. Among the questions which he was able to pose and resolve were such as:

For a very accessible account of Cantor's life (1845–1918) and work, look at the biography by Dauben [7].

In set theory it is customary to use 'natural number' to mean a non-negative integer.

1. Are there more rational numbers than natural numbers?
2. Are there more irrational real numbers than rational numbers?
3. Are there more points in the real plane than on the real line?

The obvious answer to the first of these questions is probably 'Yes', on the grounds that the set of natural numbers is a proper subset of the set of rational numbers – we make no apologies for leaping into use of the everyday language of sets! The obvious answer to the third question might similarly seem to be 'Yes'. In both cases, Cantor's answer was in fact 'No', and one aim of this book is to explain his reasons why.

However, if you are unhappy with some of the set terminology, don't worry! We shall explain it more carefully later on.

There might not be such an obvious candidate for the answer to the second question above – it all depends on how much you already know about the real numbers. Another aim of this book is to give an explanation of what the real numbers are, another aspect of Cantor's work, and part of the work of his contemporary Richard Dedekind (1831–1916). This explanation leads to the need to explain even more basic numbers, namely the natural numbers.

Cantor's answer is 'Yes'.

Cantor's work on infinity really stemmed from his and others' work on real analysis, and there are several connections between the issues of infinity and of the real numbers. First there was the drive in the 19th century to put the calculus onto a rigorous footing, giving what is now taught as real analysis: the last stage of this required a firm algebraic description of the reals, while earlier stages needed to explain away or eliminate uses of infinity, e.g. the infinitely large, like the ∞ in $\sum_{n=0}^{\infty} x^n$, and the infinitely small, namely infinitesimals. Secondly the descriptions of the reals provided by Dedekind and Cantor required the overt handling, as legitimate mathematical objects, of infinite sets. And thirdly mathematical developments in real analysis led to the desire both to study infinite subsets of the real numbers and to carry out infinite processes on such subsets.

Infinitesimals, infinitely small quantities, played a major part in the theory of calculus, until banished through the work of Weierstrass and others in the mid-19th century. Twentieth-century logic, inspired in part by Cantor's work, has resurrected infinitesimals within the context of non-standard analysis.

The key issue is whether it is legitimate to treat an infinite set as a mathematical object which one can then use in further constructions. We are so used, in the final years of the 20th century, not only to writing and talking about infinite sets like the set ℕ of natural numbers and the set ℝ of real numbers, but also to manipulating such sets using e.g. operations like intersection (∩) and union (∪), that it is easy to forget, or be unaware of, how controversial the use of infinite sets was until (and perhaps still after) Cantor's work. Mathematicians had been happy to regard, say, the natural numbers as being *potentially infinite*, meaning that however (finitely) many numbers one might list, there would always be an extra one that could be added to the list; but they were wary of treating this list as though it could be completed to give a single infinite object, what would be described as an *actually infinite* object. This rejection was very deeply rooted, going back to the work of Aristotle in the 4th century BC. Of course, over the same period there had been much discussion, and use, of both infinitely small quantities and what looked suspiciously like infinite processes (the dots '...' in so many mathematical expressions), that some sort of resolution of infinity was due. This resolution was essentially done by Cantor.

Both the history and philosophy are well described in the books by Adrian Moore [8] and Lavine [9]. Some of the pre-1900 mathematical source material on infinity and the calculus can be found in Fauvel and Gray [10].

Cantor's work, by virtue of its revolutionary nature, raised very many difficult and worrying issues. Matters were not helped by the initial imprecision of some of the concepts involved. His results, and the unresolved questions arising from them, attracted sufficient interest and enthusiasm from enough of his contemporaries that much effort was subsequently made to put his theory on more solid foundations. The aim of this book is to present this classic theory of sets as it is now understood by most mathematicians, using the framework of the Zermelo–Fraenkel axiom system (probably the most widely used of the systems for set theory). The theory is well worth studying for its own sake, as an accessible and exciting part of mathematics. It also provides a foundation for further study in modern set theory, beyond the scope of this book, but studied at postgraduate, and sometimes at advanced undergraduate, level.

The use of 'classic' in the title of this book is partly a deliberate misuse of the word, inasmuch as it can be used to mean 'to do with ancient Greece' – ancient Greeks like Aristotle might well have abhorred Cantor's use of infinity!

The development of the book is roughly as follows. Chapter 2 gives constructions of the real numbers in terms of rational numbers and shows how these ultimately stem from set constructions involving natural numbers. Chapter 3 then shows how the natural numbers themselves can be constructed from very simple sorts of set, like the empty set. Thus all these important number systems in everyday mathematics can be constructed and explained purely in terms of sets. Chapter 4 introduces axioms for set theory which underpin all these constructions. Historically, axioms were introduced after Cantor had produced his theory of infinite sets, which we describe in later chapters. Axioms were felt necessary for two reasons: to help avoid paradoxes arising from this theory of infinite sets; and to help establish the relationship between a major unresolved problem arising from the theory involving a certain sort of ordering of the real numbers and a special axiom called the axiom of choice. In Chapter 5 we discuss this latter axiom, which turned out to be a very pervasive mathematical principle. In Chapter 6 we look at some of Cantor's theory of infinite cardinal numbers, including remarkable results about the sizes of the real line, the real plane and the set of irrational numbers, all stemming from the construction of the reals in Chapter 2. In Chapter 7 we look at some of the general theory of ordered sets (a further legacy of Cantor's), and then in Chapter 8 we look at his theory of ordinal numbers, a variety of richly

Many of the crucial papers in the development of set theory and the foundations of mathematics can be found in van Heijenoort [11], in English translation and with very valuable commentary by the editor.

Many of Cantor's results about cardinal and ordinal numbers are in papers translated in [12]. These papers are not only vital source material for the whole subject, but are also very readable.

structured sets extending ideas about natural numbers into the realm of infinite sets. In Chapter 9, the final chapter of the book, we both complete Cantor's theory of infinite sets, linking cardinal and ordinal numbers via the axiom of choice, and look at two major problems concerning the real numbers which he had not resolved – in many ways the problem of the nature of the real numbers pervades the subject, so it is fitting that we both start and end the book with the reals.

1.2 Assumed knowledge

The book is written on the basis that you have already had some experience of using sets, functions and basic logic, and that you are familiar with a variety of mathematical words and notations – ideally you will have already taken a first course in real analysis so that some of the context of the material in this book is known to you. Among what we hope you will have seen before is the following material.

There are many excellent books on real analysis, for instance those by Spivak [13] and Haggarty [14].

Set Notation

A *set* X is a collection of objects called the *elements*, or *members*, of X. We sometimes use words like *family* and *collection* instead of sets, for variety and sometimes, we hope, to aid comprehension. We shall often use letters like $a, b, \ldots, y, z, A, B, \ldots, Y, Z$ to stand for sets or for elements of sets.

Later on in the book all the objects we shall consider are sets whose elements are also sets!

We write $x \in X$ to express that the object x is an element of the set X and $y \notin X$ to say that y is *not* an element of X.

When $x \in X$, we also say 'x is in X' or 'x belongs to X'.

We use curly brackets, { and }, around a list of objects to signify the set of all those objects. For instance $\{3, 8, 9\}$ is the set with elements 3, 8, 9.

The order in which the elements are listed inside the curly brackets doesn't change the set, nor does listing some element more than once. Thus $\{9, 3, 7\}$ and $\{3, 3, 7, 9\}$ both represent the same set as $\{3, 7, 9\}$. In general, two sets X and Y are equal if and only if they contain the same elements or, equivalently, if and only if every element of X is an element of Y and vice versa.

We use standard notation for the most common sets of numbers: $\mathbb{N}$ for the set of natural numbers, $\mathbb{Z}$ for the set of all integers (positive, negative and zero), $\mathbb{Q}$ for the set of rational numbers, $\mathbb{R}$ for the set of real numbers and $\mathbb{C}$ for the set of complex numbers.

Recall that in this book we take $\mathbb{N}$ to include all the positive integers *and* the number 0.

We use the notation $\varnothing$ for the *empty set*, the set which contains no elements.

We can also describe a set using curly brackets in terms of a property possessed by all its elements, as with $\{n : n \text{ is an even integer}\}$ or, equivalently, $\{n \in \mathbb{Z} : n \text{ is even}\}$ for the set of all even integers $\{\ldots, -4, -2, 0, 2, 4, 6, \ldots\}$. In general we write

$$\{x : x \text{ has property } P\}$$

The colon ':' is read as 'such that'.

for the set of all x such that x has property P (some property which may or may not be possessed by a given object x).

We shall occasionally use standard notation for intervals of the real line:

(a, b) for the *open interval* $\{x \in \mathbb{R} : a < x < b\}$;

$[a, b]$ for the *closed interval* $\{x \in \mathbb{R} : a \leq x \leq b\}$;

$(a, b]$ and $[a, b)$ for the *half open and closed* intervals $\{x \in \mathbb{R} : a < x \leq b\}$ and $\{x \in \mathbb{R} : a \leq x < b\}$ respectively;

$(-\infty, b)$ and (a, ∞) for the open intervals $\{x \in \mathbb{R} : x < b\}$ and $\{x \in \mathbb{R} : x > a\}$ respectively;

$(-\infty, b]$ and $[a, \infty)$ for the closed intervals $\{x \in \mathbb{R} : x \leq b\}$ and $\{x \in \mathbb{R} : x \geq a\}$ respectively.

Given two sets X and Y, we write

$X \cup Y$ for the *union* of X and Y, i.e. the set of elements belonging to X or Y (or both);

$X \cap Y$ for the *intersection* of X and Y, i.e. the set of elements belonging to both X and Y;

$X \setminus Y$ for the *complement* of Y in X, i.e. the set of elements of X *not* in Y.

We shall adopt the standard mathematical use of the word 'or' as allowing the 'or both' case – what's called the *inclusive* use of 'or'.

Given a family of sets $\mathscr{F}$, we write $\bigcup\{X : X \in \mathscr{F}\}$ for the union of all the sets in this family, i.e. the set $\{x : x \in X$ for some $X \in \mathscr{F}\}$. A family $\mathscr{F}$ of sets might sometimes be indexed by another set, for instance the family of all open intervals of $\mathbb{R}$ of the form $(\frac{1}{n+1}, \infty)$ for $n \in \mathbb{N}$ is effectively indexed by the set $\mathbb{N}$: in this sort of case we would write the family as $\{(\frac{1}{n+1}, \infty) : n \in \mathbb{N}\}$ and the union of the family as $\bigcup\{(\frac{1}{n+1}, \infty) : n \in \mathbb{N}\}$ (which happens to equal the set $(0, \infty)$).

X is a *subset* of the set Y means that X is a set of which every element is also an element of Y (so that, for all x, if $x \in X$ then $x \in Y$). We write $X \subseteq Y$ for 'X is a subset of Y'. A subset X of Y is said to be *proper* if $X \neq Y$. The *power set* of Y, written as $\mathscr{P}(Y)$, is the set of all subsets of Y.

We write $X \times Y$ for the *Cartesian product* of X and Y, i.e. the set of all ordered pairs (x, y) with $x \in X$ and $y \in Y$. We use X^2 as shorthand for $X \times X$, X^3 for $(X \times X) \times X$ and so on.

Function Notation

A *function* f from a set X to a set Y associates an element, $f(x)$, of Y with each element x of X. The *rule* of f, written as $x \longmapsto f(x)$, describes this process of association. The element $f(x)$ of Y is called the *image of x under f*. The *domain* of f is the set X and the *codomain* of f is the set Y. We use the standard arrow notation for such a function, combining the information of its domain, codomain and rule:

$$\begin{aligned} f : X &\longrightarrow Y \\ x &\longmapsto f(x) \end{aligned}$$

If A is a subset of the domain of this function f, the *restriction* of f to A, written as $f|_A$, is the function

$$\begin{aligned} f|_A : A &\longrightarrow Y \\ x &\longmapsto f(x) \end{aligned}$$

i.e. $f|_A$ has the same rule and codomain as f, but has its domain restricted to A.

The *image set* or *range* of $f: X \longrightarrow Y$, written as Range(f), is the set of images of f, namely $\{f(x) : x \in X\}$. For any subset A of the domain X the set $\{f(x) : x \in A\}$ is called the *image set of A under f*. We shall sometimes write $f(A)$ for this set, but we shall mostly use the notation Range($f|_A$), which represents the same set.

A function is said to be *onto* if for each $y \in Y$ there is an $x \in X$ with $f(x) = y$.

So that f is onto exactly when Range(f) $= Y$.

The function f is said to be *one-one* if for all $x, x' \in X$, if $f(x) = f(x')$ then $x = x'$ (or, equivalently, if $x \neq x'$ then $f(x) \neq f(x')$).

If f is a one–one function, then its *inverse function* f^{-1} is defined as the function

$$\begin{aligned} f^{-1}\colon \mathrm{Range}(f) &\longrightarrow X \\ y &\longmapsto \text{the unique } x \text{ such that } f(x) = y \end{aligned}$$

For any subset B of the codomain Y, its *inverse image set* under f, written as $f^{-1}(B)$, is the set $\{x \in X : f(x) \in B\}$.

The use of f^{-1} in this context does *not* mean that the inverse function f^{-1} exists – a well-known source of confusion!

If $f: X \longrightarrow Y$ and $g: Y \longrightarrow Z$ are functions then the *composite* function (or *composition of f and g*) $g \circ f$ is the function

$$\begin{aligned} g \circ f\colon X &\longrightarrow Z \\ x &\longmapsto g(f(x)) \end{aligned}$$

We define $g \circ f$ in exactly the same way when the domain of g contains the range of f as a subset, rather than requiring that the domain of g coincides exactly with the codomain of f.

If f is both one–one and onto, then f is a *bijection*. If f, g are both bijections with the codomain of f equal to the domain of g, then the composition $g \circ f$ is also a bijection.

Basic Logic

Suppose that p and q are propositions, or statements, about mathematics, of the sort where it makes sense to ask whether they are true or false under some given set of circumstances, or, equivalently, in a particular *interpretation* of the symbols or words that they involve. Then the proposition 'if p then q' is true in some interpretation unless p is true and q is false in this interpretation; i.e. it is true not only when both p and q are true, but also when p is false. And the statement 'p if and only if q' is true in an interpretation when 'if p then q' and 'if q then p' are both true in this interpretation, which boils down to both of p and q being true in this interpretation or both being false.

For example, p might be the proposition 'f is a continuous function' and q the proposition 'all continuous functions are continuous functions'. The truth of p depends on the interpretation of f – do we interpret f as a function and, if so, is it continuous? – while q is always true in normal usage of the words.

Further exercises

Exercise 1.1

Prove the following statements for all sets A, B and C.

(a) $A \cap (B \cup C) = (A \cap B) \cup (A \cap C)$

(b) $A \cup (B \cap C) = (A \cup B) \cap (A \cup C)$

(c) $A \setminus (B \cup C) = (A \setminus B) \cap (A \setminus C)$

(d) $A \setminus (B \cap C) = (A \setminus B) \cup (A \setminus C)$

Exercise 1.2

(a) If Y is a finite set with n elements, show that $\mathscr{P}(Y)$ has 2^n elements.

(b) Decide if it is true that, for every set Y, $\mathscr{P}(\mathscr{P}(Y)) = \mathscr{P}(Y)$.

(c) If A and B are sets, decide if it is true that $\mathscr{P}(A \times B) = \mathscr{P}(A) \times \mathscr{P}(B)$.

Exercise 1.3

Given a function $f: X \longrightarrow Y$, with $A_1, A_2 \subseteq X$ and $B_1, B_2 \subseteq Y$, decide which of the following statements about image sets and inverse image sets under f are *always* true.

(a) $f(A_1 \cap A_2) = f(A_1) \cap f(A_2)$

(b) $f(A_1 \cup A_2) = f(A_1) \cup f(A_2)$

(c) $f(X \setminus A_1) = Y \setminus f(A_1)$

(d) $f^{-1}(B_1 \cap B_2) = f^{-1}(B_1) \cap f^{-1}(B_2)$

(e) $f^{-1}(B_1 \cup B_2) = f^{-1}(B_1) \cup f^{-1}(B_2)$

(f) $f^{-1}(Y \setminus B_1) = X \setminus f^{-1}(B_1)$

For the remainder of this book we shall mostly use the notation Range$(f|_A)$ for the image set of a subset A of X under f, rather than the notation $f(A)$ used here. This is because when the domain X of f, which is a set, consists of elements which are also sets, the notation $f(x)$ is reserved for the image of an *element* x of X, rather than the image set of some *subset*.

Exercise 1.4

Let $f: X \longrightarrow Y$ and $g: Y \longrightarrow Z$ be functions. For each of the following statements, decide whether it is true or false, and give a proof or counterexample as appropriate.

(a) If f and g are one–one then $g \circ f$ is one–one.

(b) If $g \circ f$ is one–one then f is one–one.

(c) If $g \circ f$ is one–one then g is one–one.

(d) If f and g are onto then $g \circ f$ is onto.

(e) If $g \circ f$ is onto then f is onto.

(f) If $g \circ f$ is onto then g is onto.

2 THE REAL NUMBERS

2.1 Introduction

All modern mathematicians appreciate the vast importance, to mathematics and to a huge range of mathematically based crafts and sciences, of the differential and integral calculus. And anyone who has studied a course in real analysis should have some sense of not only a rigorous foundation for the calculus, but also why such a foundation is necessary. In the original formulations by Leibniz and Newton of the calculus, and in the extension of this work by such as the Bernoulli brothers and Euler, there were deficiencies of which contemporary natural philosophers were well aware. These deficiencies centred around use of the infinite, namely the infinitely large, for instance the '...' in

$$\frac{1}{1-x} = 1 + x + x^2 + x^3 + \dots,$$

The dots, '...', mean 'going on forever'.

and the infinitely small, i.e. the infinitesimals used to work out basic derivatives. The ramifications of the calculus for science and mathematics were so extensive and exciting that the need to remedy these and other deficiencies only began to become pressing towards the end of the 18th century. Through the work of the likes of Cauchy, Bolzano and Weierstrass up to the middle of the 19th century, the calculus acquired essentially the rigorous foundation which is passed on in the many standard undergraduate textbooks on real analysis of today. In particular, the use of the infinite was buried deep by the use of e.g. the familiar ε–δ definitions for limits, which avoid mention of infinitesimals. But there remained major problems at the heart of the subject. For instance, as Dedekind pointed out, even by the 1850s no one knew how to prove the following:

Dedekind's seminal and highly readable *Continuity and Irrational Numbers*, translated in [15], was written in 1872, but covered work he had done in 1858.

> If a magnitude x grows continually but not beyond all limits it approaches a limiting value.

To modern eyes, informed by a standard course in real analysis, this poses no problem beyond, perhaps, the archaic language: letting A stand for the set of all the relevant xs, so that the subset A of $\mathbb{R}$ is bounded above, the limit of the xs is just $\sup A$ (or $\operatorname{lub} A$, if you prefer this notation). But this viewpoint relies precisely on the work done by Dedekind (and independently by Cantor) in resolving the problem.

The root of the problem, for Dedekind, was that no one had tied down what the real numbers were in a sufficiently precise way. To be sure, there were long-established and very powerful geometric intuitions about real numbers in terms of lengths along a line, linked to ideas about continuity, e.g. continuous motion; but these were not precise enough to justify statements like the one above about limits. In this chapter we shall look in detail at two different constructions of the real numbers. That these constructions are different may sound problematic, but we shall see that, in an important sense, they give rise to the same (familiar) theory of the reals.

The language of 19th-century mathematics abounds in references to continuous motion, e.g. 'magnitude x *grows continually*' and '*approaches* a limiting value' in our problem above. Come to think of it, we still use this sort of language!

Both constructions define each real number in terms of rational numbers. By the middle of the 19th century, rational numbers and their properties were a sufficiently natural and uncontroversial part of mathematics that Dedekind and Cantor could use them without qualms to define the more difficult and controversial real numbers. Rational numbers seem in some sense simpler than the reals. But the question 'what are the rationals?' is perhaps as deserving of an answer as 'what are the reals?', so that we shall also look at a way of constructing the rationals from something even 'simpler', namely the natural numbers.

What this chapter essentially does is to provide an arithmetic or algebraic, rather than geometric, description of the real line or, equivalently, what is called the *continuum*. It is using this description to attempt further to understand the nature of the continuum that underpins the rest of this book.

A dictionary definition of 'continuum' is as a continuous body; and 'continuous' means connected or joined together.

2.2 Dedekind's construction

The central problem with the real numbers was to explain irrational numbers. Numbers like $\sqrt{2}$ have been known to be irrational for over two thousand years, and methods of approximating to $\sqrt{2}$ by rational numbers became increasingly sophisticated over that period. So it was quite natural for Dedekind and Cantor to think of defining irrational numbers by reference to rational numbers.

In any case, rational numbers were regarded as unproblematic by the 19th century.

Dedekind's definition relies on a picture of how an irrational number r sits among the rational numbers on the real line. As r is irrational, each rational lies either to the left or the right of r. So r cuts $\mathbb{Q}$ into two subsets L and R, where L consists of all the rationals to the left of r, and R consists of rationals to its right.

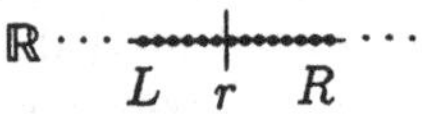

We have various intuitions about L and R. For instance, they are both non-empty; they are disjoint; any rational in L is less than any rational in R; and both L and R contain rationals arbitrarily close to r.

Of course, to define the real numbers, we cannot assume that they already exist! What Dedekind did was to define a real number to be a partition of $\mathbb{Q}$ into two non-empty subsets, L and R, with the property that every element of L is less than every element of R. This partition is called a *Dedekind cut* and $\mathbb{R}$ is defined to be the set of all such partitions. (The word 'cut' is meant to conjure up the earlier intuition of an irrational r cutting $\mathbb{Q}$ into the two subsets L and R.) Dedekind then explained how to define an order and the usual arithmetic operations on these partitions, exploiting the order and arithmetic operations of $\mathbb{Q}$.

L and R *partition* $\mathbb{Q}$ means that $L \cup R = \mathbb{Q}$ and $L \cap R = \varnothing$.

We shall take our definition of a real number to be an adaptation of Dedekind's construction. Rather than take both the L and R above, we shall just use the 'left set' L: after all, R is just $\mathbb{Q} \setminus L$, the complement of L in $\mathbb{Q}$. But this does mean that our definition has to capture the idea that every element of L is less than every element of $\mathbb{Q} \setminus L$.

Definitions

A *Dedekind left set* is a subset $\mathbf{r}$ of $\mathbb{Q}$ with the following properties:

(i) $\mathbf{r}$ is a proper, non-empty subset of $\mathbb{Q}$, so that $\varnothing \neq \mathbf{r} \neq \mathbb{Q}$;

(ii) $\mathbf{r}$ is 'closed to the left', i.e. if $q \in \mathbf{r}$ and $p <_{\mathbb{Q}} q$, then $p \in \mathbf{r}$;

(iii) $\mathbf{r}$ has no maximum element, i.e. for any $p \in \mathbf{r}$ there is some $q \in \mathbf{r}$ with $p <_{\mathbb{Q}} q$.

A *real number* is a Dedekind left set and $\mathbb{R}$ is the set of all such real numbers.

We shall use bold lower-case letters, e.g. $\mathbf{r}$,$\mathbf{s}$, for Dedekind left sets.

We use $<_{\mathbb{Q}}$ to stand for the usual order on $\mathbb{Q}$. Thus when we write $p <_{\mathbb{Q}} q$, it is implicit that p and q are rationals. The reason for the $<_{\mathbb{Q}}$ notation is that we will have to define an order $<$ on the reals, and want to avoid confusion about which order is being discussed.

Properties (i) and (ii) are the key parts of the definition: they ensure that $\mathbf{r}$ and its complement $\mathbb{Q} \setminus \mathbf{r}$ form a Dedekind cut, as you will soon check! Property (iii) deals with a relatively minor technical issue which we have avoided so far. When we pictured a real number r cutting $\mathbb{Q}$ into two subsets L and R, we had in mind the case when r was irrational – after all, it's irrational numbers that we are trying explain. But if r happened to be rational, we would have to include r in one of L and R for these to partition $\mathbb{Q}$. Property (iii) corresponds to deciding to put such an r into the R rather than the L.

Of course, our picture of the real numbers is that they include the rational numbers. But a rational number is not a Dedekind left set – the latter is a set of rationals. So we have to specify which real numbers are going to correspond to the rationals.

Definition

Let $q \in \mathbb{Q}$. Then the real number corresponding to q is

$$\mathbf{q} = \{p \in \mathbb{Q} : p <_{\mathbb{Q}} q\}.$$

The bold letter $\mathbf{q}$ is used for the real number corresponding to q.

Exercise 2.1

(a) Let $q \in \mathbb{Q}$. Check that $\mathbf{q}$ defined by

$$\mathbf{q} = \{p \in \mathbb{Q} : p <_{\mathbb{Q}} q\}$$

is a Dedekind left set.

(b) Is $\{p \in \mathbb{Q} : p \leq_{\mathbb{Q}} \frac{1}{2}\}$ a real number?

Solution

(a) Certainly $\mathbf{q}$ is a subset of $\mathbb{Q}$. It is non-empty as e.g. $q -_{\mathbb{Q}} 1 <_{\mathbb{Q}} q$, so that $q -_{\mathbb{Q}} 1 \in \mathbf{q}$; and it is a proper subset of $\mathbb{Q}$, as e.g. $q +_{\mathbb{Q}} 1 \not<_{\mathbb{Q}} q$, so that $q +_{\mathbb{Q}} 1 \notin \mathbf{q}$. Thus $\mathbf{q}$ obeys property (i) of the definition.

For property (ii), suppose that $p \in \mathbf{q}$ and that $x <_{\mathbb{Q}} p$: we must show that $x \in \mathbf{q}$. Of course $p \in \mathbf{q}$ means that $p <_{\mathbb{Q}} q$, so that as $x <_{\mathbb{Q}} p$ we have $x <_{\mathbb{Q}} q$, so that $x \in \mathbf{q}$, as required. For property (iii), we need to show that for any $p \in \mathbf{q}$, there is some $x \in \mathbf{q}$ with $p <_{\mathbb{Q}} x$. If $p \in \mathbf{q}$, then $p <_{\mathbb{Q}} q$. Take x to be $\frac{1}{2}(p +_{\mathbb{Q}} q)$. As p and q are in $\mathbb{Q}$, x is also in $\mathbb{Q}$; and

Of course, as $q \in \mathbb{Q}$, $q -_{\mathbb{Q}} 1$ is also in $\mathbb{Q}$. As with $<_{\mathbb{Q}}$, we are being careful to use the notation $-_{\mathbb{Q}}$ and $+_{\mathbb{Q}}$ to emphasize that we are using the 'known' operations on $\mathbb{Q}$. We shall soon define operations $-$ and $+$ on $\mathbb{R}$.

we also have $p <_{\mathbb{Q}} \frac{1}{2}(p +_{\mathbb{Q}} q) <_{\mathbb{Q}} q$. Thus $p <_{\mathbb{Q}} x$ and $x \in \mathbf{q}$ as required.

(b) Although the set obeys properties (i) and (ii), it fails with property (iii), as it contains a maximum element, namely $\frac{1}{2}$. Thus it isn't a real number.

We hope that you didn't find the process of checking that $\mathbf{q}$ is a real number too difficult. It is, however, rather a lengthy process, and is a warning of what will be required later on to check that various subsets of $\mathbb{Q}$ are indeed real numbers. You can also see how we have to use known properties of $\mathbb{Q}$ and $<_{\mathbb{Q}}$, $-_{\mathbb{Q}}$ and $+_{\mathbb{Q}}$ in such arguments.

Exercise 2.2

Let $\mathbf{r}$ be a Dedekind left set.

(a) Show that $\mathbf{r}$ is bounded above as a subset of $\mathbb{Q}$, i.e. there is some $x \in \mathbb{Q}$ such that $q \leq_{\mathbb{Q}} x$ for all $q \in \mathbf{r}$.

This x is then an upper bound of $\mathbf{r}$.

(b) Show that $\mathbf{r}$ and $\mathbb{Q} \setminus \mathbf{r}$ (the complement of $\mathbf{r}$ in $\mathbb{Q}$) form a Dedekind cut of $\mathbb{Q}$, i.e. that they are non-empty sets which partition $\mathbb{Q}$, such that every element of $\mathbf{r}$ is less than every element of $\mathbb{Q} \setminus \mathbf{r}$.

Solution

(a) As $\mathbf{r}$ is a proper subset of $\mathbb{Q}$, there is some $x \in \mathbb{Q} \setminus \mathbf{r}$. We claim that x is a suitable upper bound of $\mathbf{r}$. If this were not the case, there would be some $q \in \mathbf{r}$ such that $x <_{\mathbb{Q}} q$. But then, as $\mathbf{r}$ is closed to the left (property (ii)), we would have $x \in \mathbf{r}$, which would give a contradiction.

We are effectively showing that any rational not in $\mathbf{r}$ is an upper bound for $\mathbf{r}$.

Exercise 2.3

Write down a description, in terms of rational numbers and operations of $\mathbb{Q}$, of the Dedekind left sets corresponding to the following real numbers:

(a) -3;

(b) $\sqrt{2}$. (It won't do here to write $\{q \in \mathbb{Q} : q < \sqrt{2}\}$, as this assumes one already knows about the real number $\sqrt{2}$! So how might you refer to rationals less than $\sqrt{2}$ without mentioning $\sqrt{2}$?)

Solution

(a) $\{q \in \mathbb{Q} : q <_{\mathbb{Q}} -3\}$.

(b) For rationals q close to $\sqrt{2}$, if $q < \sqrt{2}$ then $q^2 < 2$. It is thus tempting to say that $\sqrt{2}$ is represented by $\{q \in \mathbb{Q} : q^2 <_{\mathbb{Q}} 2\}$. This is almost correct, except that it isn't closed to the left, as it contains e.g. -1 (because $(-1)^2 = 1 <_{\mathbb{Q}} 2$), but doesn't contain -2. So for safety we need to include all the negative rationals. Thus one representation is

$$\{q \in \mathbb{Q} : q^2 <_{\mathbb{Q}} 2 \text{ or } q \text{ is negative}\}.$$

The representation of $\sqrt{2}$ in our solution above is correct, but is it obvious that it is a Dedekind left cut? It is worth going through the process of showing this.

Exercise 2.4

Show that $\mathbf{r} = \{q \in \mathbb{Q} : q^2 <_{\mathbb{Q}} 2 \text{ or } q \text{ is negative}\}$ is a Dedekind left cut.

Solution

First of all $\mathbf{r} \neq \varnothing$, as e.g. $1 \in \mathbf{r}$; and $3 \notin \mathbf{r}$ (as $3^2 \not<_{\mathbb{Q}} 2$ and, of course, 3 is not negative), so that $\mathbf{r} \neq \mathbb{Q}$. Thus property (i) is satisfied.

It is easy, but tedious, to check that if $q \in \mathbf{r}$ and $p <_{\mathbb{Q}} q$, then $p \in \mathbf{r}$. There are two cases to consider. If p is negative, then p is automatically in $\mathbf{r}$. Otherwise we have $p \geq_{\mathbb{Q}} 0$, so that $p <_{\mathbb{Q}} q$ and $q \in \mathbf{r}$ means that $q^2 <_{\mathbb{Q}} 2$. As we can show that the function $x \longmapsto x^2$ is increasing on the set of non-negative rationals, and $0 \leq_{\mathbb{Q}} p <_{\mathbb{Q}} q$, we have $p^2 <_{\mathbb{Q}} q^2$, so that $p^2 <_{\mathbb{Q}} 2$, giving $p \in \mathbf{r}$ as required. In all cases, property (ii) is satisfied.

Property (iii), which is a technical but nevertheless necessary detail, requires more care. We need to show that for any $p \in \mathbf{r}$ there is some $q \in \mathbf{r}$ with $p <_{\mathbb{Q}} q$. If p is negative, then e.g. $q = 1$ will do perfectly well. But if $p \geq_{\mathbb{Q}} 0$, so that p is in $\mathbf{r}$ because $p^2 <_{\mathbb{Q}} 2$, we have to make an effort to specify some $q \in \mathbf{r}$ with $p <_{\mathbb{Q}} q$. Our experience of real analysis (which we are trying to underpin with our definition of $\mathbb{R}$) suggests that there should be some $n \in \mathbb{N}$ for which

$$p + \frac{1}{n} < \sqrt{2},$$

$p + \frac{1}{n}$ is a rational and will then be suitable as our q.

so that

$$\left(p + \frac{1}{n}\right)^2 <_{\mathbb{Q}} 2.$$

The + here is strictly addition, $+_{\mathbb{Q}}$, on $\mathbb{Q}$.

This last inequality, involving elements of $\mathbb{Q}$, is equivalent to

$$p^2 + \frac{2p}{n} + \frac{1}{n^2} <_{\mathbb{Q}} 2,$$

which is equivalent to

$$\frac{2p}{n} + \frac{1}{n^2} <_{\mathbb{Q}} 2 - p^2. \quad (*)$$

We are attempting, much as with standard arguments in a real analysis course, to find, for the given p, an $n \in \mathbb{N}$ such that

$$\left(p + \frac{1}{n}\right)^2 <_{\mathbb{Q}} 2$$

As $p^2 <_{\mathbb{Q}} 2$, we have $2 - p^2 >_{\mathbb{Q}} 0$. We can show within $\mathbb{Q}$ that there is some natural number $n \geq_{\mathbb{Q}} 1$ such that $n >_{\mathbb{Q}} \frac{2p+1}{2-p^2}$, so that for this n,

$$\begin{aligned}
\frac{2p}{n} + \frac{1}{n^2} &= \frac{1}{n}\left(2p + \frac{1}{n}\right) \\
&\leq_{\mathbb{Q}} \frac{1}{n}(2p+1) \quad (\text{as } n \geq_{\mathbb{Q}} 1) \\
&<_{\mathbb{Q}} \left(\frac{2-p^2}{2p+1}\right)(2p+1) \quad \left(\text{as } n >_{\mathbb{Q}} \frac{2p+1}{2-p^2}\right) \\
&= 2 - p^2,
\end{aligned}$$

so that the inequality marked $(*)$ above is satisfied. Thus for this $n \in \mathbb{N}$, we have that the rational $q = p + \frac{1}{n}$ satisfies both $p <_{\mathbb{Q}} q$ and $q \in r$.

The sort of argument above is very reminiscent of the sort of argument that one is often obliged to do at an early stage of a standard course in real analysis. The only difference is that the arithmetic and inequalities are about numbers in $\mathbb{Q}$ rather than $\mathbb{R}$. Many of the properties of $\mathbb{R}$ discussed below require proofs of a similar sort. There is a light at the end of the tunnel! Once we have shown that $\mathbb{R}$, as defined by Dedekind left sets or by any alternative definition, satisfies various properties well known from standard real analysis texts, subsequent proofs about $\mathbb{R}$ will usually follow from these properties without needing to refer to what reals numbers actually are.

We have to define, in terms of Dedekind left sets, the order $<_{\mathbb{R}}$ on the reals, and the operations $+_{\mathbb{R}}$ and $\cdot_{\mathbb{R}}$ of addition and multiplication on the reals. But first we have to say what we mean by two real numbers being equal. As each real number is given by a set of rational numbers, it is natural to define two reals as equal when they contain the same rationals. (This will mean that proving the equality of two real numbers $\mathbf{r}$ and $\mathbf{s}$ could well require proofs that $\mathbf{r} \subseteq \mathbf{s}$ and $\mathbf{s} \subseteq \mathbf{r}$.) And, given our intuition of a real number being represented by the set of all rationals less than it, it is then natural to define the order on $\mathbb{R}$, i.e. when one real is less than another, in terms of when one set of rationals is less than another.

In any construction of mathematical objects, not just in set theory, it's worth making sure that one understands what it means for two objects to be equal. In this book this will usually mean that they are equal as sets, which will be *defined* as meaning that they contain the same members.

Definitions

Let $\mathbf{r}$ and $\mathbf{s}$ be Dedekind left sets. Then $\mathbf{r}$ and $\mathbf{s}$ are equal, written as $\mathbf{r} =_{\mathbb{R}} \mathbf{s}$, if they are equal as sets, i.e. for all q,

$$q \in \mathbf{r} \text{ if and only if } q \in \mathbf{s};$$

$\mathbf{r}$ is less than or equal to $\mathbf{s}$, written as $\mathbf{r} \leq_{\mathbb{R}} \mathbf{s}$, if $\mathbf{r} \subseteq \mathbf{s}$, i.e. for all q,

$$\text{if } q \in \mathbf{r} \text{ then } q \in \mathbf{s};$$

$\mathbf{r}$ is less than $\mathbf{s}$, written as $\mathbf{r} <_{\mathbb{R}} \mathbf{s}$, if $\mathbf{r} \leq_{\mathbb{R}} \mathbf{s}$ and $\mathbf{r} \neq_{\mathbb{R}} \mathbf{s}$.

So $\mathbf{r} <_{\mathbb{R}} \mathbf{s}$ if $\mathbf{r} \subset \mathbf{s}$, i.e. $\mathbf{r}$ is a proper subset of $\mathbf{s}$.

It is easy to show that, with these definitions, $\mathbb{R}$ is *linearly ordered* by $\leq_{\mathbb{R}}$, i.e. has the properties in the following theorem.

We shall discuss linear orders in more detail in Chapter 6.

Theorem 2.1

The set $\mathbb{R}$ of Dedekind left sets has the following properties.

(O1) *Reflexive*: for all $\mathbf{r} \in \mathbb{R}$, $\mathbf{r} \leq_{\mathbb{R}} \mathbf{r}$.

(O2) *Anti-symmetric*: for all $\mathbf{r}, \mathbf{s} \in \mathbb{R}$, if $\mathbf{r} \leq_{\mathbb{R}} \mathbf{s}$ and $\mathbf{s} \leq_{\mathbb{R}} \mathbf{r}$, then $\mathbf{r} =_{\mathbb{R}} \mathbf{s}$.

(O3) *Transitive*: for all $\mathbf{r}, \mathbf{s}, \mathbf{t} \in \mathbb{R}$, if $\mathbf{r} \leq_{\mathbb{R}} \mathbf{s}$ and $\mathbf{s} \leq_{\mathbb{R}} \mathbf{t}$, then $\mathbf{r} \leq_{\mathbb{R}} \mathbf{t}$.

(O4) *Linear*: for all $\mathbf{r}, \mathbf{s} \in \mathbb{R}$, $\mathbf{r} \leq_{\mathbb{R}} \mathbf{s}$ or $\mathbf{s} \leq_{\mathbb{R}} \mathbf{r}$.

Exercise 2.5

Prove Theorem 2.1.

Solution

Properties (O1), (O2) and (O3) follow easily from the corresponding properties of the subset relation, $\subseteq$, which apply to all subsets of a set like $\mathbb{Q}$, not just Dedekind left sets. For instance, for (O2), if $\mathbf{r} \leq_{\mathbb{R}} \mathbf{s}$ and $\mathbf{s} \leq_{\mathbb{R}} \mathbf{r}$, so that $\mathbf{r} \subseteq \mathbf{s}$ and $\mathbf{s} \subseteq \mathbf{r}$, then the sets $\mathbf{r}$ and $\mathbf{s}$ contain the same elements: thus $\mathbf{r} =_{\mathbb{R}} \mathbf{s}$.

However, property (O4) requires more care, as it is not true in general for subsets A, B of $\mathbb{Q}$, that $A \subseteq B$ or $B \subseteq A$. We shall need to use the fact that $\mathbf{r}, \mathbf{s}$ are special subsets of $\mathbb{Q}$. If $\mathbf{r} =_{\mathbb{R}} \mathbf{s}$, we are done, as then $\mathbf{r} \subseteq \mathbf{s}$, so that $\mathbf{r} \leq_{\mathbb{R}} \mathbf{s}$. Otherwise there is a rational q which is an element of one of $\mathbf{r}$ and $\mathbf{s}$ but not of the other. Without loss of generality we can say that $q \in \mathbf{s}$ and $q \notin \mathbf{r}$. Then as we commented in the solution to Exercise 2.2(a), q is an upper bound for the set $\mathbf{r}$ of rationals, so that for all $p \in \mathbf{r}$, $p <_{\mathbb{Q}} q$. As $q \in \mathbf{s}$ and $\mathbf{s}$ is closed to the left, we thus have $p \in \mathbf{s}$, for all $p \in \mathbf{r}$. Hence $\mathbf{r} \subseteq \mathbf{s}$, i.e. $\mathbf{r} \leq_{\mathbb{R}} \mathbf{s}$.

Take for instance $A = \{0, 1\}$ and $B = \{-\frac{1}{2}, \frac{3}{4}\}$.

We cannot have $p = q$ for any $p \in \mathbf{r}$, as this would give $q \in \mathbf{r}$.

Actually, as $\mathbf{r} \neq_{\mathbb{R}} \mathbf{s}$, we have shown that $\mathbf{r} <_{\mathbb{R}} \mathbf{s}$.

We can now prove that $\mathbb{R}$ has the crucial completeness property. Firstly we need some definitions.

The completeness property is in some sense what makes the reals a richer system than the rationals.

Definitions

Let A be a non-empty subset of $\mathbb{R}$, and let $\mathbf{x}, \alpha$ be real numbers. Then $\mathbf{x}$ is an *upper bound* of A if $\mathbf{r} \leq_{\mathbb{R}} \mathbf{x}$, for all $\mathbf{r} \in A$.
The subset A is *bounded above* if there is some $\mathbf{x} \in \mathbb{R}$ which is an upper bound for A.
And α is the *least upper bound* of A, written as $\operatorname{lub} A$, if

(i) α is an upper bound for A; and

(ii) $\alpha \leq_{\mathbb{R}} \mathbf{x}$, for all upper bounds $\mathbf{x}$ of A.

$\operatorname{lub} A$ is often called the *supremum* of A, written as $\sup A$.

We hope that these definitions are familiar to you from real analysis. Of course we would not expect every subset of the reals to be bounded above: e.g. the set of all reals shouldn't be! But if a non-empty subset A happens to be bounded above, then it actually has a least upper bound.

This property doesn't hold for $\mathbb{Q}$. For example, if $A = \{q \in \mathbb{Q} : q^2 <_{\mathbb{Q}} 2\}$ then A doesn't have a *rational* least upper bound, as $\sqrt{2}$ isn't rational.

Theorem 2.2 Completeness property of $\mathbb{R}$

Suppose that the non-empty subset A of $\mathbb{R}$ is bounded above. Then A has a least upper bound in $\mathbb{R}$, i.e. $\operatorname{lub} A$ exists.

Proof

First note that the requirement that $\mathbf{r} \leq_{\mathbb{R}} \alpha$ for each $\mathbf{r} \in A$, is equivalent to requiring that $\mathbf{r} \subseteq \alpha$ for each such $\mathbf{r}$, when we regard real numbers as subsets of $\mathbb{Q}$. Thus we need $\bigcup\{\mathbf{r} : \mathbf{r} \in A\}$, or equivalently $\bigcup A$, to be a subset of α. Thus α is an upper bound for A in terms of $\subseteq$. And in these terms of $\subseteq$, the *least* upper bound of A would be $\bigcup A$, as this is the smallest set containing each $\mathbf{r} \in A$ as a subset. If $\alpha = \bigcup A$ is also a real number, then, translating back into terms of $\leq_{\mathbb{R}}$, α must be the least upper bound of A.

$\bigcup A$ is an abbreviation for $\{q : q \in \mathbf{r}$ for some $\mathbf{r} \in A\}$, which in everyday maths is usually written as $\bigcup\{\mathbf{r} : \mathbf{r} \in A\}$.

So let's show that $\bigcup A$ is indeed a real number, in which case it is lub A as required.

Taking the union, $\bigcup A$, of a set A, turns out to be a construction of major importance within set theory.

First of all, as A is non-empty and each real in A is a non-empty set of rationals, $\bigcup A$ is also a non-empty set of rationals. To show that $\bigcup A \neq \mathbb{Q}$, we use the fact that A is bounded above. This means that there is a real number $\mathbf{s}$ such that $\mathbf{r} \leq_{\mathbb{R}} \mathbf{s}$ for all $\mathbf{r} \in A$. As $\mathbf{s} \in \mathbb{R}$ there is some rational q such that $q \notin \mathbf{s}$. If q were in $\bigcup A$, then we would have $q \in \mathbf{r}$, for some $\mathbf{r} \in A$; but then, as $\mathbf{r} \subseteq \mathbf{s}$, we would have $q \in \mathbf{s}$, contradicting $q \notin \mathbf{s}$. Thus $q \notin \bigcup A$, so that $\bigcup A \neq \mathbb{Q}$.

Secondly we must show that $\bigcup A$ is closed to the left. Suppose that $p <_{\mathbb{Q}} q$, where $q \in \bigcup A$. Then $q \in \mathbf{r}$, for some $\mathbf{r} \in A$, so that as $\mathbf{r}$ is closed to the left we have $p \in \mathbf{r}$. Thus $p \in \bigcup A$. Hence $\bigcup A$ is closed to the left.

Lastly we need to show that $\bigcup A$ has no maximum element. Suppose that $p \in \bigcup A$. Then $p \in \mathbf{r}$, for some $\mathbf{r} \in A$. As $\mathbf{r}$ is a real number, there is some $q \in \mathbf{r}$ with $p <_{\mathbb{Q}} q$; and as $\mathbf{r} \in A$, we have $q \in \bigcup A$. Hence no element of $\bigcup A$ can be a maximum element.

Thus $\bigcup A$ is a real number. Hence lub A exists. ■

Exercise 2.6

Show that $\mathbb{R}$ is not bounded above, i.e. there is no real number $\mathbf{s}$ such that $\mathbf{r} \leq_{\mathbb{R}} \mathbf{s}$ for all $\mathbf{r} \in \mathbb{R}$. [Hint: suppose that there was such an $\mathbf{s}$. Then there is a rational q with $q \notin \mathbf{s}$. Let $p = q + 1$. Show that the corresponding real number $\mathbf{p}$ does not satisfy $\mathbf{p} \leq_{\mathbb{R}} \mathbf{s}$.]

Let us now define the arithmetic operations on $\mathbb{R}$. Our aim is to define $\mathbf{r} +_{\mathbb{R}} \mathbf{s}$ and $\mathbf{r} \cdot_{\mathbb{R}} \mathbf{s}$ as real numbers, so that they are closed to the left and so on. And we shall want the definitions to work the right way on our representation of the rationals within $\mathbb{R}$, e.g. $\mathbf{2} +_{\mathbb{R}} \mathbf{3} = \mathbf{5}$ and $\mathbf{2} \cdot_{\mathbb{R}} \mathbf{3} = \mathbf{6}$. We might guess that suitable definitions were

$$\mathbf{r} +_{\mathbb{R}} \mathbf{s} = \{p +_{\mathbb{Q}} q : p \in \mathbf{r} \text{ and } q \in \mathbf{s}\}$$
$$\mathbf{r} \cdot_{\mathbb{R}} \mathbf{s} = \{p \cdot_{\mathbb{Q}} q : p \in \mathbf{r} \text{ and } q \in \mathbf{s}\}.$$

The first of these works, but the second, for $\cdot_{\mathbb{R}}$, doesn't.

Exercise 2.7

Why won't

$$\mathbf{r} \cdot_{\mathbb{R}} \mathbf{s} = \{p \cdot_{\mathbb{Q}} q : p \in \mathbf{r} \text{ and } q \in \mathbf{s}\}$$

be suitable as a definition?

Solution

It has problems with the large, negative rationals that must be in each of $\mathbf{r}$ and $\mathbf{s}$. For instance, $-11 \in \mathbf{2}$ and $-12 \in \mathbf{3}$, so that with this attempt at a definition we would get

$$-11 \cdot_{\mathbb{Q}} -12 = 132$$

as an element of what we wanted to be $\mathbf{6}$. Another problem, which we would encounter if both $\mathbf{r}$ and $\mathbf{s}$ were negative, would be to ensure that their product is closed to the left: for such $\mathbf{r}, \mathbf{s}$, multiplying $p \in \mathbf{r}$ with $q \in \mathbf{s}$ would give only positive rationals $p \cdot_{\mathbf{Q}} q$.

Our definition of multiplication thus has to be a bit fiddly, depending on whether $\mathbf{r} >_{\mathbf{R}} \mathbf{0}$ and $\mathbf{s} >_{\mathbf{R}} \mathbf{0}$. We shall give the definition of $\mathbf{r} \cdot_{\mathbf{R}} \mathbf{s}$ for $\mathbf{r} >_{\mathbf{R}} \mathbf{0}$ and leave the definition for $\mathbf{r} \leq_{\mathbf{R}} \mathbf{0}$ as an exercise.

Definitions

Let $\mathbf{r}, \mathbf{s} \in \mathbb{R}$. Then their sum is defined as follows:

$$\mathbf{r} +_{\mathbf{R}} \mathbf{s} = \{p +_{\mathbf{Q}} q : p \in \mathbf{r} \text{ and } q \in \mathbf{s}\}.$$

If $\mathbf{r} >_{\mathbf{R}} \mathbf{0}$, then the product of $\mathbf{r}$ and $\mathbf{s}$ is defined as follows:

$$\mathbf{r} \cdot_{\mathbf{R}} \mathbf{s} = \begin{cases} \{p \cdot_{\mathbf{Q}} q : p \in \mathbf{r},\ p >_{\mathbf{Q}} 0 \text{ and } q \in \mathbf{s},\ q >_{\mathbf{Q}} 0\} \\ \qquad\qquad \cup \{q \in \mathbb{Q} : q \leq_{\mathbf{Q}} 0\}, & \text{if } \mathbf{s} >_{\mathbf{R}} \mathbf{0}, \\ \{p \cdot_{\mathbf{Q}} q : p \notin \mathbf{r} \text{ and } q \in \mathbf{s}\}, & \text{if } \mathbf{s} \leq_{\mathbf{R}} \mathbf{0}. \end{cases}$$

$p \notin \mathbf{r}$ means that p is greater than all the rationals in $\mathbf{r}$, e.g. p is large and positive.

The definition for $\mathbf{r} >_{\mathbf{R}} \mathbf{0}$ and $\mathbf{s} >_{\mathbf{R}} \mathbf{0}$ matches our intuition that rationals close to, but less than, the product of two positive reals r, s arise as products pq, where p and q are positive rationals a bit less than r and s. Likewise the definition for $\mathbf{r} >_{\mathbf{R}} \mathbf{0}$ and $\mathbf{s} \leq_{\mathbf{R}} \mathbf{0}$ exploits the intuition that rationals close to, but less than, the product of $r > 0$ with $s \leq 0$ arise as pq, where p is a rational a bit larger than r and q is a rational a bit less than s.

Exercise 2.8

Give the definition of $\mathbf{r} \cdot_{\mathbf{R}} \mathbf{s}$ for $\mathbf{r} \leq_{\mathbf{R}} \mathbf{0}$ and any $\mathbf{s}$.

Checking that these definitions give real numbers and using them to prove various arithmetic identities is something of a chore!

Exercise 2.9

Show that $\mathbf{r} +_{\mathbf{R}} \mathbf{s}$ and $\mathbf{r} \cdot_{\mathbf{R}} \mathbf{s}$ are real numbers.

Exercise 2.10

Show that $\mathbf{2} +_{\mathbf{R}} \mathbf{3} = \mathbf{5}$ and $\mathbf{2} \cdot_{\mathbf{R}} \mathbf{3} = \mathbf{6}$.

You will need to prove that the sets $\mathbf{2} +_{\mathbf{R}} \mathbf{3}$ and $\mathbf{5}$ are equal etc.

Exercise 2.11

Show that $(\sqrt{2})^2 = 2$, i.e. putting

$$\mathbf{r} = \{q \in \mathbb{Q} : q^2 <_{\mathbf{Q}} 2 \text{ or } q \text{ is negative}\},$$

show that $\mathbf{r} \cdot_{\mathbf{R}} \mathbf{r} = \mathbf{2}$.

With these definitions, it can be shown that $\mathbb{R}$, given by Dedekind left sets, satisfies the properties in the following theorem, which are usually given as the axioms for the reals in a modern real analysis text.

Theorem 2.3

Consider the following properties, expressed in terms of a set X with operations $+$ and $\cdot$, relations $=$ (of equality) and $\leq$, with special elements written as 0 and 1, where 0 does not equal 1. They are often described as the axioms for a *complete ordered field*.

1. For all x, $x \leq x$.
2. For all x, y, if $x \leq y$ and $y \leq x$, then $x = y$.
3. For all x, y, z, if $x \leq y$ and $y \leq z$, then $x \leq z$.
4. For all x, y, $x \leq y$ or $y \leq x$.
5. For all x, y, z, if $x \leq y$, then $x + z \leq y + z$.
6. For all x, y, z, if $x \leq y$ and $0 \leq z$, then $x \cdot z \leq y \cdot z$.
7. For all x, y, z, $x + (y + z) = (x + y) + z$.
8. For all x, $x + 0 = 0 + x = x$.
9. For each x, there is an element y such that $x + y = y + x = 0$.
10. For all x, y, $x + y = y + x$.
11. For all x, y, z, $x \cdot (y \cdot z) = (x \cdot y) \cdot z$.
12. For all x, $x \cdot 1 = 1 \cdot x = x$.
13. For all $x \neq 0$, there is an element y such that $x \cdot y = y \cdot x = 1$.
14. For all x, y, $x \cdot y = y \cdot x$.
15. For all x, y, z, $x \cdot (y + z) = (x \cdot y) + (x \cdot z)$.
16. Any non-empty subset A of X which is bounded above has a least upper bound in X.

The set $\mathbb{R}$ of all Dedekind left sets has all of these properties, when the operations $+$ and $\cdot$ are interpreted by $+_{\mathbb{R}}$ and $\cdot_{\mathbb{R}}$, the relations $=$ and $\leq$ by $=_{\mathbb{R}}$ and $\leq_{\mathbb{R}}$, and the special elements 0 and 1 of X by $\mathbf{0}$ and $\mathbf{1}$.

We are writing these properties in terms of a general set X because we shall be looking, in the next section, at other sets, besides that of all Dedekind left sets, that have all these properties.

$\mathbf{0}$ and $\mathbf{1}$ are the Dedekind left sets corresponding to the rational numbers 0 and 1.

We have already discussed Properties 1 to 4, in Theorem 2.1 above; and Property 16 is the completeness property, also proved earlier. We shall not prove that $\mathbb{R}$ has the remaining properties, as the proofs are very tedious, especially where $\cdot_{\mathbb{R}}$ is involved! Some parts are recommended as exercises.

Once we have established that the set $\mathbb{R}$ satisfies the properties in Theorem 2.3, we can proceed to prove all the usual results about the reals by the methods found in standard real analysis texts. These methods use only the fact that the reals obey these properties: they never 'look inside' any real number, so that the fact that a real number has been defined as a set of rationals ceases to be relevant. Nevertheless, we have achieved the aim of explaining what a real number *is*. It is relevant to note at what cost we have defined the real numbers. First, we have defined reals in terms of rational numbers. Although these are, in terms of the mathematics of, say, the last 300 years, relatively unproblematic objects, it might be as reasonable to ask what is a rational number as ask what is a real number. Secondly, the definition of an individual real number is as an *infinite* set of rationals. Use of the infinite in mathematics has been a matter of controversy for a good 2000

For example, we can prove the intermediate value theorem and use it to deduce that there is a real number x such that $x^2 = 2$, i.e. $\sqrt{2}$ exists.

Rational numbers were certainly more problematic to ancient Greek mathematicians, who dealt with them as ratios of quantities.

years. Arguably mathematicians of the 19th century were confident with what is called a *potentially* infinite set, one for which, however (finitely) many elements you have, there is always another available. But treating an *actually* infinite set, like a Dedekind left set of rationals, as a legitimate mathematical object suitable for all sorts of manipulation, seemed somewhat dubious.

For instance, for any finite set of natural numbers, there is always a natural number bigger than any of those in the set.

In the next section we shall look at some alternative constructions of the reals. In the final section of the chapter we shall look at how to construct the rational numbers from something 'simpler'. And the issue of infinity is one which will be a major topic in the rest of the book.

Further exercises

Exercise 2.12

Show that $\mathbf{r} +_{\mathbb{R}} \mathbf{0} =_{\mathbb{R}} \mathbf{r}$, for any $\mathbf{r} \in \mathbb{R}$.

Exercise 2.13

For each $\mathbf{r} \in \mathbb{R}$ there is an $\mathbf{s} \in \mathbb{R}$ such that $\mathbf{r} +_{\mathbb{R}} \mathbf{s} = \mathbf{0}$, i.e. $\mathbf{s}$ is $-\mathbf{r}$. Describe $\mathbf{s}$ as a Dedekind left set, giving your answer in the form $\mathbf{s} = \{q \in \mathbb{Q} : \ldots\}$, in terms of the rationals in $\mathbf{r}$.

2.3 Alternative constructions

Although Dedekind dreamt up his construction in 1858, he only published it in 1872, when he received a paper of Cantor's with an alternative construction. This latter is the first such alternative which we shall discuss in this section.

Cantor's construction exploits *Cauchy sequences* of rationals. You may have encountered the idea of a Cauchy sequence in real analysis. It is a sequence for which you might not have been told, or have yet guessed, a limit, but whose members become arbitrarily close to each other. (Actually any convergent sequence of real numbers is a Cauchy sequence, but this turns out to be a dull observation!) Cauchy sequences often arise from constructions like recurrence relations or iterative processes; and there's a standard theorem of real analysis which then ensures that the sequence has a limit. As an example, define a sequence $\langle a_n \rangle$ by

If this seems like the vague sort of mumbo-jumbo which real analysis was developed to make more precise, bear with us! We shall soon give a proper definition of a Cauchy sequence.

$$\begin{aligned} a_0 &= 0, \\ a_1 &= 1, \\ a_n &= \tfrac{1}{2}(a_{n-1} + a_{n-2}), \text{ for all } n \geq 2. \end{aligned}$$

We shall write sequences using the notation $\langle a_n \rangle$ rather than the more common notation $\{a_n\}$, because the latter notation might easily be confused with that for the set whose only element is a_n – this is a penalty of having a book on set theory!

It is easy to show that successive terms of this sequence get progressively closer to each other: for all $n \geq 2$

$$\begin{aligned} |a_n - a_{n-1}| &= \left|\tfrac{1}{2}(a_{n-1} + a_{n-2}) - a_{n-1}\right| \\ &= \left|\tfrac{1}{2}a_{n-2} - \tfrac{1}{2}a_{n-1}\right| \\ &= \tfrac{1}{2}|a_{n-1} - a_{n-2}|, \end{aligned}$$

so that a simple induction shows that

$$\begin{aligned}|a_n - a_{n-1}| &= (\tfrac{1}{2})^{n-1}|a_1 - a_0| \\ &= (\tfrac{1}{2})^{n-1}.\end{aligned}$$

Although the terms of this sequence become very close to each other, it may not be at all obvious what the limit of the sequence is.

The limit is $\frac{2}{3}$.

Cantor's idea was based on the idea that any irrational could be regarded as the limit of a Cauchy sequence of rationals. Let us take $\sqrt{2}$ as an example. We shall construct a sequence $\langle x_n \rangle$ of rationals approximating to $\sqrt{2}$ by using a standard numerical analysis technique, the bisection method. We define three sequences, $\langle a_n \rangle$, $\langle b_n \rangle$ and $\langle x_n \rangle$, as follows:

$$a_0 = 1, \quad b_0 = 2,$$

and for each $n \geq 1$,

$$x_n = \tfrac{1}{2}(a_n + b_n);$$

$$a_{n+1} = \begin{cases} x_n, & \text{if } x_n^2 < 2, \\ a_n, & \text{otherwise;} \end{cases}$$

$$b_{n+1} = \begin{cases} b_n, & \text{if } x_n^2 < 2, \\ x_n, & \text{otherwise.} \end{cases}$$

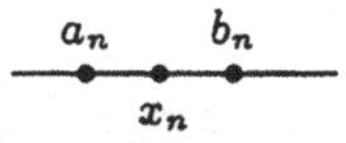

Easy induction arguments show that each of a_n, b_n, x_n is rational,that $a_n \leq a_{n+1} \leq b_{n+1} \leq b_n$ and $a_n^2 < 2 < b_n^2$ for each $n \geq 0$; and, as x_n is halfway between a_n and b_n,

$$\begin{aligned}b_{n+1} - a_{n+1} &= \tfrac{1}{2}(b_n - a_n) \\ &= (\tfrac{1}{2})^{n+1}.\end{aligned}$$

For each n, the closed interval $[a_n, b_n]$ contains $\sqrt{2}$ and $[a_{n+1}, b_{n+1}] \subseteq [a_n, b_n]$. Also the length of $[a_n, b_n]$, i.e. $b_n - a_n$, is $(\frac{1}{2})^n$. So for each of the sequences $\langle a_n \rangle$, $\langle b_n \rangle$ and $\langle x_n \rangle$ of rationals, the terms of the sequence are getting closer and closer to each other, and closer to $\sqrt{2}$. But suppose now that we didn't know all about the reals, and that all we knew about were the rationals. Then all we could say was that each of these sequences appeared to be converging, but not to a limit within the set $\mathbb{Q}$ with which we were familiar. Cantor's idea was essentially to *define* a real number to be such a sequence – the example above would define $\sqrt{2}$. But there's a complication! We want $\sqrt{2}$ to be defined as a unique object, whereas we have obtained three different sequences which intuitively converge to it. So Cantor defined a real not just as a single sequence of rationals, but as the set of all sequences of rationals whose terms get arbitrarily close to the terms of this sequence. Let's now give the proper definitions, starting with that of a Cauchy sequence.

Dedekind's construction gives a unique cut of the rationals corresponding to each real. In Cantor's construction, a real is intuitively represented by the set of all Cauchy sequences of rationals converging to it.

> ***Definition***
>
> The sequence $\langle q_n \rangle$ of rationals is a *Cauchy sequence* if for each $\varepsilon >_{\mathbb{Q}} 0$ (where $\varepsilon \in \mathbb{Q}$) there is an $N \in \mathbb{N}$ such that
>
> $$|q_i -_{\mathbb{Q}} q_j| <_{\mathbb{Q}} \varepsilon, \quad \text{for all } i, j \geq_{\mathbb{N}} N.$$

As in the last section, we shall occasionally emphasize that we are exploiting the operations, relations and properties of $\mathbb{Q}$ and $\mathbb{N}$.

So the terms of a Cauchy sequence get arbitrarily close to each other. As with our earlier example involving $\sqrt{2}$, there is the likelihood that such a sequence will not converge to any limit in $\mathbb{Q}$. But some Cauchy sequences will converge quite happily in $\mathbb{Q}$, for instance $\langle \frac{n-1}{n} \rangle$ is a Cauchy sequence and converges to 1.

Next we shall capture the idea of two such sequences getting arbitrarily close to each other, so that they are in some sense equivalent.

Definition

Let $\langle a_n \rangle$ and $\langle b_n \rangle$ be Cauchy sequences of rationals. We shall say that they are *equivalent* and write $\langle a_n \rangle \sim \langle b_n \rangle$ if for each $\varepsilon >_{\mathbb{Q}} 0$ there is an $N \in \mathbb{N}$ such that

$$|a_n - b_n| <_{\mathbb{Q}} \varepsilon, \quad \text{for all } n \geq_{\mathbb{N}} N.$$

This is equivalent to saying that the sequence $\langle a_n - b_n \rangle$ converges to 0.

What fleshes out the word 'equivalent' is that $\sim$ is an equivalence relation.

Exercise 2.14

Show that $\sim$ is an *equivalence relation* on the set of all Cauchy sequences of rationals, i.e. it has the following properties:

(a) *reflexive*: for all Cauchy sequences $\langle a_n \rangle$, $\langle a_n \rangle \sim \langle a_n \rangle$;

(b) *symmetric*: for all Cauchy sequences $\langle a_n \rangle$ and $\langle b_n \rangle$, if $\langle a_n \rangle \sim \langle b_n \rangle$ then $\langle b_n \rangle \sim \langle a_n \rangle$;

(c) *transitive*: for all Cauchy sequences $\langle a_n \rangle$, $\langle b_n \rangle$ and $\langle c_n \rangle$, if $\langle a_n \rangle \sim \langle b_n \rangle$ and $\langle b_n \rangle \sim \langle c_n \rangle$ then $\langle a_n \rangle \sim \langle c_n \rangle$.

Solution

Reflexivity follows from the fact that the constant sequence $\langle 0, 0, 0, \ldots, 0, \ldots \rangle$ converges to 0. And showing symmetry is likewise straightforward, as $|a_n - b_n| = |b_n - a_n|$ for all n. The argument for transitivity, however, requires some analytic skulduggery, as follows.

We have $a_n - a_n = 0$ for all n.

Suppose that $\langle a_n \rangle \sim \langle b_n \rangle$ and $\langle b_n \rangle \sim \langle c_n \rangle$. To show that $\langle a_n \rangle \sim \langle c_n \rangle$, we need to show that for any $\varepsilon >_{\mathbb{Q}} 0$, there is an $N \in \mathbb{N}$ such that for all $n \geq_{\mathbb{N}} N$, $|a_n - c_n| <_{\mathbb{Q}} \varepsilon$. So take any rational $\varepsilon >_{\mathbb{Q}} 0$. As $\langle a_n \rangle \sim \langle b_n \rangle$ there is some $N_1 \in \mathbb{N}$ such that for all $n \geq_{\mathbb{N}} N_1$,

$$|a_n - b_n| <_{\mathbb{Q}} \frac{\varepsilon}{2}.$$

Likewise as $\langle b_n \rangle \sim \langle c_n \rangle$ there is some $N_2 \in \mathbb{N}$ such that for all $n \geq_{\mathbb{N}} N_2$,

$$|b_n - c_n| <_{\mathbb{Q}} \frac{\varepsilon}{2}.$$

Put $N = \max\{N_1, N_2\}$. Then for all $n \geq_{\mathbb{Q}} N$,

$$\begin{aligned} |a_n - c_n| &= |(a_n - b_n) + (b_n - c_n)| \\ &\leq_{\mathbb{Q}} |a_n - b_n| + |b_n - c_n| \quad \text{(by the triangle inequality for } \mathbb{Q}) \\ &<_{\mathbb{Q}} \frac{\varepsilon}{2} + \frac{\varepsilon}{2} \quad (\text{as } n \geq_{\mathbb{N}} N = \max\{N_1, N_2\}) \\ &= \varepsilon, \end{aligned}$$

as required to show the transitivity of $\sim$.

Any equivalence relation on a set can be used to partition the set into equivalence classes. Cantor defined a real number to be any equivalence class arising from $\sim$, as follows.

A *partition* of a set X is a set of subsets of X such that no two of the subsets have any elements in common, and the union of the subsets is all of X.

Definitions

A *Cantor real number* is any *equivalence class* under the relation $\sim$, i.e. any set of the form

$$\{\langle b_n\rangle : \langle b_n\rangle \sim \langle a_n\rangle\},$$

where $\langle a_n\rangle$ is a Cauchy sequence. We shall write such a class as $[\![\langle a_n\rangle]\!]$.

We shall use $\mathbb{R}_C$ to stand for the set of all Cantor real numbers.

Given a rational number q, the corresponding Cantor real number, $\mathbf{q}_C$ is defined by

$$\mathbf{q}_C = [\![\langle q_n\rangle]\!],$$

where $\langle q_n\rangle$ is the constant sequence defined by $q_n = q$ for all n (which is clearly a Cauchy sequence).

Throughout we take $\langle a_n\rangle$, $\langle b_n\rangle$ etc. to be Cauchy sequences of rationals.

For this section we shall use C, for Cantor, as a subscript to distinguish between operations etc. involving Cantor reals from those involving the set $\mathbb{R}$ of Dedekind left sets.

We must now define the standard relations and arithmetic operations on C. Equality is just straightforward equality as sets. Each Cantor real is an equivalence class under $\sim$ and two such are classes are either equal or disjoint. How about the order relation $<$? Our intuition of how this construction of $\mathbb{R}_C$ relates to the reals as we really know them is that a real number is represented by the set of all Cauchy sequences of rationals which converge to it. Given two reals r and s, and sequences $\langle a_n\rangle$ converging to r and $\langle b_n\rangle$ converging to s, if $r < s$ then we would expect not just that $a_n < b_n$ for all large enough n, but that there's some $\varepsilon > 0$ such that $a_n \leq b_n - \varepsilon$ for all large enough n. (In standard real analysis one would take ε to be something like $\frac{1}{2}(s-r)$.) Turning this around the other way, as we are trying to define $<$ for Cantor reals, it is tempting to define $[\![\langle a_n\rangle]\!] <_C [\![\langle b_n\rangle]\!]$ by

> there are an $\varepsilon \in \mathbb{Q}$ with $\varepsilon >_{\mathbb{Q}} 0$ and an $N \in \mathbb{N}$
> such that $a_n \leq_{\mathbb{Q}} b_n -_{\mathbb{Q}} \varepsilon$ for all $n \geq_{\mathbb{N}} N$.

There is, however, a potential problem with this: the definition seems to depend on the particular sequences $\langle a_n\rangle$ and $\langle b_n\rangle$ taken out of the two equivalence classes. If we were to take different Cauchy sequences $\langle a'_n\rangle$ and $\langle b'_n\rangle$ out of the equivalence classes, is it always the case that the a'_ns are eventually smaller than the b'_ns? If not, we could not be said to have defined anything: the test whether one class is less than the other has to be independent of the representative sequences which we pull out of them. Luckily the test is independent of the representative sequences, as the following result shows.

Theorem 2.4

Suppose that $\langle a_n\rangle \sim \langle a'_n\rangle$ and $\langle b_n\rangle \sim \langle b'_n\rangle$, where all the sequences involved are Cauchy sequences of rationals; and that there are an $\varepsilon \in \mathbb{Q}$ with $\varepsilon >_{\mathbb{Q}} 0$ and an $N \in \mathbb{N}$ such that $a_n \leq_{\mathbb{Q}} b_n -_{\mathbb{Q}} \varepsilon$ for all $n \geq_{\mathbb{N}} N$. Then there are an $\varepsilon' >_{\mathbb{Q}} 0$ and an $N' \in \mathbb{N}$ such that $a'_n \leq_{\mathbb{Q}} b'_n -_{\mathbb{Q}} \varepsilon'$ for all $n \geq_{\mathbb{N}} N'$.

Proof

The general idea of the proof is that as $\langle a_n \rangle \sim \langle a'_n \rangle$ and $\langle b_n \rangle \sim \langle b'_n \rangle$, then for all large enough n the a'_ns are within $\frac{\varepsilon}{4}$ of the corresponding a_ns and the b'_ns are within $\frac{\varepsilon}{4}$ of the b_ns. (As ever with such arguments, there is nothing very special about taking $\frac{\varepsilon}{4}$. Taking e.g. $\frac{\varepsilon}{3}$ or $\frac{\varepsilon}{5}$ would have done just as well.)

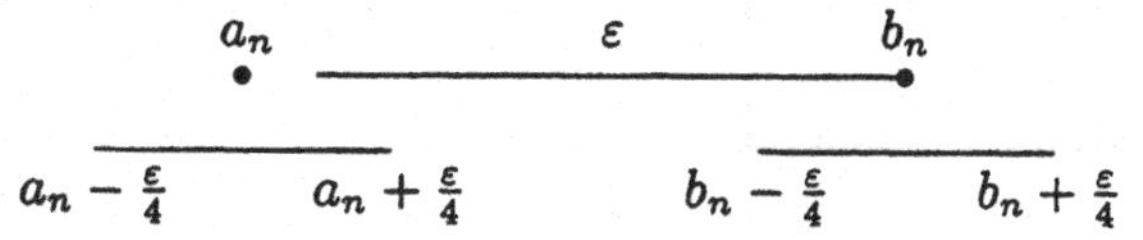

So that even in the worst case of all the a'_ns being bigger than the a_ns and the b'_ns being smaller than the b_ns, the a'_ns are still at least $\varepsilon - (\frac{\varepsilon}{4} + \frac{\varepsilon}{4}) = \frac{\varepsilon}{2}$ less than the b'_ns.

As $\langle a_n \rangle \sim \langle a'_n \rangle$ there is some $N_1 \in \mathbb{N}$ such that $|a_n -_{\mathbb{Q}} a'_n| <_{\mathbb{Q}} \frac{\varepsilon}{4}$ for all $n \geq_{\mathbb{N}} N_1$: in particular $a_n +_{\mathbb{Q}} \frac{\varepsilon}{4} >_{\mathbb{Q}} a'_n$ for all $n \geq_{\mathbb{N}} N_1$.

Likewise as $\langle b_n \rangle \sim \langle b'_n \rangle$ there is some $N_2 \in \mathbb{N}$ such that $|b_n -_{\mathbb{Q}} b'_n| <_{\mathbb{Q}} \frac{\varepsilon}{4}$ for all $n \geq_{\mathbb{N}} N_2$: in particular $b'_n >_{\mathbb{Q}} b_n -_{\mathbb{Q}} \frac{\varepsilon}{4}$ for all $n \geq_{\mathbb{N}} N_2$.

Put $N' = \max\{N, N_1, N_2\}$ so that for all $n \geq_{\mathbb{N}} N'$ the above inequalities hold, as does the inequality

$$b_n -_{\mathbb{Q}} \frac{3\varepsilon}{4} \geq_{\mathbb{Q}} a_n +_{\mathbb{Q}} \frac{\varepsilon}{4},$$

which follows from the assumption that $b_n -_{\mathbb{Q}} \varepsilon \geq_{\mathbb{Q}} a_n$ for all $n \geq_{\mathbb{N}} N$.

Finally put $\varepsilon' = \frac{\varepsilon}{2}$, so that $\varepsilon' >_{\mathbb{Q}} 0$. Then for all $n \geq_{\mathbb{N}} N'$, we have

$$\begin{aligned} b'_n -_{\mathbb{Q}} \varepsilon' &= b'_n -_{\mathbb{Q}} \frac{\varepsilon}{2} \\ &>_{\mathbb{Q}} b_n -_{\mathbb{Q}} \frac{3\varepsilon}{4} \quad (\text{as } b'_n >_{\mathbb{Q}} b_n -_{\mathbb{Q}} \frac{\varepsilon}{4}) \\ &\geq_{\mathbb{Q}} a_n +_{\mathbb{Q}} \frac{\varepsilon}{4} \\ &>_{\mathbb{Q}} a'_n, \end{aligned}$$

as required. ■

Thanks to this result, we can now define an order relation $<$ on the equivalence classes as suggested earlier, safe in the knowledge that our definition is unambiguous. (You may wonder why we are defining a $<$ relation first, rather than a $\leq$ relation. For these equivalence classes of Cauchy sequences, it just seems to turn out to be easier this way!)

Definitions

Given two Cantor reals $[\![\langle a_n\rangle]\!]$ and $[\![\langle b_n\rangle]\!]$, we define $[\![\langle a_n\rangle]\!] <_C [\![\langle b_n\rangle]\!]$ by

there are an $\varepsilon \in \mathbb{Q}$ with $\varepsilon >_{\mathbb{Q}} 0$ and an $N \in \mathbb{N}$
such that $a_n \leq_{\mathbb{Q}} b_n -_{\mathbb{Q}} \varepsilon$ for all $n \geq_{\mathbb{N}} N$.

And we say that $[\![\langle a_n\rangle]\!] \leq_C [\![\langle b_n\rangle]\!]$ when $[\![\langle a_n\rangle]\!] <_C [\![\langle b_n\rangle]\!]$ or $[\![\langle a_n\rangle]\!] = [\![\langle b_n\rangle]\!]$.

Exercise 2.15

Show that $\mathbb{R}_C$ is linearly ordered by $<_C$, i.e. has the following properties:

(a) *Irreflexive*: for all $[\![\langle a_n\rangle]\!] \in \mathbb{R}_C$, it is not the case that $[\![\langle a_n\rangle]\!] <_C [\![\langle a_n\rangle]\!]$ (which we write as $[\![\langle a_n\rangle]\!] \not<_C [\![\langle a_n\rangle]\!]$);

(b) *Transitive*: for all $[\![\langle a_n\rangle]\!], [\![\langle b_n\rangle]\!], [\![\langle c_n\rangle]\!] \in \mathbb{R}_C$, if $[\![\langle a_n\rangle]\!] <_C [\![\langle b_n\rangle]\!]$ and $[\![\langle b_n\rangle]\!] <_C [\![\langle c_n\rangle]\!]$, then $[\![\langle a_n\rangle]\!] <_C [\![\langle c_n\rangle]\!]$;

(c) *Linear*: for all $[\![\langle a_n\rangle]\!], [\![\langle b_n\rangle]\!] \in \mathbb{R}_C$, $[\![\langle a_n\rangle]\!] <_C [\![\langle b_n\rangle]\!]$ or $[\![\langle a_n\rangle]\!] = [\![\langle b_n\rangle]\!]$ or $[\![\langle b_n\rangle]\!] <_C [\![\langle a_n\rangle]\!]$.

Solution

We leave the details for you: these are very similar to standard arguments involving sequences that you find in real analysis, except that all the numbers involved are rationals. Of course, in the light of Theorem 2.4 above, it is enough to argue using representative Cauchy sequences out of each equivalence class.

The definitions of addition and multiplication on Cantor reals are relatively straightforward. The standard results from real analysis about the addition and multiplication of convergent sequences $\langle a_n\rangle$ and $\langle b_n\rangle$ are that if

The definition of multiplication for Cantor reals is indeed much more natural than the corresponding definition for Dedekind left sets.

$$\lim_{n\to\infty} a_n = l \text{ and } \lim_{n\to\infty} b_n = m,$$

then

$$\lim_{n\to\infty} a_n + b_n = l + m \text{ and } \lim_{n\to\infty} a_n b_n = lm.$$

So that if we think of l and m as the numbers approximated by Cauchy sequences $\langle a_n\rangle$ and $\langle b_n\rangle$, the sensible definitions of $l+m$ and $l \cdot m$ are in terms of the sequences $\langle a_n + b_n\rangle$ and $\langle a_n b_n\rangle$. Before we leap to the corresponding definition in terms of equivalence classes of Cauchy sequences, we need a couple of key results to ensure that the definition makes sense. First, are $\langle a_n + b_n\rangle$ and $\langle a_n \cdot b_n\rangle$ Cauchy sequences? (They are clearly sequences of rationals if the a_ns and b_ns are.) And second, will the equivalence class resulting as the sum (or product) of two equivalence classes be the same regardless of which representative sequences $\langle a_n\rangle$ and $\langle b_n\rangle$ are chosen from the latter? These results are given as the next exercises and their proofs are left to you.

Exercise 2.16

Suppose that $\langle a_n \rangle$ and $\langle b_n \rangle$ are Cauchy sequences (of rationals). Show that $\langle a_n + b_n \rangle$ and $\langle a_n b_n \rangle$ are also Cauchy sequences.

Exercise 2.17

Suppose that $\langle a_n \rangle \sim \langle b_n \rangle$ and $\langle a'_n \rangle \sim \langle b'_n \rangle$, where all the sequences are Cauchy sequences of rationals. Show that:

(a) $\langle a_n + b_n \rangle \sim \langle a'_n + b'_n \rangle$;

(b) $\langle a_n b_n \rangle \sim \langle a'_n b'_n \rangle$.

Definitions

Let $[\![\langle a_n \rangle]\!]$ and $[\![\langle b_n \rangle]\!]$ be Cantor reals. Then their sum and product are defined as follows:

$$[\![\langle a_n \rangle]\!] +_C [\![\langle b_n \rangle]\!] = [\![\langle a_n +_{\mathbb{Q}} b_n \rangle]\!];$$
$$[\![\langle a_n \rangle]\!] \cdot_C [\![\langle b_n \rangle]\!] = [\![\langle a_n \cdot_{\mathbb{Q}} b_n \rangle]\!].$$

This completes Cantor's definition of the real numbers, building them up from the rational numbers. Clearly Cantor reals look very different from Dedekind left sets. Shouldn't this worry us, given that we are trying to capture *the* real numbers, i.e. shouldn't we expect just a single definition of the reals? Fortunately, both sets of 'real numbers' have exactly the same properties, in the sense of how we discuss real numbers in e.g. real analysis. The proof of this is usually split into two stages, as follows.

Firstly recall that the set $\mathbb{R}$ of all Dedekind left sets satisfies all the axioms for a complete ordered field, listed in Theorem 2.3 in the previous section. The Cantor reals also satisfy these axioms, which we state as the next theorem.

Theorem 2.5

The set $\mathbb{R}_C$ of all Cantor reals has all 16 of the properties listed in Theorem 2.3 as axioms for a complete ordered field, when the operations $+$ and $\cdot$ are interpreted by $+_C$ and $\cdot_C$, the relations $=$ and $\leq$ by set equality and $\leq_C$, and the special elements 0 and 1 by $\mathbf{0}_C$ and $\mathbf{1}_C$.

$\mathbf{0}_C$ and $\mathbf{1}_C$ are the equivalence classes of the constant sequences $\langle 0 \rangle$ and $\langle 1 \rangle$.

Proof

The details are left to you! However, the proofs are, by and large, much more straightforward than for Dedekind left sets, because the definitions of addition and multiplication are so much simpler. Consider for instance the property

Usually described as the distributive law.

$$\text{for all } x, y, z, \quad x \cdot (y + z) = (x \cdot y) + (x \cdot z).$$

For $\mathbb{R}$ one essentially has to verify that for any Dedekind left sets x, y, z, every rational which belongs to $x \cdot (y + z)$ also belongs to $(x \cdot y) + (x \cdot z)$, and vice versa: there's the further complication of allowing for the different definitions of the product of two left sets, depending on whether they are positive, negative or $\mathbf{0}$. But for $\mathbb{R}_C$ the result follows very quickly from the corresponding result for $\mathbb{Q}$, as follows.

We need to show that for all Cauchy sequences of rationals $\langle a_n\rangle$, $\langle b_n\rangle$ and $\langle c_n\rangle$,

$$[\![\langle a_n\rangle]\!] \cdot_C ([\![\langle b_n\rangle]\!] +_C [\![\langle c_n\rangle]\!]) = ([\![\langle a_n\rangle]\!] \cdot_C [\![\langle b_n\rangle]\!]) +_C ([\![\langle a_n\rangle]\!] \cdot_C [\![\langle c_n\rangle]\!]).$$

Using the definitions of $+_C$ and $\cdot_C$

$$\begin{aligned}
[\![\langle a_n\rangle]\!] \cdot_C ([\![\langle b_n\rangle]\!] +_C [\![\langle c_n\rangle]\!]) &= [\![\langle a_n\rangle]\!] \cdot_C [\![\langle b_n +_\mathbb{Q} c_n\rangle]\!] \\
&= [\![\langle a_n \cdot_\mathbb{Q} (b_n +_\mathbb{Q} c_n)\rangle]\!] \\
&= [\![\langle (a_n \cdot_\mathbb{Q} b_n) +_\mathbb{Q} (a_n \cdot_\mathbb{Q} c_n)\rangle]\!] \quad \text{(by the distributive property for } \mathbb{Q}) \\
&= [\![\langle a_n \cdot_\mathbb{Q} b_n\rangle]\!] +_C [\![\langle a_n \cdot_\mathbb{Q} c_n\rangle]\!] \\
&= ([\![\langle a_n\rangle]\!] \cdot_C [\![\langle b_n\rangle]\!]) +_C ([\![\langle a_n\rangle]\!] \cdot_C [\![\langle c_n\rangle]\!]),
\end{aligned}$$

as required. ■

Inasmuch as real analysis textbooks take these properties of the reals as the starting point of the subject, rather than any explanation of what real numbers are, Cantor's definition seems on a par with Dedekind's: both give a set which has these properties. But there is a deeper reason why both definitions are actually of equal status: they give sets with exactly the same arithmetic and order structure. Indeed the same is true for *any* two sets equipped with an order relation $\leq$ and operations corresponding to $+$ and $\cdot$ which satisfy the 16 properties of Theorem 2.3. Two such sets are *isomorphic*, meaning that it is possible to match the elements of one set with those of the other so that the effects of adding and multiplying in each set also match, as do the pairs of elements in the order relation. The formal definition is as follows.

Definitions

Let A be a set with two binary operations $+_A$ and $\cdot_A$, a binary relation $\leq_A$ and specially labelled elements $\mathbf{0}_A$ and $\mathbf{1}_A$ (so $\mathbf{0}_A, \mathbf{1}_A \in A$). Let B be a set similarly equipped with binary operations $+_B$ and $\cdot_B$, a binary relation $\leq_B$ and special elements $\mathbf{0}_B, \mathbf{1}_B \in B$.

This means that for each $a, a' \in A$, both $a +_A a'$ and $a \cdot_A a'$ are uniquely defined elements of A. The relation $\leq_A$ could be specified by the pairs of elements (a, a') for which $a \leq_A a'$ is the case.

Then A and B are *isomorphic* if there is a function $\theta\colon A \longrightarrow B$ (called an *isomorphism*) such that

1. θ is a bijection (i.e. one–one and onto),
2. for all $a, a' \in A$, $\theta(a +_A a') = \theta(a) +_B \theta(a')$,
3. for all $a, a' \in A$, $\theta(a \cdot_A a') = \theta(a) \cdot_B \theta(a')$,
4. for all $a, a' \in A$, $a \leq_A a'$ if and only if $\theta(a) \leq_B \theta(a')$,
5. $\theta(\mathbf{0}_A) = \mathbf{0}_B$ and $\theta(\mathbf{1}_A) = \mathbf{1}_B$.

It is a straightforward exercise to show that the relation 'A is isomorphic to B' on sets equipped with the operations, relation and special elements as above is an equivalence relation. (For instance, to show symmetry you would need to show that if $\theta\colon A \longrightarrow B$ is an isomorphism, then so is $\theta^{-1}\colon B \longrightarrow A$.)

You may well have seen similar arguments for e.g. vector spaces or groups.

Isomorphic sets are essentially the same: one set is merely the other with the names of the elements changed.

The key reason why Dedekind left sets and Cantor reals are of equal status as definitions of the real numbers is given by the following theorem.

> ***Theorem 2.6***
>
> Any two sets which satisfy the axioms of a complete ordered field (the 16 properties in Theorem 2.3) are isomorphic.

This can be rephrased as 'any two complete ordered fields are isomorphic'.

Proof

The proof is left as a (long) exercise for you! The essence of the proof is as follows. For any complete ordered field A, take the subset generated from $\mathbf{1}_A$ by repeated use of the operations $+_A$, $\cdot_A$ and additive and multiplicative inverses (which stem from axioms 9 and 13). This subset can be shown to be isomorphic to the set of rationals with its usual structure, so it's appropriate to label it as $\mathbb{Q}_A$. Thus given two complete ordered fields A and B, start the definition of an isomorphism θ from A to B by matching the 'rationals' in $\mathbb{Q}_A$ with the corresponding elements of $\mathbb{Q}_B$. Then for each $a \in A$ define a set $L(a) = \{q \in \mathbb{Q}_A : q <_A a\}$, the Dedekind left set corresponding to a. The corresponding subset $\{\theta(q) : q \in L(a)\}$ of B can be shown to be bounded above in B, so has a least upper bound in B (by property 16, the completeness property). Defining $\theta(a)$ to be this least upper bound gives the required isomorphism. ∎

For example, the multiplicative inverse of $\mathbf{1}_A +_A \mathbf{1}_A$ corresponds to $\frac{1}{2} \in \mathbb{Q}$.

Essentially, thanks to the completeness property the rationals in a complete ordered field 'fix' all the other elements of the field.

Let us outline yet another way of constructing the reals from the rationals, one that may well be closest to the algebraic picture of the reals with which you began reading this book, namely by decimal expansions. We are quite accustomed to writing numbers by their decimal expansions, e.g.

We could call these the reals defined by *decimal expansions*.

$$\sqrt{2} = 1.41421356237309504\ldots.$$

An expansion of this sort is really an infinite series of the form

$$\sum_{n=0}^{\infty} \frac{a_n}{10^n},$$

where a_0 is an integer and, for $n \geq 1$, each a_n is an integer in the set $\{0, 1, 2, \ldots, 9\}$. The definition of an infinite series says that this is the limit of the sequence of its partial sums $\langle s_N \rangle$, where

$$s_N = \sum_{n=0}^{N} \frac{a_n}{10^n}.$$

It is a straightforward piece of real analysis to show that such a sequence $\langle s_N \rangle$ is a Cauchy sequence of rationals, which connects decimal expansions to Cantor reals – each equivalence class in Cantor's definition contains such a sequence $\langle s_N \rangle$.

It is possible to define the reals in terms of such expansions, although the definitions of $+$ and $\cdot$ are fairly unpleasant, even by comparison to those we have already met with our earlier constructions. First of all we define the reals to be the set D of all sequences $\langle a_n \rangle$, also written as $\langle a_0, a_1, a_2, \ldots \rangle$, such that $a_0 \in \mathbb{Z}$ and $a_n \in \{0, 1, 2, \ldots, 9\}$ for all $n \geq 1$, with the extra proviso that the sequence does not eventually consist of 9s. This last proviso avoids counting some numbers twice, e.g. 1.2 as both $\langle 1, 2, 0, 0, 0, 0, 0, 0, \ldots \rangle$ and $\langle 1, 1, 9, 9, 9, 9, 9, 9, \ldots \rangle$ (corresponding to the decimal expansions 1.2000000... and 1.1999999...). The rational numbers can then be represented by those sequences which eventually cycle, corresponding to their decimal expansions, e.g.

Corresponding to the decimal expansion $\sum_{n=0}^{\infty} \frac{a_n}{10^n}$ above.

$1 = \langle 1, 0, 0, 0 \ldots \rangle$ ends with the repeated cycle of 0,

and

$$\frac{21}{22} = \langle 0, 9, 5, 4, 5, 4, 5, 4, \ldots \rangle \text{ ends with the repeated cycle of } 5, 4.$$

$\frac{21}{22} = 0.9545454\ldots$.

Define a strict order on these sequences by

$$\langle a_n \rangle <_D \langle b_n \rangle$$

if

there is some $k \in \mathbb{N}$ such that $a_k <_{\mathbb{Z}} b_k$ but $a_i = b_i$ for all $i <_{\mathbb{N}} k$.

So for instance

$$\langle -11, 3, 0, 0, 0, 0, \ldots \rangle <_D \langle 2, 1, 7, 5, 0, 0, \ldots \rangle <_D \langle 2, 1, 8, 3, 0, 0, \ldots \rangle.$$

Corresponding to $-11.3 < 2.175 < 2.183$.

One can verify that $<_D$ is indeed a strict linear order on D and that the completeness property is satisfied.

Defining addition on D is pretty tricky. One might hope to find a rule for adding two infinite sequences directly, but the carrying of digits arbitrarily far down the sequence causes problems, not to mention avoiding ending in a sequence of 9s. The way round this is to look at partial sums of the corresponding decimal expansions, as follows. Given $\langle a_n \rangle, \langle b_n \rangle \in D$, define sequences of rational numbers $\langle A_n \rangle, \langle B_n \rangle$ by

Imagine trying to cope with $\langle 3, 3, 3, 3, \ldots \rangle +_D \langle 6, 6, 6, 6, \ldots \rangle$!

$$A_n = \sum_{i=0}^{n} a_i \text{ and } B_n = \sum_{i=0}^{n} b_i.$$

For each n add the rationals A_n and B_n in the usual way, and construct the corresponding sequence in D. Then define $\langle a_n \rangle +_D \langle b_n \rangle$ to be the least upper bound of these sequences (known to exist as one has already verified the completeness property for D!).

One defines multiplication on D in a similar way. Verifying that D is a complete ordered field is fairly gruelling! Nevertheless this version of the reals is of great importance, precisely because it is one that we use in everyday maths. Likewise in everyday maths we are aware of other representations of the real numbers similar to D, differing only in the number base chosen, e.g. 2 or 3 rather than 10. In general, working with number base $M \geq 2$, we could represent the reals by sequences $\langle a_n \rangle$ corresponding to the infinite series

We could call these the reals defined by *M-ary expansions*. Again, we'd need a convention about sequences ending in recurring $(M-1)$s.

$$\sum_{n=0}^{\infty} \frac{a_n}{M^n}.$$

Just as we did in the previous section with Dedekind left sets, let us review some of the costs of these alternative definitions. Again they involve the manipulation of infinite objects, e.g. sequences and equivalence classes, as single objects. They also require particular care that one has actually *defined* something, e.g. when defining the product of two Cantor reals in terms of representative Cauchy sequences in them. There's a message here of taking care with definitions. And all the constructions assume that we know all about the rationals. How we might define the rationals in terms of something simpler is the subject of the next section.

A Cauchy sequence codes infinitely many rationals. And an equivalence class $[\![\langle a_n \rangle]\!]$ contains infinitely many Cauchy sequences, e.g. the sequences $\langle a_n + \frac{k}{n} \rangle$ for each fixed $k \in \mathbb{N}$.

Further exercises

Exercise 2.18

At the beginning of the section, we defined three sequences of rationals $\langle a_n \rangle$, $\langle b_n \rangle$ and $\langle x_n \rangle$ by:

$$a_0 = 1, \quad b_0 = 2,$$

and for each $n \geq 1$,

$$x_n = \tfrac{1}{2}(a_n + b_n);$$

$$a_{n+1} = \begin{cases} x_n, & \text{if } x_n^2 < 2, \\ a_n, & \text{otherwise;} \end{cases}$$

$$b_{n+1} = \begin{cases} b_n, & \text{if } x_n^2 < 2, \\ x_n, & \text{otherwise.} \end{cases}$$

(a) Verify that these are all Cauchy sequences.

(b) Show that $\langle a_n \rangle \sim \langle b_n \rangle \sim \langle x_n \rangle$.

(c) Show that $[\![\langle a_n \rangle]\!] \cdot_C [\![\langle a_n \rangle]\!] = \mathbf{2}_C$.

Exercise 2.19

Show that any non-zero Cantor real has a multiplicative inverse in $\mathbb{R}_C$, i.e. given any Cauchy sequence of rationals $\langle a_n \rangle$ such that $[\![\langle a_n \rangle]\!] \neq \mathbf{0}_C$, there is a Cauchy sequence $\langle b_n \rangle$ of rationals such that $[\![\langle a_n \rangle]\!] \cdot_C [\![\langle b_n \rangle]\!] = \mathbf{1}_C$.

Exercise 2.20

Explain why one cannot define a strict order $<$ on Cantor reals by $[\![\langle a_n \rangle]\!] < [\![\langle b_n \rangle]\!]$ if

there is some $N \in \mathbb{N}$ such that $a_n <_{\mathbb{Q}} b_n$ for all $n \geq_{\mathbb{N}} N$.

Would the above give a suitable definition of a weak order $\leq$ on $\mathbb{R}_C$?

Exercise 2.21

Define a weak order $\leq$ on Cantor reals as directly as you can using an ε–N definition (as opposed to what we have done above, defining it in terms of $<$).

2.4 The rational numbers

The question 'What are the real numbers?' has been answered by giving a variety of constructions exploiting the rational numbers. It seems reasonable then to ask the question 'What are the rationals?' and in this section we shall outline an answer. The approach is essentially the same as for the reals: we explain the rationals in terms of something more 'basic', namely the set $\mathbb{Z}$ of integers. This then suggests the question 'What are the integers?' and we answer this in terms of the set $\mathbb{N}$ of natural numbers.

You may well be able to guess the next question we should be asking! Its answer will occupy the next chapter.

There is an alternative way of approaching this issue. Just as modern real analysis texts tend to dodge the question 'What are the reals?' by building the subject on the axioms for a complete ordered field (as in Theorem 2.3), one could present axioms which in some sense describe the rationals. For instance one might take all properties in Theorem 2.3 except the completeness property (Property 16), which axiomatize an *ordered field*. Any set A which satisfies all these properties contains a subset, generated by $\mathbf{0}_A$ and $\mathbf{1}_A$ and closed under addition, multiplication and inverses, which serves as a version of $\mathbb{Q}$. The reasons for putting this mathematics onto an axiomatic basis are historically bound up with the foundational consequences of Dedekind's and Cantor's constructions of $\mathbb{R}$ from $\mathbb{Q}$, and these consequences are the meat of set theory. So we shall persist in giving a construction of $\mathbb{Q}$ rather than leaning on axioms.

Such a set might be $\mathbb{R}$ itself!

When we are reasoning about rationals, we are accustomed to writing rationals in the form $\frac{a}{b}$ where a and b are integers, with b positive. Thus, to explain the rationals in terms of the integers, we might describe a rational number as given by an ordered pair (a, b) of integers, with b positive, corresponding to the fraction $\frac{a}{b}$. This would create the problem of representing a rational by several distinct pairs of integers in the set $\mathbb{Z} \times \mathbb{Z}^+$. For instance, as

We write $\mathbb{Z}^+$ for the set $\{b \in \mathbb{Z} : b >_{\mathbb{Z}} 0\}$.

$$\tfrac{1}{2} = \tfrac{2}{4} = \tfrac{7}{14} = \tfrac{13}{26},$$

the rational number $\frac{1}{2}$ would be represented by each of the ordered pairs

$$(1,2),\ (2,4),\ (7,14),\ (13,26).$$

This is the same sort of problem that we encountered when representing a real number by a Cauchy sequence of rationals, and we resolve it in the same way, via an equivalence relation. We need to find a way of saying why e.g. $\frac{1}{2}$ and $\frac{7}{14}$ are the same, without using the properties of $\mathbb{Q}$ – we are, after all, trying to define $\mathbb{Q}$! But we can exploit the properties of $\mathbb{Z}$, and the key to explaining, in terms of $\mathbb{Z}$ and its operations, that

We shall write $<_{\mathbb{Z}}$ for the order on $\mathbb{Z}$ and $+_{\mathbb{Z}}$, $\cdot_{\mathbb{Z}}$ for its arithmetic operations.

$$\frac{a}{b} = \frac{c}{d}$$

is by cross-multiplying the terms in this equation to get

$$ad = bc.$$

This neatly avoids the problem that $\mathbb{Z}$ has no division operation and recasts everything in terms of the multiplication which $\mathbb{Z}$ does have. We shall exploit this last equation to specify when ordered pairs of integers are equivalent.

> ***Definition***
>
> For any $a, b, c, d \in \mathbb{Z}$ with $b, d >_{\mathbb{Z}} 0$, we shall write $(a, b) \sim (c, d)$ when $a \cdot_{\mathbb{Z}} d = b \cdot_{\mathbb{Z}} c$.

The symbol $\sim$ is being used with a different meaning from the previous section.

Exercise 2.22

Show that $\sim$ is an equivalence relation on the set $\mathbb{Z} \times \mathbb{Z}^+$ of all ordered pairs (a, b) of integers with $b >_{\mathbb{Z}} 0$, i.e. it is reflexive, symmetric and transitive.

> ***Definitions***
>
> Let $[\![(a, b)]\!]$ be the equivalence class of the ordered pair (a, b) of integers under the equivalence relation $\sim$, i.e. the set
>
> $$\{(c, d) \in \mathbb{Z} \times \mathbb{Z}^+ : (a, b) \sim (c, d)\}.$$
>
> A *rational number* is such an equivalence class and $\mathbb{Q}$ is the set of all these equivalence classes.

So, for instance, the rational $-\frac{1}{3}$ is represented by the set

$$\{(-k, 3k) : k \in \mathbb{Z}^+\}.$$

We must now describe the order relation on $\mathbb{Q}$ and its arithmetic operations in terms of these equivalence classes and the order and arithmetic operations of $\mathbb{Z}$. We shall exploit various known results about the (desired!) set $\mathbb{Q}$, which will help us translate terms involving fractions $\frac{x}{y}$, with x, y integers, into ones involving ordered pairs (x, y). For integers a, b, c, d we have

$$\frac{a}{b} < \frac{c}{d} \text{ if and only if } ad < bc, \text{ for } b, d > 0,$$
$$\frac{a}{b} + \frac{c}{d} = \frac{ad + bc}{bd},$$
$$\frac{a}{b} \cdot \frac{c}{d} = \frac{ac}{bd}.$$

These will help us define $<_{\mathbb{Q}}$, $+_{\mathbb{Q}}$ and $\cdot_{\mathbb{Q}}$ respectively. For example, the ordered pair $(ad + bc, bd)$ could represent $\frac{a}{b} + \frac{c}{d}$.

As rational numbers are represented by equivalence classes, we shall have to make sure that any construction involving them yields the same result regardless of which representatives we take from the classes. So our definitions need to be preceded by the results given in the next exercise.

Exercise 2.23

Suppose that $(a, b), (a', b'), (c, d), (c', d') \in \mathbb{Z} \times \mathbb{Z}^+$ and that $(a, b) \sim (a', b')$, $(c, d) \sim (c', d')$. Use the standard properties of $\mathbb{Z}$ to show the following.

(a) $a \cdot_{\mathbb{Z}} d <_{\mathbb{Z}} b \cdot_{\mathbb{Z}} c$ if and only if $a' \cdot_{\mathbb{Z}} d' <_{\mathbb{Z}} b' \cdot_{\mathbb{Z}} c'$.

(b) $\big((a \cdot_{\mathbb{Z}} d) +_{\mathbb{Z}} (b \cdot_{\mathbb{Z}} c),\ b \cdot_{\mathbb{Z}} d\big) \sim \big((a' \cdot_{\mathbb{Z}} d') +_{\mathbb{Z}} (b' \cdot_{\mathbb{Z}} c'),\ b' \cdot_{\mathbb{Z}} d'\big)$.

(c) $(a \cdot_{\mathbb{Z}} c,\ b \cdot_{\mathbb{Z}} d) \sim (a' \cdot_{\mathbb{Z}} c',\ b' \cdot_{\mathbb{Z}} d')$.

Solution

We shall do (c) and leave the others to you.

As $(a,b) \sim (a',b')$ and $(c,d) \sim (c',d')$, we have

$$a \cdot_{\mathbb{Z}} b' = b \cdot_{\mathbb{Z}} a' \text{ and } c \cdot_{\mathbb{Z}} d' = d \cdot_{\mathbb{Z}} c'.$$

Thus

$$\begin{aligned}(a \cdot_{\mathbb{Z}} c) \cdot_{\mathbb{Z}} (b' \cdot_{\mathbb{Z}} d') &= (a \cdot_{\mathbb{Z}} b') \cdot_{\mathbb{Z}} (c \cdot_{\mathbb{Z}} d') && \text{(by the commutativity and associativity of } \cdot_{\mathbb{Z}}) \\ &= (b \cdot_{\mathbb{Z}} a') \cdot_{\mathbb{Z}} (d \cdot_{\mathbb{Z}} c') && \text{(as } (a,b) \sim (a',b') \text{ and } (c,d) \sim (c',d')) \\ &= (b \cdot_{\mathbb{Z}} d) \cdot_{\mathbb{Z}} (a' \cdot_{\mathbb{Z}} c') && \text{(by the commutativity and associativity of } \cdot_{\mathbb{Z}}),\end{aligned}$$

which is what is required to show that $(a \cdot_{\mathbb{Z}} c,\ b \cdot_{\mathbb{Z}} d) \sim (a' \cdot_{\mathbb{Z}} c',\ b' \cdot_{\mathbb{Z}} d')$.

Definitions

Let $[\![(a,b)]\!]$ and $[\![(c,d)]\!]$ be any two rationals. Then

$[\![(a,b)]\!] <_{\mathbf{Q}} [\![(c,d)]\!]$ if $a \cdot_{\mathbb{Z}} d <_{\mathbb{Z}} b \cdot_{\mathbb{Z}} c$;

$[\![(a,b)]\!] +_{\mathbf{Q}} [\![(c,d)]\!] = [\![((a \cdot_{\mathbb{Z}} d) +_{\mathbb{Z}} (b \cdot_{\mathbb{Z}} c),\ b \cdot_{\mathbb{Z}} d)]\!]$;

$[\![(a,b)]\!] \cdot_{\mathbf{Q}} [\![(c,d)]\!] = [\![(a \cdot_{\mathbb{Z}} c,\ b \cdot_{\mathbb{Z}} d)]\!]$.

Given an integer k, the corresponding rational number is $\mathbf{k}_{\mathbf{Q}}$ defined by

$$\mathbf{k}_{\mathbf{Q}} = [\![(k,1)]\!].$$

Exercise 2.24

Show that $[\![(-1,2)]\!] \cdot_{\mathbf{Q}} -\mathbf{2}_{\mathbf{Q}} <_{\mathbf{Q}} -\mathbf{1}_{\mathbf{Q}} +_{\mathbf{Q}} [\![(14,6)]\!]$.

Solution

$$\begin{aligned}[\![(-1,2)]\!] \cdot_{\mathbf{Q}} -\mathbf{2}_{\mathbf{Q}} &= [\![(-1,2)]\!] \cdot_{\mathbf{Q}} [\![(-2,1)]\!] \\ &= [\![(-1 \cdot_{\mathbb{Z}} -2,\ 2 \cdot_{\mathbb{Z}} 1)]\!] \\ &= [\![(2,2)]\!].\end{aligned}$$

This matches our expectation that $-\frac{1}{2} \cdot -2 = 1$ as $[\![(2,2)]\!] = \mathbf{1}_{\mathbf{Q}}$.

$$\begin{aligned}-\mathbf{1}_{\mathbf{Q}} +_{\mathbf{Q}} [\![(14,6)]\!] &= [\![(-1,1)]\!] +_{\mathbf{Q}} [\![(14,6)]\!] \\ &= [\![((-1 \cdot_{\mathbb{Z}} 6) +_{\mathbb{Z}} (1 \cdot_{\mathbb{Z}} 14),\ 1 \cdot_{\mathbb{Z}} 6)]\!] \\ &= [\![(8,6)]\!].\end{aligned}$$

Again, we expect that $-1 + \frac{14}{6} = \frac{8}{6} (= \frac{4}{3})$ and $[\![(8,6)]\!]$ represents $\frac{4}{3}$.

To test whether $[\![(2,2)]\!] <_{\mathbf{Q}} [\![(8,6)]\!]$ we need to investigate whether

$$2 \cdot_{\mathbb{Z}} 6 <_{\mathbb{Z}} 2 \cdot_{\mathbb{Z}} 8.$$

As $2 \cdot_{\mathbb{Z}} 6 = 12 <_{\mathbb{Z}} 16 = 2 \cdot_{\mathbb{Z}} 8$, the inequality holds, as required.

Exercise 2.25

Suppose that $[\![(a,b)]\!]$ and $[\![(c,d)]\!]$ are rationals such that $[\![(a,b)]\!] <_{\mathbf{Q}} [\![(c,d)]\!]$. Show, by giving a construction, that there is a rational $[\![(x,y)]\!]$ such that $[\![(a,b)]\!] <_{\mathbf{Q}} [\![(x,y)]\!] <_{\mathbf{Q}} [\![(c,d)]\!]$.

This shows that $<_{\mathbf{Q}}$ is a *dense* order.

Solution

With ordinary rationals $\frac{a}{b}$ and $\frac{c}{d}$, we would expect the rational

$$\frac{1}{2}\left(\frac{a}{b}+\frac{c}{d}\right)=\frac{ad+bc}{2bd}$$

to be halfway between them. So $[\![((a \cdot_{\mathbb{Z}} d) +_{\mathbb{Z}} (b \cdot_{\mathbb{Z}} c),\ 2 \cdot_{\mathbb{Z}} b \cdot_{\mathbb{Z}} d)]\!]$ ought to be suitable as an $[\![(x, y)]\!]$. This is indeed the case and we leave you to verify the details.

The result of the previous exercise also follows from the axioms for an ordered field. We state without proof the following theorem, that $\mathbb{Q}$ is such a field. You might like to give the proof yourself. (The proof is much more straightforward than the corresponding proofs for $\mathbb{R}$ and $\mathbb{R}_C$.)

> ***Theorem 2.7***
>
> $\mathbb{Q}$ as constructed by equivalence classes of pairs of integers is an ordered field, i.e. satisfies Properties 1 to 15 of Theorem 2.3.

We have defined the rationals in terms of integers. But what are the integers? We shall define them in terms of the set of natural numbers $\mathbb{N} = \{0, 1, 2, 3, \ldots\}$ and the order and arithmetic of $\mathbb{N}$. The problem is, of course, how to represent the negative integers without using subtraction, which is not a closed operation on $\mathbb{N}$. One trick is to represent the integer n by a pair (a, b) of natural numbers such that, in $\mathbb{Z}$, $a - b = n$. So for instance -3 could be represented by $(1, 4)$ or $(7, 10)$, and the integer 2 by $(2, 0)$ or $(7, 5)$. As ever, we have the problem of representing each integer by a single object; and as ever we resolve it via a suitable equivalence relation. This time, the key idea is that one can express an equation involving $-$,

It should come as no surprise that we shall write $<_{\mathbb{N}}$ for the order on $\mathbb{N}$ and $+_{\mathbb{N}}$, $\cdot_{\mathbb{N}}$ for its arithmetic operations.

$$a - b = c - d,$$

in an equivalent way avoiding use of $-$, by

$$a + d = b + c.$$

> ***Definition***
>
> For any $a, b, c, d \in \mathbb{N}$ we shall write $(a, b) \sim (c, d)$ if $a +_{\mathbb{N}} d = b +_{\mathbb{N}} c$.

Yet another meaning for $\sim$! The precise details of the constructions of $\mathbb{Q}$ from $\mathbb{Z}$ and of $\mathbb{Z}$ from $\mathbb{N}$ are not going to be of great significance in the rest of the book, so it's not worth inventing a special notation for each of these equivalence relations.

We leave you to check that $\sim$ is an equivalence relation in the next exercise.

Exercise 2.26

Show that $\sim$ as just defined is an equivalence relation on the set $\mathbb{N} \times \mathbb{N}$ of all pairs of natural numbers.

Definitions

Let $[\![(a,b)]\!]$ be the equivalence class of the ordered pair (a,b) of natural numbers under the equivalence relation $\sim$, i.e. the set $\{(c,d) \in \mathbb{N} \times \mathbb{N} : (a,b) \sim (c,d)\}$. An *integer* is such an equivalence class and $\mathbb{Z}$ is the set of all these equivalence classes.

To define $<_{\mathbb{Z}}$, $+_{\mathbb{Z}}$ and $\cdot_{\mathbb{Z}}$ in terms of $\mathbb{N}$, we are again guided by results that we expect to hold for $\mathbb{Z}$, involving natural numbers a, b, c, d:

$a - b < c - d$ if and only if $a + d < b + c$;

$(a - b) + (c - d) = (a + c) - (b + d)$;

$(a - b)(c - d) = (ac + bd) - (ad + bc)$.

E.g. the ordered pair $(a + c, b + d)$ could represent the sum of $a - b$ and $c - d$.

And to prepare the ground for our definitions, we have to check that they will make sense for whatever representatives we take from the equivalence classes. The results that we need are left for you as the next exercise.

Exercise 2.27

Suppose that $(a,b), (a',b'), (c,d), (c',d') \in \mathbb{N} \times \mathbb{N}$ and that $(a,b) \sim (a',b')$, $(c,d) \sim (c',d')$. Use the standard properties of $\mathbb{N}$ to show the following.

(a) $a +_{\mathbb{N}} d <_{\mathbb{N}} b +_{\mathbb{N}} c$ if and only if $a' +_{\mathbb{N}} d' <_{\mathbb{N}} b' +_{\mathbb{N}} c'$.

(b) $(a +_{\mathbb{N}} c, b +_{\mathbb{N}} d) \sim (a' +_{\mathbb{N}} c', b' +_{\mathbb{N}} d')$.

(c) $((a \cdot_{\mathbb{N}} c) +_{\mathbb{N}} (b \cdot_{\mathbb{N}} d),\ (a \cdot_{\mathbb{N}} d) +_{\mathbb{N}} (b \cdot_{\mathbb{N}} c))$
$\sim ((a' \cdot_{\mathbb{N}} c') +_{\mathbb{N}} (b' \cdot_{\mathbb{N}} d'),\ (a' \cdot_{\mathbb{N}} d') +_{\mathbb{N}} (b' \cdot_{\mathbb{N}} c'))$.

Definitions

Let $[\![(a,b)]\!]$ and $[\![(c,d)]\!]$ be any two integers. Then

$[\![(a,b)]\!] <_{\mathbb{Z}} [\![(c,d)]\!]$ if $a +_{\mathbb{N}} d <_{\mathbb{N}} b +_{\mathbb{N}} c$;

$[\![(a,b)]\!] +_{\mathbb{Z}} [\![(c,d)]\!] = [\![(a +_{\mathbb{N}} c, b +_{\mathbb{N}} d)]\!]$;

$[\![(a,b)]\!] \cdot_{\mathbb{Z}} [\![(c,d)]\!] = [\![((a \cdot_{\mathbb{N}} c) +_{\mathbb{N}} (b \cdot_{\mathbb{N}} d), (b \cdot_{\mathbb{N}} c) +_{\mathbb{N}} (a \cdot_{\mathbb{N}} d))]\!]$.

Given a natural number n, the corresponding integer is $\mathbf{n}_{\mathbb{Z}}$ defined by

$$\mathbf{n}_{\mathbb{Z}} = [\![(n,0)]\!].$$

We shall not investigate the properties satisfied by the integers – not because they are of no interest, but because they take us too far from the point of this section, which is how one might construct the rationals, and thus the reals, starting from simpler objects. As in the earlier sections of this chapter, we note that our constructions involve treating infinite objects, equivalence classes containing infinitely many pairs, as single objects; and this use of infinity will need investigation. We have also reduced everything using quite sophisticated set constructions to terms involving natural numbers. But what are the natural numbers? This question, like that of 'What are the reals?', is of sufficient significance to merit a chapter to itself.

The idea of representing one set of objects in terms of simpler objects is of importance in computer science, where complicated data held on a computer are actually represented, ultimately, by electrical charges, or their absence.

For example, $-\mathbf{2}_{\mathbb{Z}} = \{(n, n+2) : n \in \mathbb{N}\}$, which has infinitely many elements.

3 THE NATURAL NUMBERS

3.1 Introduction

In the previous chapter we saw a way of constructing the real numbers from the rational numbers. The rationals can be constructed from the integers, and these can in turn be constructed from the set of natural numbers, $\mathbb{N} = \{0, 1, 2, 3, \ldots\}$. In this chapter we shall continue this process of reduction by defining the natural numbers in terms of sets.

One might reasonably argue on historical and psychological grounds that only the positive integers (which are the counting numbers) should be termed 'natural', not 0. But logicians include 0 in the set, mainly because of the importance of 0 in defining arithmetic within set theory.

Why sets? One answer to this is that set constructions, in particular ones involving infinite sets, were a vital part of the definition of the reals. Thus set ideas, as well as the natural numbers, underlie our work so far. If we can express natural numbers in terms of sets, then we have a single foundation for our theory. Using sets in this way will, however, require us to be, or to become, clear about the ways in which we might legitimately use sets. When we were constructing reals in terms of rationals, rationals in terms of integers, etc., we were reasonably relaxed about using familiar properties of the 'known' number system when proving statements about the number system being constructed from it. (Strictly speaking, we should perhaps have defined $\mathbb{N}$ and proved all of its standard properties, then used these to define $\mathbb{Z}$ and prove all its standard properties, and then done the same for $\mathbb{Q}$ and then $\mathbb{R}$.) But the use of sets to do much more than take the odd union or intersection is less familiar; and the development of the theory of sets which stemmed from e.g. Dedekind's and Cantor's work, turned out to be trickier than anticipated! The impact for this chapter is that we shall have to start the process of formalizing set theory, so that it supports both the special sets which we shall take as the natural numbers and will also provide a framework for establishing their familiar properties.

Also, as in the previous chapter, there is a different approach to explaining the natural numbers, namely by giving axioms for them. Ideally such axioms should be satisfied by essentially just one structure, as is the case with the axioms for a complete ordered field: any two sets satisfying the properties should be isomorphic. Such axioms were devised by Peano and are as follows, expressed in terms of a set X.

The Italian mathematician Giuseppe Peano (1858–1932) introduced these axioms in his 1889 paper, which you can find in [11].

Peano's axioms for the natural numbers

X is a set with a special element $0_X \in X$ and a function $S\colon X \longrightarrow X$ such that the following also hold:

1. the function S is one–one, i.e. for all $x, y \in X$, if $S(x) = S(y)$ then $x = y$;
2. for all $x \in X$, $0_X \neq S(x)$;
3. for all subsets $A \subseteq X$, if A contains 0_X and contains $S(x)$ whenever $x \in A$, then A is all of X.

Such an X is often called a *Peano system*.

The idea of S is that it will be the *successor* function, $S(x) = x + 1$.

This is called the *induction principle*.

If we take X to be the set $\mathbb{N} = \{0, 1, 2, 3, \ldots\}$, and interpret 0_X by 0 and the function S by the function $n \longmapsto n + 1$, then it's clear that $\mathbb{N}$ satisfies all these properties – at least inasmuch as we've always known that the principle of induction holds for ℕ! Perhaps this principle seems more familiar in the guise:

> for all properties P, if 0_X has property P, and $S(x)$ has property P whenever x has it, then every element of X has property P.

Exercise 3.1

What is the connection between the sets A and properties P in the two versions of the induction principle above?

Solution

Each property P satisfied by (some or all of) the elements of X corresponds to a subset of X, namely

$$\{x \in X : x \text{ has the property } P\}.$$

Likewise a subset A of X corresponds to the property 'x belongs to A'.

The idea of a property P giving rise to a set is very important, but will be seen to require care.

Peano's axioms may not, at first sight, seem so powerful. For instance, they don't refer to the standard arithmetic operations of ℕ, let alone give the impression that a result like the fundamental theorem of arithmetic must hold. But, as you will see, one can construct sum and product operations on such a set X so that the full theory of ℕ can then be developed.

The fundamental theorem of arithmetic states that every positive integer greater than 1 can be expressed uniquely as a product of primes.

The following theorem is an example of the power of the axioms.

Theorem 3.1

Suppose that the set X satisfies Peano's axioms. Then every $x \in X$, other than 0_X, is $S(y)$ for some $y \in X$.

We have to exclude 0_X because of property 2 of the axioms.

Proof

We shall exploit the induction principle, which is where the real power of the axioms resides. Define a subset A of X by

$$A = \{x \in X : x = 0_X \text{ or } x = S(y) \text{ for some } y \in X\}.$$

Then $0_X \in A$, by definition. And if $x \in A$, then $S(x) \in A$ – by definition $S(x)$ is in A, regardless of whether x is! So by the induction principle, $A = X$, meaning that every $x \neq 0_X$ is $S(y)$ for some y. ■

Why must induction be used?

A typical example of a set defined by a property of its members.

Exercise 3.2

Give a similar proof from the axioms that for all $x \in X$, $S(x) \neq x$. [Hints: define $A = \{x \in X : S(x) \neq x\}$. You will need to use the fact that S is one–one.]

We shall show how to construct arithmetic operations like addition later in the chapter, for a specific set X. A key tool will be the definition of a function f by *recursion*. In the context of a set X satisfying Peano's axioms, this means giving $f(0_X)$ some value and explaining how to define $f(S(x))$ assuming one knew the value of $f(x)$. For example, define f on $\{0, 1, 2, 3, \ldots\}$ by

$$\begin{aligned} f(0) &= 1, \\ f(n+1) &= (n+1)f(n) \text{ for } n > 0. \end{aligned}$$

Then to work out $f(m)$ for some specific m, use the second part of the definition to relate $f(m)$ to the value of $f(m-1)$, then relate $f(m-1)$ to $f(m-2)$, and so on, until you eventually hit $f(0)$, which is defined here to equal 1. For instance, to compute $f(3)$, we have

$$\begin{aligned} f(3) = f(2+1) &= 3f(2) \\ = 3f(1+1) &= 3 \cdot 2f(1) \\ = 6f(0+1) &= 6 \cdot 1f(0) \\ &= 6 \cdot 1 = 6. \end{aligned}$$

In fact this f is just the factorial function, defined by $f(n) = n! = n \cdot (n-1) \cdot (n-2) \cdot \ldots \cdot 2 \cdot 1$.

A general result about defining a function by recursion on a set X satisfying Peano's axioms is as follows.

Theorem 3.2 Definition by recursion

Let X satisfy Peano's axioms. Let Y be any set, y_0 any element of Y and $h: X \times Y \longrightarrow Y$ a function on pairs $(x, y) \in X \times Y$. Then there exists a unique function $f: X \longrightarrow Y$ such that

$$\begin{aligned} f(0_X) &= y_0, \\ f(S(x)) &= h(x, f(x)), \text{ for all } x. \end{aligned}$$

For the factorial example above, we could take both X and Y to be the set of natural numbers, $y_0 = 1$ and h the function defined by $h(x, y) = (x+1) \cdot y$.

We shall delay a proof of this theorem until we have established a sensible framework for such a proof. For the moment, let's use Theorem 3.2 to show that any two sets satisfying Peano's axioms are isomorphic.

Theorem 3.3

Let X be a set with a special element 0_X and a function $S_X: X \longrightarrow X$, such that X satisfies Peano's axioms. And let Y, with an element 0_Y and a function $S_Y: Y \longrightarrow Y$, similarly satisfy these axioms. Then X and Y are isomorphic, i.e. there is a function $f: X \longrightarrow Y$ such that

1. f is a bijection;
2. $f(0_X) = 0_Y$;
3. $f(S_X(x)) = S_Y(f(x))$, for all $x \in X$.

f is an *isomorphism*.

f matches the elements of X with those of Y so that the effects of the functions S_X and S_Y also match. The theorem can be read as 'any two Peano systems are isomorphic'.

Proof

As X satisfies Peano's axioms, we can apply Theorem 3.2, taking $y_0 = 0_Y$ and the function $h\colon X \times Y \longrightarrow Y$ defined by $h(x, y) = S_Y(y)$. This defines a function f of the form

$$f\colon X \longrightarrow Y$$
$$f(0_X) = 0_Y,$$
$$f(S_X(x)) = S_Y(f(x)), \text{ for all } x \in X.$$

As f, by definition, satisfies requirements 2 and 3 for an isomorphism, we need only show that f is a bijection. First, we show that f is one–one.

We need to show that for all $x, x' \in X$, if $f(x) = f(x')$ then $x = x'$. We shall exploit the induction principle for X. Define a subset A of X by

You will find it quite common to use induction to prove things about a function defined by recursion.

$$A = \{x \in X : \text{ for all } x' \in X, \text{ if } f(x) = f(x') \text{ then } x = x'\},$$

or, equivalently,

$$A = \{x \in X : \text{ for all } x' \in X, \text{ if } x \neq x' \text{ then } f(x) \neq f(x')\}.$$

So A consists of those x for which $f(x)$ isn't the image under f of another $x' \neq x$. We shall show that $A = X$ by induction, and this will show that f is one–one.

Do we have $0_X \in A$? Take any $x' \in X$ such that $x' \neq 0_X$. We shall show that this forces $f(x') \neq f(0_X)$. As $x' \neq 0_X$, then by Theorem 3.1 $x' = S_X(x)$ for some $x \in X$. This means that $f(x') = f(S_X(x))$ which, by the definition of f, equals $S_Y(f(x))$. But Y satisfies Peano's axioms, so that for any $y \in Y$ we have $S_Y(y) \neq 0_Y$. In particular $S_Y(f(x)) \neq 0_Y$. As the definition of f gives that $0_Y = f(0_X)$, this means that $f(x') \neq f(0_X)$, as required. Thus $0_X \in A$.

Now we suppose that $x \in A$ and show that $S_X(x) \in A$. Again we suppose that $x' \neq S_X(x)$ and show that $f(x') \neq f(S_X(x))$. There are two cases: when $x' = 0_X$ and when $x' \neq 0_X$. In the case that $x' = 0_X$, we have already shown that $0_X \in A$, so that if $x' = 0_X \neq S_X(x)$ then $f(x') = f(0_X) \neq f(S_X(x))$. In the case that $x' \neq 0_X$, then by Theorem 3.1 we have $x' = S_X(x'')$, for some $x'' \in X$. The condition that

$$x' \neq S_X(x)$$

becomes

$$S(x'') \neq S_X(x),$$

which, as S_X is one–one, means that

$$x'' \neq x.$$

As $x \in A$, this means that

$$f(x'') \neq f(x),$$

so that, as S_Y is one–one,

$$S_Y(f(x'')) \neq S_Y(f(x)).$$

But by definition of f, $S_Y(f(x'')) = f(S_X(x'')) = f(x')$, while $S_Y(f(x)) = f(S_X(x))$, so that

$$f(x') \neq f(S_X(x)),$$

as required. Thus by the induction principle for X, we have $A = X$, so that f is indeed one–one.

Now let us show that f is onto. This time we shall use the induction principle for Y to show that the subset B of Y defined by

$$B = \{y \in Y : y = f(x) \text{ for some } x \in X\}$$

is all of Y.

B is the image set, or range, of the function f.

We have $0_Y \in B$, because $0_Y = f(0_X)$. Now suppose that $y \in B$ and show that $S_Y(y) \in B$. As $y \in B$, there is some $x \in X$ such that $f(x) = y$. Then by definition of f we have

$$\begin{aligned} f(S_X(x)) &= S_Y(f(x)) \\ &= S_Y(y), \end{aligned}$$

so that $S_Y(y) \in B$, as required. Thus by the induction principle for Y, we have $B = Y$, so that f is onto, completing the proof that f is a bijection. ■

Now we have established that any sets which satisfy Peano's axioms are essentially the same, let us return to the issue of defining one such set, which we shall designate as *the* set $\mathbb{N}$ of natural numbers, in terms of something more basic. The something more basic will be very simple sorts of sets. As with definitions of the real numbers in terms of $\mathbb{Q}$, there are likely to be several reasonable candidates. The first of these was proposed by the German mathematician and philosopher Gottlob Frege (1848–1925), defining the natural number 0 and, given the number n, using it to define its successor $S(n)$. The representation of 0 was by the set $\underline{0}$, where

$$\underline{0} = \{\varnothing\},$$

which is the set of all sets containing 0 elements – the empty set $\varnothing$ being the only such set. In general, the set $\underline{n}$ representing the natural number n also consists of all sets with n elements. A circular definition is avoided by using formal logic, as well simple ideas about sets, to define $\underline{n+1}$ in terms of $\underline{n}$; and as $\underline{0}$ has been defined, this gives a recursive definition of $\underline{n}$ in general. Informally, the idea is to exploit a definition of when a set C has exactly one element, by

We shall look at a suitable framework of formal logic in the next chapter.

$$(\text{there exists } x)\big(x \in C \text{ and } (\text{for all } y)(\text{if } y \in C \text{ then } y = x)\big).$$

Then given $\underline{n}$, define its successor by

$$\begin{aligned} \underline{n+1} = \{B : (\text{ there exists } A)(A \in \underline{n} \text{ and } A \subseteq B \\ \text{and } B \setminus A \text{ has one element})\}. \end{aligned}$$

Imagining $\underline{n}$ to consist of all sets with the same number, n, of elements, the set $\underline{n+1}$ should then consist of all sets with one more element.

Alas! This idea has a major flaw, because treating $\underline{n}$ as a set can lead to a contradiction, as we shall discuss in the next chapter. This is a shame, as Frege's idea fits in well with the principle behind several of the definitions in the previous chapter, in that $\underline{n}$ can be regarded as an equivalence class arising from a very important equivalence relation. The idea behind this relation is that one way of judging that two sets X and Y have the same number of elements is that there is a bijection from one to the other. So define a relation $\approx$ by $X \approx Y$ when there is a bijection $f: X \longrightarrow Y$. It's easy to show that $\approx$ is an equivalence relation and that each $\underline{n}$ as above is one of its equivalence classes. As treating $\underline{n}$ can lead to a contradiction, clearly this plausible definition of $\approx$ needs more careful treatment.

We shall return to this important relation in Chapter 6.

For the rest of this chapter we shall concentrate on a different definition of natural numbers in terms of sets and logic. Our definition will in fact represent each n by a specific set in the equivalence class $\underline{n}$. To put the definition on a sound footing, we shall keep an eye on what properties we shall expect of sets, but we won't firm up on these properties until the next chapter.

3.2 The construction of the natural numbers

To match Peano's axioms, we must not only describe the set $\mathbb{N}$ but also explain which of its elements is the special element 0 and how to define the successor function on it. There are many possible definitions. Ours, which is now accepted as the standard one, represents natural numbers by sets whose construction and relationship with each other depend on very basic properties of sets. A major example of what we mean by a 'basic property' is the membership relation $\in$, i.e. the property of an object x being a 'member of' or 'element of' a set y. Furthermore, we shall take all objects, like the x here, to be sets themselves, to fit in with our aim of constructing natural numbers purely from sets. But we shall not seek to explain here what we mean by a 'set' or what $x \in y$ means: for our purposes in this chapter, these are notions informally understood from everyday mathematics.

As ever, we write $x \in y$ for 'x is an element of y', and $x \notin y$ for 'x is not an element of y'.

The attitude is similar to our definition of the reals in terms of the rationals, putting to one side the issue of what the rationals are.

Perhaps the simplest set we encounter in normal mathematics is the empty set $\varnothing$. We shall take $\varnothing$ to represent the natural number 0. We represent the successor function as follows.

One way of explaining $\varnothing$ is as the set y such that

$$(\text{for all } x)(x \notin y).$$

> ***Definition***
>
> Given a set x, the *successor* of x, written as x^+, is the set
>
> $$x^+ = x \cup \{x\}.$$

To support this definition we need $\{x\}$ to be a set whenever x is a set, and $A \cup B$ to be a set whenever A and B are.

So

$$\varnothing^+ = \varnothing \cup \{\varnothing\} = \{\varnothing\},$$
$$\varnothing^{++} = (\varnothing^+)^+ = \{\varnothing\} \cup \{\{\varnothing\}\} = \{\varnothing, \{\varnothing\}\},$$
$$\varnothing^{+++} = (\varnothing^{++})^+ = \{\varnothing, \{\varnothing\}\} \cup \{\{\varnothing, \{\varnothing\}\}\} = \{\varnothing, \{\varnothing\}, \{\varnothing, \{\varnothing\}\}\}.$$

$\varnothing^+, \varnothing^{++}, \varnothing^{+++}$ will represent $1, 2, 3$ respectively.

It is an immediate consequence of the definition that $x \subseteq x^+$, for all x. And it looks as though x^+ is a set with one element more than x.

We shall make frequent use of the facts that $x \in x^+$ and $x \subseteq x^+$.

Exercise 3.3

Is it true that x^+ has one element more than the set x?

Solution

As $x^+ = x \cup \{x\}$, certainly x^+ contains x as a subset, but would appear to contain also the extra element x (the single element of the $\{x\}$ in the union). But for this element to be an extra, we need that $x \notin x$. If $x \in x$ then $\{x\}$ is a subset of x, so that $x \cup \{x\} = x$.

In general, if $y \in z$, then $\{y\} \subseteq z$.

As we are aiming for the set x representing n to have, at least intuitively, n elements, we shall want to show that for this sort of set it is not the case that $x \in x$.

It isn't obvious that there is no set X which belongs to itself. Indeed, the set of all sets, if there were such a thing, would have to belong to itself.

This definition turns out to have more technical advantages than may be immediately apparent. For instance, as we shall show in this section, it will allow us to order $\mathbb{N}$ by the $\in$ relation, which is about as basic a notion in set theory as you can get. But first we use it to define sets which will turn out to include $\mathbb{N}$.

For instance, $\varnothing^+ \in \varnothing^{+++}$, corresponding to $1 < 3$.

> ***Definition***
>
> The set y is *inductive* if $\varnothing \in y$ and $x^+ \in y$ whenever $x \in y$.

Exercise 3.4

Show that if y and z are both inductive sets, then their intersection $y \cap z$ is also inductive.

Solution

We need to show that $\varnothing \in y \cap z$ and that whenever $x \in y \cap z$, then $x^+ \in y \cap z$.

As y and z are both inductive, $\varnothing$ belongs to both of them, and hence to their intersection. Likewise if $x \in y \cap z$, then both $x \in y$ and $x \in z$, so that as y and z are inductive, x^+ belongs to both of them, and hence to $y \cap z$.

The significance of inductive sets is that we are going to define $\mathbb{N}$ as the intersection of all inductive sets, so that it will be the smallest inductive set. There's only one snag with this: are there *any* inductive sets in the first place? *We shall assume that there is at least one*, y say. (This is a major assumption! From it flows the existence of infinite sets, from which most of the interest in set theory derives. It will figure as one of the axioms of set theory.) Then the intersection, $\mathbb{N}$, of all inductive sets, which is a subset of each of them, must be a subset of y in particular. And if z is any other inductive set, then $\mathbb{N}$ is also a subset of z and y, and hence of $y \cap z$, which is inductive by the result of Exercise 3.4. Thus, given our assumption that there is an inductive set y and our desire to define $\mathbb{N}$ as the intersection of all inductive sets, it is sufficient to define $\mathbb{N}$ in terms of y and its inductive subsets, as follows.

> ***Definitions***
>
> The set of natural numbers $\mathbb{N}$ is the intersection of all inductive subsets of any inductive set y, i.e.
>
> $$\begin{aligned}\mathbb{N} &= \bigcap\{z : z \text{ is an inductive subset of } y\}\\ &= \{x : x \in z \text{ for all inductive } z \subseteq y\}.\end{aligned}$$
>
> A *natural number* is a member of $\mathbb{N}$. We write $x = y$ for $x, y \in \mathbb{N}$ when x and y are equal as sets.

If A is a set (of sets), then $\bigcap A$ is the set $\{x : x \in z \text{ for all } z \in A\}$.

So $x = y$ means that every member of x is a member of y, and vice versa.

Given that we want $\mathbb{N}$ to satisfy Peano's axioms and that we are going to take the map $S\colon x \longmapsto x^+$ as the successor function on $\mathbb{N}$, we require $\mathbb{N}$ to be an inductive set. The next theorem shows this.

For S to be a function from $\mathbb{N}$ to itself, we need that $x^+ \in \mathbb{N}$ whenever $x \in \mathbb{N}$.

> ***Theorem 3.4***
>
> The set $\mathbb{N}$ is inductive.

Proof

The proof is very similar to the solution to Exercise 3.4. First, $\varnothing$ belongs to any inductive set, hence to all inductive subsets of y, and hence to their intersection $\mathbb{N}$. Now suppose that $x \in \mathbb{N}$, so that $x \in z$ for all inductive subsets z of y. As each such z is inductive, x^+ also belongs to z. Thus x^+ belongs to the intersection of these zs, i.e. $x^+ \in \mathbb{N}$. Hence $\mathbb{N}$ is inductive. ■

As $\mathbb{N}$ is inductive, we can now define the successor function on $\mathbb{N}$.

> ***Definitions***
>
> The *successor* function S on $\mathbb{N}$ is the function
>
> $$\begin{aligned} S\colon \mathbb{N} &\longrightarrow \mathbb{N}\\ x &\longmapsto x^+.\end{aligned}$$
>
> We shall write $\mathbf{0}$ for $\varnothing$ in its role as the 0 of $\mathbb{N}$, and likewise $\mathbf{1}$ for $\mathbf{0}^+$, $\mathbf{2}$ for $\mathbf{1}^+$, i.e. $\mathbf{0}^{++}$, etc. In general we shall use bold numbers and letters, like $\mathbf{n}$, for elements of $\mathbb{N}$.

As we are trying to build $\mathbb{N}$ on a basis of just sets and logic, the question arises of what is a function. We shall later define functions in terms of sets.

And we can show that $\mathbb{N}$ satisfies the inductive principle of Peano's axioms.

> ***Theorem 3.5 Proof by induction on*** $\mathbb{N}$
>
> For all subsets $A \subseteq \mathbb{N}$, if A contains $\mathbf{0}$ and contains $\mathbf{n}^+$ whenever $\mathbf{n} \in A$, then A is all of $\mathbb{N}$.

Proof

Any such A is an inductive set. $\mathbb{N}$ is the intersection of all inductive sets and is therefore a subset of any one of them. Thus $\mathbb{N} \subseteq A$. But $A \subseteq \mathbb{N}$. So $A = \mathbb{N}$. ■

Theorem 3.5 provides the basis for almost every proof about the natural numbers. Obviously we will encounter results about $\mathbb{N}$ whose proofs exploit previously proved results about $\mathbb{N}$, but somewhere in the history of these results a proof by induction will almost inevitably lurk.

There are a number of tempting pictures of what $\mathbb{N}$ looks like. One is that, because $\mathbb{N}$ is the smallest set containing $\mathbf{0}$ $(= \varnothing)$ and closed under the successor function,

$$\mathbb{N} = \{\mathbf{0}, \mathbf{0}^+, \mathbf{0}^{++}, \mathbf{0}^{+++}, \mathbf{0}^{++++}, \ldots\},$$

with every element of $\mathbb{N}$ looking like

$$\mathbf{0}^{\overbrace{++ + \cdots +}},$$

where the superscript consists of finitely many +s. Although this is a helpful picture, the use of dots '...' in these expressions, standing for 'and so on', fails to give a finite description of the set – and because statements about infinite sets turn out to require some care, we shall try to avoid, where possible, infinite descriptions of mathematical objects (even though we may use them as a guide to our intuition). A finite description or definition of something is, in principle, something which people can communicate to each other within a finite time.

Another tempting picture is that not only is every element of a natural number also a natural number, but also each natural number is the set of all its predecessors:

It is indeed true that if $x \in \mathbf{n}$, where $\mathbf{n} \in \mathbb{N}$, then x is a natural number. See Exercise 3.13.

$$\mathbf{1} = \mathbf{0}^+ = \mathbf{0} \cup \{\mathbf{0}\} = \{\mathbf{0}\} \quad (\text{as } \mathbf{0} = \varnothing),$$
$$\mathbf{2} = \mathbf{1}^+ = \mathbf{1} \cup \{\mathbf{1}\} = \{\mathbf{0}\} \cup \{\mathbf{1}\} = \{\mathbf{0}, \mathbf{1}\},$$
$$\mathbf{3} = \mathbf{2}^+ = \mathbf{2} \cup \{\mathbf{2}\} = \{\mathbf{0}, \mathbf{1}\} \cup \{\mathbf{2}\} = \{\mathbf{0}, \mathbf{1}, \mathbf{2}\},$$

and so on. Furthermore, the set $\mathbf{n}$ contains, intuitively, n elements. But given that we are trying to define the natural numbers, this latter observation involves a circularity. However, the intuition that each natural number is the set of all its predecessors is one that we can put on a firm foundation, by defining an order $<$ on $\mathbb{N}$ in such a way that this becomes true.

For a given everyday number n, we use the bold version of the same letter, $\mathbf{n}$, for the set formally representing it. Intuitively

$$\mathbf{n} = \mathbf{0}^{\overbrace{++ + \cdots +}^{n}}.$$

> ***Definitions***
>
> For all $\mathbf{m}, \mathbf{n} \in \mathbb{N}$ we write $\mathbf{m} < \mathbf{n}$ when $\mathbf{m} \in \mathbf{n}$.
>
> We write $\mathbf{m} \leq \mathbf{n}$ when $\mathbf{m} < \mathbf{n}$ or $\mathbf{m} = \mathbf{n}$.

This defines the $<_{\mathbb{N}}$ used in the previous chapter. We shan't use this notation from now on as there will be no other order relations around to confuse us!

We shall prove that $<$ is a linear order on $\mathbb{N}$. This is worth doing in its own right, but will also help us finish showing that $\mathbb{N}$ satisfies Peano's axioms. We shall state the results in terms of $\in$ rather than $<$, as their proof will rely heavily on the properties of $\in$.

Theorem 3.6

The relation $\in$ linearly orders $\mathbb{N}$, i.e. it has the following properties:

(i) *irreflexive*: for all $\mathbf{n} \in \mathbb{N}$, $\mathbf{n} \notin \mathbf{n}$;

(ii) *transitive*: for all $\mathbf{m}, \mathbf{n}, \mathbf{p} \in \mathbb{N}$, if $\mathbf{m} \in \mathbf{n}$ and $\mathbf{n} \in \mathbf{p}$, then $\mathbf{m} \in \mathbf{p}$;

(iii) *linear*: for all $\mathbf{m}, \mathbf{n} \in \mathbb{N}$, $\mathbf{m} \in \mathbf{n}$ or $\mathbf{m} = \mathbf{n}$ or $\mathbf{n} \in \mathbf{m}$.

Proof

We shall do the proof in stages, leaving some parts as exercises for you. Note that $\in$ doesn't linearly order every set of sets. For instance, as $\mathbf{0} \in \{\mathbf{0}\}$ and $\{\mathbf{0}\} \in \{\{\mathbf{0}\}\}$ but $\mathbf{0} \notin \{\{\mathbf{0}\}\}$, $\in$ is not in general transitive. So the result depends on the particular properties of $\mathbb{N}$. The picture that $\mathbb{N} = \{\mathbf{0}, \mathbf{1}, \mathbf{2}, \mathbf{3}, \ldots\}$, where $\mathbf{1} = \{\mathbf{0}\}$, $\mathbf{2} = \{\mathbf{0}, \mathbf{1}\}$, $\mathbf{3} = \{\mathbf{0}, \mathbf{1}, \mathbf{2}\}$ and so on, makes the theorem plausible, but does not constitute a general proof for all natural numbers. We must instead make heavy use of Theorem 3.5, i.e. proof by induction on $\mathbb{N}$. We shall start with the reflexive property, as we shall need this as a lemma for our proofs of the other properties.

Strictly speaking, we have shown that $\in$ is not transitive on the set $\left\{\mathbf{0}, \{\mathbf{0}\}, \{\{\mathbf{0}\}\}\right\}$.

A direct proof that if $\mathbf{m} \in \mathbf{n}$ and $\mathbf{n} \in \mathbf{p}$ then $\mathbf{m} \in \mathbf{p}$ gets nowhere. Instead we use induction on $\mathbf{p}$ for fixed $\mathbf{m}$ and $\mathbf{n}$. In terms of Theorem 3.5, we define a subset A of $\mathbb{N}$ by

$$A = \{\mathbf{p} \in \mathbb{N} : \text{ if } \mathbf{m} \in \mathbf{n} \in \mathbf{p} \text{ then } \mathbf{m} \in \mathbf{p}\},$$

and shall show that A is inductive.

First, as $\mathbf{0} = \varnothing$ and $\varnothing$ contains no elements, it cannot be true that $\mathbf{m} \in \mathbf{n} \in \mathbf{0}$, so that the statement 'if $\mathbf{m} \in \mathbf{n} \in \mathbf{0}$ then $\mathbf{m} \in \mathbf{0}$' is vacuously true. Thus $\mathbf{0} \in A$.

A statement of the form 'if P then Q' is false only when P is true and Q is false. So when P is false, as here, the statement is true.

Now suppose that $\mathbf{p} \in A$ and show that $\mathbf{p}^+ \in A$. We need to show that if $\mathbf{m} \in \mathbf{n} \in \mathbf{p}^+$ then $\mathbf{m} \in \mathbf{p}^+$ (i.e. $\mathbf{p}^+ \in A$), where we can exploit the information that if $\mathbf{m} \in \mathbf{n} \in \mathbf{p}$ then $\mathbf{m} \in \mathbf{p}$ (i.e. $\mathbf{p} \in A$). So suppose that

$$\mathbf{m} \in \mathbf{n} \in \mathbf{p}^+.$$

As $\mathbf{n} \in \mathbf{p}^+$ and $\mathbf{p}^+ = \mathbf{p} \cup \{\mathbf{p}\}$, we have two possibilities: $\mathbf{n} \in \mathbf{p}$; or $\mathbf{n} \in \{\mathbf{p}\}$, i.e. $\mathbf{n} = \mathbf{p}$. In the case that $\mathbf{n} \in \mathbf{p}$ we then have $\mathbf{m} \in \mathbf{n} \in \mathbf{p}$, so that, as $\mathbf{p} \in A$, we have $\mathbf{m} \in \mathbf{p}$. But $\mathbf{p} \subseteq \mathbf{p}^+$, so that $\mathbf{m} \in \mathbf{p}^+$, as required. In the case that $\mathbf{n} = \mathbf{p}$, then $\mathbf{m} \in \mathbf{n}$ is the same as $\mathbf{m} \in \mathbf{p}$, so that, as in the first case, $\mathbf{m} \in \mathbf{p}^+$. In both cases we have obtained $\mathbf{m} \in \mathbf{p}^+$, which is what is required to show that $\mathbf{p}^+ \in A$.

The only element of $\{\mathbf{p}\}$ is $\mathbf{p}$, so if $\mathbf{n} \in \{\mathbf{p}\}$ then $\mathbf{n} = \mathbf{p}$.

Thus A is inductive, so that by Theorem 3.5 $A = \mathbb{N}$, which shows that $\in$ is transitive on $\mathbb{N}$. ■

Exercise 3.5

Give an example to show that $\in$ is not in general linear on a set of sets.

Exercise 3.6

Show that $\in$ is irreflexive on $\mathbb{N}$. [Hints: put $A = \{\mathbf{n} \in \mathbb{N} : \mathbf{n} \notin \mathbf{n}\}$ and use induction. You should find that the transitive property of $\in$ on $\mathbb{N}$ is needed at some stage of the argument.]

Solution

Define the subset A of $\mathbb{N}$ by

$$A = \{\mathbf{n} \in \mathbb{N} : \mathbf{n} \notin \mathbf{n}\}.$$

First, as $\mathbf{0} = \varnothing$ and $\varnothing$ contains no elements, then we have, in particular, that $\mathbf{0}$ is not an element of $\mathbf{0}$, i.e. $\mathbf{0} \notin \mathbf{0}$. Thus $\mathbf{0} \in A$.

Now suppose that $\mathbf{n} \in A$ and show that $\mathbf{n}^+ \in A$, i.e. $\mathbf{n}^+ \notin \mathbf{n}^+$. We shall suppose that $\mathbf{n}^+ \in \mathbf{n}^+$ and try to derive a contradiction. If $\mathbf{n}^+ \in \mathbf{n}^+$, then

$$\mathbf{n}^+ \in \mathbf{n} \cup \{\mathbf{n}\}.$$

This gives two possibilities, that $\mathbf{n}^+ \in \mathbf{n}$ or $\mathbf{n}^+ = \mathbf{n}$.

In the first case, that $\mathbf{n}^+ \in \mathbf{n}$, we have, as $\mathbf{n} \in \mathbf{n} \cup \{\mathbf{n}\} = \mathbf{n}^+$,

$$\mathbf{n} \in \mathbf{n}^+ \in \mathbf{n}.$$

But $\in$ is transitive on $\mathbb{N}$, so that $\mathbf{n} \in \mathbf{n}$, contradicting that $\mathbf{n} \in A$.

In the other case, that $\mathbf{n}^+ = \mathbf{n}$, we have

$$\mathbf{n} \in \mathbf{n} \cup \{\mathbf{n}\} = \mathbf{n}^+ = \mathbf{n},$$

i.e. $\mathbf{n} \in \mathbf{n}$, again contradicting that $\mathbf{n} \in A$.

In both cases we have a contradiction, so that in fact $\mathbf{n}^+ \notin \mathbf{n}^+$, showing that $\mathbf{n}^+ \in A$. Thus A is inductive and by Theorem 3.5 $A = \mathbb{N}$. Hence $\in$ is irreflexive on $\mathbb{N}$.

So for natural numbers $\mathbf{n}$, the set $\mathbf{n}^+$ does contain one more element than $\mathbf{n}$.

Before we tackle the linear property of $\in$ on $\mathbb{N}$, we need some results about the successor function, whose proofs we shall leave as exercises for you, to give you practice in using induction with sets.

> ***Theorem 3.7***
>
> For all $\mathbf{m}, \mathbf{n} \in \mathbb{N}$,
>
> (i) $\mathbf{0} \neq \mathbf{n}^+$;
>
> (ii) if $\mathbf{m} \in \mathbf{n}$ then $\mathbf{m}^+ \in \mathbf{n}^+$;
>
> (iii) if $\mathbf{m}^+ = \mathbf{n}^+$ then $\mathbf{m} = \mathbf{n}$.

Theorem 3.7(i) and (iii), along with Theorem 3.5, show that $\mathbb{N}$ satisfies Peano's axioms.

Exercise 3.7

Prove Theorem 3.7. [Hints: (i) is very straightforward! Then prove (ii), using induction. It should then be possible to prove (iii) without needing to use induction. Our solution to (iii) uses the result of (ii) and the fact that $\in$ is irreflexive on $\mathbb{N}$, i.e. $\mathbf{n} \notin \mathbf{n}$, for all $\mathbf{n} \in \mathbb{N}$.]

Solution

(i) Note that if $\mathbf{0} = \mathbf{n}^+$ for some $\mathbf{n}$, then

$$\mathbf{n} \in \mathbf{n} \cup \{\mathbf{n}\} = \mathbf{n}^+ = \mathbf{0} = \varnothing,$$

i.e.

$$\mathbf{n} \in \varnothing.$$

This of course contradicts that $\varnothing$ has no members. Thus for all $\mathbf{n} \in \mathbb{N}$, $\mathbf{0} \neq \mathbf{n}^+$.

(ii) We shall prove this by induction on $\mathbf{n}$ for fixed $\mathbf{m}$, and we shall present the induction proof in a more familiar mathematical style. Rather than show that some subset A of $\mathbb{N}$ is inductive, we shall work with a property $P(\mathbf{n})$ of natural numbers $\mathbf{n}$, and show that

$P(\mathbf{0})$ holds and $P(\mathbf{n}^+)$ holds whenever $P(\mathbf{n})$ does,

For the connection between A and P, see Exercise 3.1 in the previous section.

from which it follows that $P(\mathbf{n})$ holds for every natural number $\mathbf{n}$. Here our A is the subset

$$A = \{\mathbf{n} \in \mathbb{N} : \text{ if } \mathbf{m} \in \mathbf{n} \text{ then } \mathbf{m}^+ \in \mathbf{n}^+\},$$

and the corresponding property $P(\mathbf{n})$ is

if $\mathbf{m} \in \mathbf{n}$ then $\mathbf{m}^+ \in \mathbf{n}^+$,

where $\mathbf{m}$ is a fixed natural number.

For $\mathbf{n} = \mathbf{0}$, $\mathbf{m} \in \mathbf{0}$ is false (as $\mathbf{0} = \varnothing$), so that

if $\mathbf{m} \in \mathbf{0}$ then $\mathbf{m}^+ \in \mathbf{0}^+$

This is the basis of the induction, equivalent to showing $\mathbf{0} \in A$.

is vacuously true (i.e. $P(\mathbf{0})$ holds).

Suppose that $P(\mathbf{n})$ holds, i.e.

if $\mathbf{m} \in \mathbf{n}$ then $\mathbf{m}^+ \in \mathbf{n}^+$,

This is the inductive step, equivalent to showing that $\mathbf{n}^+ \in A$ whenever $\mathbf{n} \in A$.

and that $\mathbf{m} \in \mathbf{n}^+$. We shall show that $\mathbf{m}^+ \in \mathbf{n}^{++}$, which then establishes that $P(\mathbf{n}^+)$ holds. As $\mathbf{m} \in \mathbf{n}^+$ and $\mathbf{n}^+ = \mathbf{n} \cup \{\mathbf{n}\}$, we have

$$\mathbf{m} \in \mathbf{n} \cup \{\mathbf{n}\},$$

so that either $\mathbf{m} \in \mathbf{n}$ or $\mathbf{m} = \mathbf{n}$.

In the case that $\mathbf{m} \in \mathbf{n}$, the inductive hypothesis gives that $\mathbf{m}^+ \in \mathbf{n}^+$. As $\mathbf{n}^+ \subseteq \mathbf{n}^{++}$, this gives $\mathbf{m}^+ \in \mathbf{n}^{++}$. In the case that $\mathbf{m} = \mathbf{n}$, then $\mathbf{m}^+ = \mathbf{n}^+ \in \mathbf{n}^{++}$. So in either case $\mathbf{m}^+ \in \mathbf{n}^{++}$, as required to show that $P(\mathbf{n}^+)$ holds. The result follows by induction (i.e. Theorem 3.5, but we'll stop mentioning this now!)

We've shown the equivalent of $A = \mathbb{N}$, i.e. $P(\mathbf{n})$ holds for all $\mathbf{n} \in \mathbb{N}$.

(iii) This can be done directly, without induction. Suppose that $\mathbf{m}^+ = \mathbf{n}^+$. This can be rewritten as

$$\mathbf{m} \cup \{\mathbf{m}\} = \mathbf{n} \cup \{\mathbf{n}\},$$

so that $\mathbf{m}$, which is an element of $\mathbf{m} \cup \{\mathbf{m}\}$, is an element of $\mathbf{n} \cup \{\mathbf{n}\}$. There are two possibilities: $\mathbf{m} \in \mathbf{n}$ and $\mathbf{m} \in \{\mathbf{n}\}$, i.e. $\mathbf{m} = \mathbf{n}$.

In the case that $\mathbf{m} \in \mathbf{n}$, part (ii) of the theorem gives that $\mathbf{m}^+ \in \mathbf{n}^+$. But $\mathbf{m}^+ = \mathbf{n}^+$, so that this means that $\mathbf{m}^+ \in \mathbf{m}^+$, contradicting $\in$ being irreflexive on $\mathbb{N}$. This leaves only the case that $\mathbf{m} = \mathbf{n}$, the desired result.

As remarked above, parts (i) and (iii) of Theorem 3.7, along with Theorem 3.5, show that $\mathbb{N}$ satisfies Peano's axioms. As this was our main objective in this chapter, it's worth recording as a theorem.

> ***Theorem 3.8***
>
> $\mathbb{N}$, with special element $\mathbf{0}$ and the successor function $S\colon \mathbf{n} \longmapsto \mathbf{n}^+$, satisfies Peano's axioms.

This means that any consequence of Peano's axioms holds for $\mathbb{N}$, like Theorem 3.1 of the previous section.

Exercise 3.8

Suppose that $\mathbf{n} \in \mathbb{N}$ with $\mathbf{n} \neq \mathbf{0}$. Show that $\mathbf{0} \in \mathbf{n}$. [Hint: use Theorem 3.1 of the previous section and induction.]

You should now have enough machinery to conclude the proof of Theorem 3.6 by proving that $\in$ is linear on $\mathbb{N}$.

Exercise 3.9

(a) Show that for all $\mathbf{m}, \mathbf{n} \in \mathbb{N}$, $\mathbf{m} \in \mathbf{n}$ or $\mathbf{m} = \mathbf{n}$ or $\mathbf{n} \in \mathbf{m}$. [Hint: use induction on $\mathbf{n}$ for fixed $\mathbf{m}$.]

(b) Show further that for all $\mathbf{m}, \mathbf{n} \in \mathbb{N}$, *exactly one* of the above holds.

Solution

(a) We shall use induction on $\mathbf{n}$ for a fixed $\mathbf{m}$ to show that $\mathbf{m} \in \mathbf{n}$ or $\mathbf{m} = \mathbf{n}$ or $\mathbf{n} \in \mathbf{m}$ for all $\mathbf{n} \in \mathbb{N}$.

For $\mathbf{n} = \mathbf{0}$ we cannot have $\mathbf{m} \in \mathbf{0}$ (as $\mathbf{0} = \varnothing$). If $\mathbf{m}$ happens to be $\mathbf{0}$ we are done, as then $\mathbf{m} = \mathbf{0} = \mathbf{n}$. And if $\mathbf{m} \neq \mathbf{0}$ then, by Exercise 3.8, $\mathbf{0} \in \mathbf{m}$. Thus the result holds for $\mathbf{n} = \mathbf{0}$.

Now suppose that the result holds for $\mathbf{n}$, so that $\mathbf{m} \in \mathbf{n}$ or $\mathbf{m} = \mathbf{n}$ or $\mathbf{n} \in \mathbf{m}$, and show that the result holds for $\mathbf{n}^+$, i.e. $\mathbf{m} \in \mathbf{n}^+$ or $\mathbf{m} = \mathbf{n}^+$ or $\mathbf{n}^+ \in \mathbf{m}$.

In the case that $\mathbf{m} \in \mathbf{n}$, then as $\mathbf{n} \subseteq \mathbf{n}^+$ we have $\mathbf{m} \in \mathbf{n}^+$. And in the case that $\mathbf{m} = \mathbf{n}$, then as $\mathbf{n} \in \mathbf{n}^+$ we have $\mathbf{m} \in \mathbf{n}^+$.

What about the case that $\mathbf{n} \in \mathbf{m}$? This means that $\mathbf{m}$ is non-empty, so that $\mathbf{m} \neq \mathbf{0}$. Then by Theorem 3.1 in the previous section (which applies as $\mathbb{N}$ satisfies Peano's axioms), $\mathbf{m} = \mathbf{k}^+$ for some $\mathbf{k} \in \mathbb{N}$. This gives $\mathbf{n} \in \mathbf{k} \cup \{\mathbf{k}\}$, so that either $\mathbf{n} \in \mathbf{k}$ or $\mathbf{n} = \mathbf{k}$. If $\mathbf{n} \in \mathbf{k}$ then, by Theorem 3.7(ii), $\mathbf{n}^+ \in \mathbf{k}^+$, i.e. $\mathbf{n}^+ \in \mathbf{m}$; and if $\mathbf{n} = \mathbf{k}$ then $\mathbf{n}^+ = \mathbf{k}^+$, i.e. $\mathbf{n}^+ = \mathbf{m}$.

In all cases we have $\mathbf{m} \in \mathbf{n}^+$ or $\mathbf{m} = \mathbf{n}^+$ or $\mathbf{n}^+ \in \mathbf{m}$, as required.

(b) This one ought to be straightforward and is left to you.

The order on $\mathbb{N}$ has one extra property which is of great importance, in everyday mathematics and within set theory, and that is being well-ordered.

Definition

A linearly ordered set X with (weak) order $\leq$ is *well-ordered* if every non-empty subset of X has a least element, i.e. for all non-empty $B \subseteq X$,

$$\text{there is an element } b_0 \in B \text{ such that } b_0 \leq b, \text{ for all } b \in B.$$

Theorem 3.9

$\mathbb{N}$ is well-ordered by $\in$.

Proof

Let B be a non-empty subset of $\mathbb{N}$, which we have shown to be linearly ordered by $<$, where $<$ is $\in$. We shall show that B has a least element by assuming that it doesn't have one and deriving a contradiction. So assume that B doesn't have a least element. Define a subset A of $\mathbb{N}$ by

$\in$ is a strict order $<$ on $\mathbb{N}$. The corresponding weak order $\leq$ is defined by $\mathbf{m} \leq \mathbf{n}$ if $\mathbf{m} \in \mathbf{n}$ or $\mathbf{m} = \mathbf{n}$.

$$A = \{\mathbf{n} \in \mathbb{N} : \mathbf{m} \notin B \text{ for all } \mathbf{m} \leq \mathbf{n}\}.$$

We shall show by induction that $A = \mathbb{N}$, so that B must be empty, contradicting that B is non-empty.

First, $\mathbf{0}$ cannot be in B, as otherwise $\mathbf{0}$ would automatically be the least element of B. Thus $\mathbf{0} \in A$.

As $\mathbf{0} = \varnothing$, there can be no $\mathbf{m}$ with $\mathbf{m} \in \mathbf{0}$ (or equivalently $\mathbf{m} < \mathbf{0}$). Thus $\mathbf{0}$ is the least element of $\mathbb{N}$.

Now suppose that $\mathbf{n} \in A$ and show that $\mathbf{n}^+ \in A$. As $\mathbf{n} \in A$ we have $\mathbf{m} \notin B$ for all $\mathbf{m} \leq \mathbf{n}$. For $\mathbf{n}^+$ to be in A, all we need to show is that $\mathbf{n}^+ \notin B$, because if $\mathbf{m} \leq \mathbf{n}^+$ then either $\mathbf{m} \leq \mathbf{n}$ (and we already know that $\mathbf{m} \notin B$ for such $\mathbf{m}$) or $\mathbf{m} = \mathbf{n}^+$. If it were the case that $\mathbf{n}^+ \in B$, then, as $\mathbf{m} \notin B$ for all $\mathbf{m} < \mathbf{n}^+$, $\mathbf{n}^+$ would be the least element of B: this would contradict that B has no least element, so we conclude that $\mathbf{n}^+ \notin B$. Hence $\mathbf{n}^+ \in A$, as required.

This needs a bit of justification and is left as an exercise.

Then, by induction, $A = \mathbb{N}$. This means that B is empty, contradicting that B is non-empty. Thus B must indeed have a least element. ■

Exercise 3.10

Fill in the missing details of the proof above by showing that for all $\mathbf{m}, \mathbf{n} \in \mathbb{N}$, $\mathbf{m} < \mathbf{n}^+$ if and only if $\mathbf{m} \leq \mathbf{n}$.

A consequence of this exercise and of Theorem 3.9 is that the least number greater than $\mathbf{n}$ is its successor $\mathbf{n}^+$. This is worth stating as a theorem.

Theorem 3.10

For all $\mathbf{n} \in \mathbb{N}$, $\mathbf{n}^+ = \min\{\mathbf{k} \in \mathbb{N} : \mathbf{n} < \mathbf{k}\}$.

We write $\min B$ for the least element of a non-empty subset of the well-ordered set $\mathbb{N}$.

Proof

First of all, note that the set $\{\mathbf{k} \in \mathbb{N} : \mathbf{n} < \mathbf{k}\}$ is non-empty, as it contains $\mathbf{n}^+$. So as $\mathbb{N}$ is well-ordered by $<$ this set does indeed have a least element. To show that this element is $\mathbf{n}^+$, note that by Exercise 3.10, if $\mathbf{m} < \mathbf{n}^+$ then $\mathbf{m} \leq \mathbf{n}$. This means that there can be no $\mathbf{m}$ with $\mathbf{n} < \mathbf{m} < \mathbf{n}^+$. Thus $\mathbf{n}^+$ is the least element of $\{\mathbf{k} \in \mathbb{N} : \mathbf{n} < \mathbf{k}\}$, as required. ■

In a later chapter we shall extend many of the features of $\mathbb{N}$ to give a theory of infinite numbers; and well-ordered sets will be the basis of this theory.

These numbers are called ordinals, and the natural numbers turn out to be the finite ordinals.

In the next section we shall look at how to do arithmetic in $\mathbb{N}$.

Further exercises

Exercise 3.11

Show that for all $\mathbf{m}, \mathbf{n} \in \mathbb{N}$, $\mathbf{m} < \mathbf{n}$ if and only if $\mathbf{m}$ is a proper subset of $\mathbf{n}$.

Exercise 3.12

Show that for all $\mathbf{m}, \mathbf{n} \in \mathbb{N}$, the least of $\mathbf{m}$ and $\mathbf{n}$ in the linear order on $\mathbb{N}$, $\min\{\mathbf{m}, \mathbf{n}\}$, equals $\mathbf{m} \cap \mathbf{n}$.

Exercise 3.13

Show that for all $\mathbf{n} \in \mathbb{N}$, if $x \in \mathbf{n}$ then $x \in \mathbb{N}$.

So that each natural number is a set of natural numbers.

Exercise 3.14

Show that for all $\mathbf{m}, \mathbf{n} \in \mathbb{N}$, if $\mathbf{m}^+ \in \mathbf{n}^+$ then $\mathbf{m} \in \mathbf{n}$.

Exercise 3.15

Show that for all $\mathbf{n} \in \mathbb{N}$, $\mathbf{n}^{++} \neq \mathbf{n}$.

Exercise 3.16

Show that for all $\mathbf{m}, \mathbf{n} \in \mathbb{N}$, if $\mathbf{m} < \mathbf{n}$ then $\mathbf{m}^+ \leq \mathbf{n}$.

Exercise 3.17

Let X be a set well-ordered (so also linearly ordered) by $\leq$ (a weak order). Suppose that X has a least element 0 and has the further property that for each $x \in X$ the set

So that $0 \leq x$, for all $x \in X$.

$$\{y \in X : x < y\}$$

$x < y$ meaning $x \leq y$ and $x \neq y$.

is non-empty. For each x let x^+ be the least element of this set.

Show that X is inductive.

3.3 Arithmetic

In this section we shall define operations of addition, multiplication and exponentiation on the set of natural numbers, and show that these operations have the properties that we expect from everyday mathematics. Although we shall frame the definitions in terms of our special set $\mathbb{N}$, the definitions could have been done for any set X satisfying Peano's axioms. Our major tool for defining the operations will be definition by recursion (Theorem 3.2); and our major tool for establishing their properties will be induction.

For instance, we expect addition to be commutative, namely $x + y = y + x$, for all x, y.

We shall use the following special case of definition by recursion obtained by taking $X = Y = \mathbb{N}$ in Theorem 3.2.

As $\mathbb{N}$ satisfies Peano's axioms, we can indeed apply Theorem 3.2 to it.

Theorem 3.11 Recursion on $\mathbb{N}$

Let $\mathbf{y_0}$ be any element of $\mathbb{N}$ and $h\colon \mathbb{N} \times \mathbb{N} \longrightarrow \mathbb{N}$ a function on pairs $(x, y) \in \mathbb{N} \times \mathbb{N}$. Then there exists a unique function $f\colon \mathbb{N} \longrightarrow \mathbb{N}$ such that

$$f(\mathbf{0}) = \mathbf{y_0},$$
$$f(\mathbf{n}^+) = h(\mathbf{n}, f(\mathbf{n})), \text{ for all } \mathbf{n} \in \mathbb{N}.$$

First of all, we shall define addition. The trick is to use recursion to define $\mathbf{m} + \mathbf{n}$ for a *fixed* $\mathbf{m}$ and *all* $\mathbf{n}$. The f in Theorem 3.11 will be defined so that $f(\mathbf{n})$ is to be regarded as $\mathbf{m} + \mathbf{n}$. To emphasize this, we shall refer to this f as $f_{\mathbf{m}}$. We choose $\mathbf{y_0}$ to give us the desired value of $f_{\mathbf{m}}(\mathbf{0})$, which is $\mathbf{m}$ itself: so put $\mathbf{y_0}$ equal to $\mathbf{m}$. How about the choice of h? It is h that has to get us the value of $f_{\mathbf{m}}(\mathbf{n}^+)$ from the values of $\mathbf{n}$ and $f_{\mathbf{m}}(\mathbf{n})$, or, equivalently, of $\mathbf{m} + \mathbf{n}^+$ from $\mathbf{n}$ and $\mathbf{m} + \mathbf{n}$.

Of course we want $\mathbf{m} + \mathbf{0}$ to equal $\mathbf{m}$.

The definition could also use the fixed value $\mathbf{m}$.

Exercise 3.18

Define $h(x, y)$ so that $f_{\mathbf{m}}(\mathbf{n}^+) = h(\mathbf{m}, f_{\mathbf{m}}(\mathbf{n}))$.

Solution

Our intention is that for any natural number $\mathbf{k}$, its successor $\mathbf{k}^+$ ought to be the same as $\mathbf{k} + \mathbf{1}$. So our definition of $+$ should give that $\mathbf{m} + \mathbf{n}^+$ is the same as $\mathbf{m} + (\mathbf{n} + \mathbf{1})$, which in turn ought to equal $(\mathbf{m} + \mathbf{n}) + \mathbf{1}$, which is the same as $(\mathbf{m} + \mathbf{n})^+$. We exploit this desired feature of $+$ to define $\mathbf{m} + \mathbf{n}^+$ to be $(\mathbf{m} + \mathbf{n})^+$, i.e. define $f_{\mathbf{m}}(\mathbf{n}^+)$ as $(f_{\mathbf{m}}(\mathbf{n}))^+$.

Recall that we use $\mathbf{1}$ as a shorthand for $\mathbf{0}^+$.

Thus we define h by

$$h(x, y) = y^+.$$

An important aspect of our definition of h above is that it exploits only a function we assume that we already know about, namely the successor function $\mathbf{n} \longmapsto \mathbf{n}^+$ on $\mathbb{N}$. This is a vital part of defining a function by recursion, that it is being defined in terms of known functions, although the importance of this might not be obvious from the wording of Theorem 3.11. Bear in mind that we are seeking to represent natural numbers by sets and that Theorem 3.11 will need proof within whatever framework we establish for sets. It will

You might also notice that in this case the value of $h(x, y)$ doesn't depend on x.

transpire that functions are themselves sets and that, along with other sets, one has to justify their status as sets, given that there are dangerous non-sets which look suspiciously like sets – all will be explained in the next chapter! For the moment, note that by defining a function by recursion in terms of other functions already known to be well-behaved sets, we are guaranteed that the new function is also a well-behaved set.

To summarize, Theorem 3.11 will guarantee that there is a function $f_{\mathbf{m}} : \mathbb{N} \longrightarrow \mathbb{N}$ defined by

$$f_{\mathbf{m}}(\mathbf{0}) = \mathbf{m},$$
$$f_{\mathbf{m}}(\mathbf{n}^+) = (f_{\mathbf{m}}(\mathbf{n}))^+.$$

Writing $f_{\mathbf{m}}(\mathbf{n})$ as $\mathbf{m} + \mathbf{n}$, this becomes

$$\mathbf{m} + \mathbf{0} = \mathbf{m},$$
$$\mathbf{m} + \mathbf{n}^+ = (\mathbf{m} + \mathbf{n})^+.$$

We shall normally use this latter form, mainly because it is customary to represent addition by +. But *be very careful* when working through the rest of the section to remember that our definition of $\mathbf{m} + \mathbf{n}$ is for a fixed $\mathbf{m}$ and variable $\mathbf{n}$.

The definition of + by recursion appears in the work of both Dedekind and Peano. Dedekind appreciated the need to establish results like Theorem 3.2 to validate the construction.

This gives a definition for the $+_{\mathbb{N}}$ used in the previous chapter.

Exercise 3.19

Use this definition to compute the following:

(a) $\mathbf{3} + \mathbf{2}$;

(b) $\mathbf{2} + \mathbf{3}$.

Recall that $\mathbf{2} = \mathbf{1}^+ = \mathbf{0}^{++}$ and $\mathbf{3} = \mathbf{2}^+ =$ etc.

Solution

(a) In terms of the $f_{\mathbf{m}}$ notation above, $\mathbf{3} + \mathbf{2}$ is the same as $f_{\mathbf{3}}(\mathbf{2})$. To be able to evaluate this using the definition of $f_{\mathbf{3}}$, we need to express the $\mathbf{2}$ as $\mathbf{1}^+$, giving

$$f_{\mathbf{3}}(\mathbf{2}) = f_{\mathbf{3}}(\mathbf{1}^+)$$
$$= (f_{\mathbf{3}}(\mathbf{1}))^+.$$

This in turn requires us to compute $f_{\mathbf{3}}(\mathbf{1})$, which, as $\mathbf{1} = \mathbf{0}^+$, the definition gives as

$$f_{\mathbf{3}}(\mathbf{1}) = f_{\mathbf{3}}(\mathbf{0}^+)$$
$$= (f_{\mathbf{3}}(\mathbf{0}))^+.$$

The definition specifically gives the value of $f_{\mathbf{3}}(\mathbf{0})$ to be $\mathbf{3}$. We can now amalgamate these calculations and do so using the + notation:

$$\mathbf{3} + \mathbf{2} = \mathbf{3} + \mathbf{1}^+$$
$$= (\mathbf{3} + \mathbf{1})^+$$
$$= (\mathbf{3} + \mathbf{0}^+)^+$$
$$= ((\mathbf{3} + \mathbf{0})^+)^+$$
$$= ((\mathbf{3})^+)^+$$
$$= \mathbf{3}^{++}$$
$$= \mathbf{5}.$$

(b)
$$\begin{aligned} 2+3 &= 2+2^+ \\ &= (2+2)^+ \\ &= (2+1^+)^+ \\ &= ((2+1)^+)^+ \\ &= ((2+0^+)^+)^+ \\ &= (((2+0)^+)^+)^+ \\ &= (((2)^+)^+)^+ \\ &= 2^{+++} \\ &= 5 \end{aligned}$$

Exercise 3.20

Show that for all $n \in \mathbb{N}$, $n+1 = n^+$.

Solution

$$\begin{aligned} n+1 &= n+0^+ \\ &= (n+0)^+ \\ &= n^+. \end{aligned}$$

The definition of multiplication is handled similarly, by using recursion as in Theorem 3.11 and defining $m \cdot n$ for fixed m and all natural numbers n. We could again write f_m instead of the f in Theorem 3.11, to emphasize that $f(n)$ will really mean $m \cdot n$. We first define $f_m(0)$ by

This defines the $m \cdot_{\mathbb{N}} n$ used in the previous chapter.

$$f_m(0) = 0.$$

How might we define $f_m(n^+)$ in terms of n and the previous value $f_m(n)$ (and the fixed value m itself)? What we would expect from normal arithmetic is that $m \cdot n^+$ is the same as $m \cdot (n+1)$ which ought to equal $(m \cdot n) + m$. This could be used to define $f_m(n^+)$, which is $m \cdot n^+$, in terms of $f_m(n)$ and m and $+$. One of the points of recursion is to define a function in terms of previously constructed functions, and we have just constructed $+$. So we can legitimately define $f_m(n^+)$ by

$$f_m(n^+) = f_m(n) + m.$$

In terms of the more usual notation for multiplication, this becomes

$$\begin{aligned} m \cdot 0 &= 0, \\ m \cdot n^+ &= (m \cdot n) + m. \end{aligned}$$

Exercise 3.21

Use a similar method to define exponentiation, i.e. m^n. (As $+$ and $\cdot$ have already been defined, these functions can be exploited in the definition.)

Solution

We can use Theorem 3.11 to define m^n for fixed m and all n, by the following:

We could in fact have defined $m+n$ and $m \cdot n$ for fixed n and all m. But we could not do this for m^n. Why not?

$$\begin{aligned} m^0 &= 1, \\ m^{n^+} &= m^n \cdot m. \end{aligned}$$

Let us summarize the definitions of the arithmetic operations on $\mathbb{N}$, as follows.

Definitions

Addition, multiplication and exponentiation on $\mathbb{N}$ are defined, for fixed $\mathbf{m}$ and all $\mathbf{n}$, by:

$$\begin{aligned} \mathbf{m}+\mathbf{0} &= \mathbf{m}, \\ \mathbf{m}+\mathbf{n}^+ &= (\mathbf{m}+\mathbf{n})^+; \end{aligned}$$

$$\begin{aligned} \mathbf{m}\cdot\mathbf{0} &= \mathbf{0}, \\ \mathbf{m}\cdot\mathbf{n}^+ &= (\mathbf{m}\cdot\mathbf{n})+\mathbf{m}; \end{aligned}$$

$$\begin{aligned} \mathbf{m}^{\mathbf{0}} &= \mathbf{1}, \\ \mathbf{m}^{\mathbf{n}^+} &= \mathbf{m}^{\mathbf{n}}\cdot\mathbf{m}. \end{aligned}$$

Exercise 3.22

Use the definitions above to compute the following.

(a) $\mathbf{0}\cdot\mathbf{3}$

(b) $\mathbf{2^1}$

Solution

(a) Multiplication is the more complicated operation, as it is defined in terms of the more basic operation of addition. When we make a use of the definition of multiplication that results in an expression involving $+$, it makes sense to then simplify the latter as much as possible using the definition of addition, before going back to the more complicated multiplication operation.

$$\begin{aligned} \mathbf{0}\cdot\mathbf{3} &= \mathbf{0}\cdot\mathbf{2}^+ \\ &= (\mathbf{0}\cdot\mathbf{2})+\mathbf{0} \quad (\text{definition of } \mathbf{m}\cdot\mathbf{n}^+) \\ &= \mathbf{0}\cdot\mathbf{2} \quad (\text{definition of } \mathbf{m}+\mathbf{0}) \\ &= \mathbf{0}\cdot\mathbf{1}^+ \\ &= (\mathbf{0}\cdot\mathbf{1})+\mathbf{0} \quad (\text{definition of } \mathbf{m}\cdot\mathbf{n}^+) \\ &= \mathbf{0}\cdot\mathbf{1} \quad (\text{definition of } \mathbf{m}+\mathbf{0}) \\ &= \mathbf{0}\cdot\mathbf{0}^+ \\ &= (\mathbf{0}\cdot\mathbf{0})+\mathbf{0} \quad (\text{definition of } \mathbf{m}\cdot\mathbf{n}^+) \\ &= \mathbf{0}\cdot\mathbf{0} \quad (\text{definition of } \mathbf{m}+\mathbf{0}) \\ &= \mathbf{0} \quad (\text{definition of } \mathbf{m}\cdot\mathbf{0}) \end{aligned}$$

(b) This time the most complicated operation is exponentiation, and we use its definition to reduce the computation to the use of the definitions of the more basic operations of multiplication and then of addition.

$$
\begin{aligned}
2^1 &= 2^{0^+} \\
&= (2^0)\cdot 2 \quad \text{(definition of } m^{n^+}) \\
&= 1\cdot 2 \quad \text{(definition of } m^0) \\
&= 1\cdot 1^+ \\
&= (1\cdot 1)+1 \quad \text{(definition of } m\cdot n^+) \\
&= (1\cdot 0^+)+1 \\
&= ((1\cdot 0)+1)+1 \quad \text{(definition of } m\cdot n^+) \\
&= (0+1)+1 \quad \text{(definition of } m\cdot 0) \\
&= (0+0^+)+1 \\
&= (0+0)^+ + 1 \quad \text{(definition of } m+n^+) \\
&= 0^+ + 1 \quad \text{(definition of } m+0) \\
&= 0^+ + 0^+ \\
&= (0^+ + 0)^+ \quad \text{(definition of } m+n^+) \\
&= 0^{++} \quad \text{(definition of } m+0) \\
&= 2
\end{aligned}
$$

Exercise 3.23

Use the definitions to prove the following identities.

(a) $3 = 0 + (2+1) = (0+2)+1$

(b) $2\cdot 3 = 6 = 3\cdot 2$

(c) $3^{1+1} = 9 = 3^1\cdot 3^1$

(d) $2^{1\cdot 2} = 4$

(e) $2^{2^2} = 16$

We hope that you found these exercises restful, albeit a bit tedious! Of course the results obtained were entirely to be expected, assuming that we have indeed correctly defined the arithmetic operations. We expect, for instance, that addition is associative, i.e.

$$m + (n+p) = (m+n)+p, \text{ for all } m,n,p \in \mathbb{N},$$

and that

$$m^{n+p} = m^n\cdot m^p, \text{ for all } m,n,p \in \mathbb{N}.$$

We can verify these properties for small, specific, values of m, n and p, but this process does not prove that they hold in general. For a general proof, there is the one major proof technique, namely induction. This should not be surprising, as induction is a key part of the definition of the natural numbers!

We shall list a whole host of standard properties which we desire $\mathbb{N}$ to have, to match our everyday notions. We shall work through the proofs of some of them and leave the rest for you to do as exercises. Most of the proofs require induction. We give some of these properties below in the form of a theorem.

Theorem 3.12 Properties of arithmetic on $\mathbb{N}$

The following hold for all $\mathbf{m}, \mathbf{n}, \mathbf{p} \in \mathbb{N}$ (as appropriate).

1. $\mathbf{m} + (\mathbf{n} + \mathbf{p}) = (\mathbf{m} + \mathbf{n}) + \mathbf{p}$ (associativity of addition);
2. $\mathbf{m} + \mathbf{n} = \mathbf{n} + \mathbf{m}$ (commutativity of addition);
3. $\mathbf{m} \cdot (\mathbf{n} + \mathbf{p}) = (\mathbf{m} \cdot \mathbf{n}) + (\mathbf{m} \cdot \mathbf{p})$ (distributivity of multiplication over addition);
4. $\mathbf{m} \cdot (\mathbf{n} \cdot \mathbf{p}) = (\mathbf{m} \cdot \mathbf{n}) \cdot \mathbf{p}$ (associativity of multiplication);
5. $\mathbf{m} \cdot \mathbf{n} = \mathbf{n} \cdot \mathbf{m}$ (commutativity of multiplication);
6. $\mathbf{m}^{\mathbf{n}+\mathbf{p}} = \mathbf{m}^{\mathbf{n}} \cdot \mathbf{m}^{\mathbf{p}}$;
7. $(\mathbf{m}^{\mathbf{n}})^{\mathbf{p}} = \mathbf{m}^{\mathbf{n} \cdot \mathbf{p}}$;
8. $(\mathbf{m} \cdot \mathbf{n})^{\mathbf{p}} = \mathbf{m}^{\mathbf{p}} \cdot \mathbf{n}^{\mathbf{p}}$ (distributivity of exponentiation over multiplication).

These properties will hold for any set satisfying Peano's axioms.

Exercise 3.24

Among these properties are the commutativity of multiplication,

$$\mathbf{m} \cdot \mathbf{n} = \mathbf{n} \cdot \mathbf{m}, \quad \text{for all } \mathbf{m}, \mathbf{n} \in \mathbb{N},$$

and the associativity of addition. Will it make any difference which we prove first?

Solution

As the definition of multiplication is in terms of addition, it might be advisable to establish properties of addition first, before tackling those involving multiplication. Likewise, we should probably establish properties of multiplication before those of exponentiation, given that the definition of exponentiation exploits multiplication.

There are some curious constraints besides these on the order in which one proves the properties in Theorem 3.12. For instance, one needs the distributivity of $\cdot$ over $+$ to prove the commutativity of $\cdot$.

First let us prove the associativity of addition, $\mathbf{m} + (\mathbf{n} + \mathbf{p}) = (\mathbf{m} + \mathbf{n}) + \mathbf{p}$ for all $\mathbf{m}, \mathbf{n}, \mathbf{p}$. We shall use induction on one of $\mathbf{m}, \mathbf{n}, \mathbf{p}$, leaving the other two fixed.

Exercise 3.25

Which of $\mathbf{m}$, $\mathbf{n}$ and $\mathbf{p}$ shall be the induction variable?

Solution

For some of these proofs it doesn't matter which variable one chooses for the induction. But it does matter here! Suppose that we chose $\mathbf{n}$ as the induction variable. Then for the inductive step we would assume that the result, $\mathbf{m} + (\mathbf{n} + \mathbf{p}) = (\mathbf{m} + \mathbf{n}) + \mathbf{p}$, holds for $\mathbf{n}$, and would try to show the result for $\mathbf{n}^+$, namely $\mathbf{m} + (\mathbf{n}^+ + \mathbf{p}) = (\mathbf{m} + \mathbf{n}^+) + \mathbf{p}$. We will have a problem trying to relate the sides of the desired equation with the terms, $\mathbf{m} + (\mathbf{n} + \mathbf{p})$ and $(\mathbf{m} + \mathbf{n}) + \mathbf{p}$, about which we know something. Although the definition of addition turns $(\mathbf{m} + \mathbf{n}^+) + \mathbf{p}$ into $(\mathbf{m} + \mathbf{n})^+ + \mathbf{p}$, it doesn't then help us turn this into anything to do with e.g. $(\mathbf{m} + \mathbf{n}) + \mathbf{p}$.

Induction on $\mathbf{m}$ will hit a similar difficulty.

The problem is that the definition of addition, for $m + n$, is in terms of fixed m, the term on the *left* of the $+$, and variable n, the term on the *right* of the $+$. So our best bet for any proof by induction of something about $+$ is to use the variable on the *right* of the $+$ for the induction. For the sums of three numbers involved in the associative law, this means doing the induction on the rightmost variable, p, for fixed m and n.

Similar situations will arise for inductions involving multiplication or exponentiation. When in doubt, do induction on the rightmost variable, with any others fixed.

So we fix m and n, and use induction on p. For $p = 0$ the definition of $n + 0$ gives that $m + (n + 0) = m + n$, while also by definition $(m + n) + 0 = m + n$. Thus $m + (n + p) = (m + n) + p$ holds for $p = 0$.

For the inductive step, suppose that $m + (n + p) = (m + n) + p$ holds for p. Then

$$\begin{aligned} m + (n + p^+) &= m + (n + p)^+ && \text{(by definition of addition)} \\ &= \left(m + (n + p)\right)^+ && \text{(by definition of addition)} \\ &= \left((m + n) + p\right)^+ && \text{(by the inductive hypothesis)} \\ &= (m + n) + p^+ && \text{(by definition of addition),} \end{aligned}$$

so that the hypothesis also holds for p^+. The result follows by induction.

Our argument is essentially showing that

$$\{p \in \mathbb{N} : m + (n + p) = (m + n) + p\}$$

is inductive.

Exercise 3.26

Use induction to prove each of the following for all $n \in \mathbb{N}$.

(a) $0 + n = n$

(b) $1 + n = n^+$

(c) $0 \cdot n = 0$

(d) $1 \cdot n = n$

(e) $1^n = 1$

(f) $n^1 = n$

Solution

We shall show by induction that $1 \cdot n = n$ for all $n \in \mathbb{N}$ and leave the other parts with no solution.

For $n = 0$ we have $1 \cdot 0 = 0$, by the definition of multiplication. So $1 \cdot n = n$ holds for $n = 0$.

For the inductive step, suppose that $1 \cdot n = n$ holds for n. Then

$$\begin{aligned} 1 \cdot n^+ &= (1 \cdot n) + 1 && \text{(by definition of multiplication)} \\ &= n + 1 && \text{(by the inductive hypothesis)} \\ &= n^+ && \text{(by Exercise 3.20),} \end{aligned}$$

so that the hypothesis also holds for n^+. The result follows by induction.

We shall now show that addition is commutative, i.e. $m + n = n + m$ for all $m, n \in \mathbb{N}$. We shall use induction on n for fixed m – because of the symmetry of the formula we are trying to prove, we could have equally reversed the roles of m and n.

For $\mathbf{n} = \mathbf{0}$, we have $\mathbf{m} + \mathbf{0} = \mathbf{m}$, by definition of addition. And how about $\mathbf{0} + \mathbf{m}$? By the first part of Exercise 3.26, this also equals $\mathbf{m}$, so that $\mathbf{m} + \mathbf{n} = \mathbf{n} + \mathbf{m}$ holds for $\mathbf{n} = \mathbf{0}$. There's a message here. When you do this sort of thing on your own, you might come across an expression like $\mathbf{0} + \mathbf{n}$, the simplification of which requires its own induction proof. And this won't always have been done conveniently for you beforehand, as in Exercise 3.26!

For the inductive step, suppose that $\mathbf{m} + \mathbf{n} = \mathbf{n} + \mathbf{m}$ holds for $\mathbf{n}$ and try to show that $\mathbf{m} + \mathbf{n}^+ = \mathbf{n}^+ + \mathbf{m}$. There's no problem in getting started, as we can write

$$\begin{aligned} \mathbf{m} + \mathbf{n}^+ &= (\mathbf{m} + \mathbf{n})^+ \\ &= (\mathbf{n} + \mathbf{m})^+ \quad \text{(by the inductive hypothesis);} \end{aligned}$$

but the next obvious manipulation that we can do gives

$$(\mathbf{n} + \mathbf{m})^+ = \mathbf{n} + \mathbf{m}^+,$$

rather than the $\mathbf{n}^+ + \mathbf{m}$ which we want. Instead of giving up in despair, we can actually show that $\mathbf{n} + \mathbf{m}^+$ does equal $\mathbf{n}^+ + \mathbf{m}$.

Exercise 3.27

Use induction on $\mathbf{a}$ for fixed $\mathbf{b}$ to show that $\mathbf{a} + \mathbf{b}^+ = \mathbf{a}^+ + \mathbf{b}$, for all $\mathbf{a}, \mathbf{b} \in \mathbb{N}$.

It seems unlikely, though true in this case, that when one gets stuck during one induction proof, one proves the recalcitrant step by another induction!

Solution

For $\mathbf{b} = \mathbf{0}$,

$$\begin{aligned} \mathbf{a} + \mathbf{0}^+ &= (\mathbf{a} + \mathbf{0})^+ \\ &= \mathbf{a}^+ \\ &= \mathbf{a}^+ + \mathbf{0}, \end{aligned}$$

so the base step holds.

Suppose that $\mathbf{a} + \mathbf{b}^+ = \mathbf{a}^+ + \mathbf{b}$ holds for $\mathbf{b}$ and try to prove the corresponding result for $\mathbf{b}^+$, i.e. $\mathbf{a} + \mathbf{b}^{++} = \mathbf{a}^+ + \mathbf{b}^+$.

$$\begin{aligned} \mathbf{a} + \mathbf{b}^{++} &= (\mathbf{a} + \mathbf{b}^+)^+ \\ &= (\mathbf{a}^+ + \mathbf{b})^+ \quad \text{(by the inductive hypothesis)} \\ &= \mathbf{a}^+ + \mathbf{b}^+, \end{aligned}$$

as required. The result follows by induction.

As a consequence of the result of this exercise, we can complete the inductive step of our original argument. With the supposition that $\mathbf{m} + \mathbf{n} = \mathbf{n} + \mathbf{m}$, we can now deduce that

$$\begin{aligned} \mathbf{m} + \mathbf{n}^+ &= \mathbf{n} + \mathbf{m}^+ \quad \text{(we had got this far)} \\ &= \mathbf{n}^+ + \mathbf{m} \quad \text{(by the last exercise).} \end{aligned}$$

Thus the result, that addition is commutative, follows by induction.

We shall leave the remaining results of Theorem 3.12 as exercises for you without providing our solutions. They all involve very therapeutic inductions, except perhaps the commutativity of multiplication, for which one needs a lemma in the middle, rather as we needed in our proof of the commutativity of addition. To some extent the later results below need earlier results as lemmas.

Exercise 3.28

Prove that the following hold for all $\mathbf{m}, \mathbf{n}, \mathbf{p} \in \mathbb{N}$ (as appropriate).

(a) $\mathbf{m} \cdot (\mathbf{n} + \mathbf{p}) = (\mathbf{m} \cdot \mathbf{n}) + (\mathbf{m} \cdot \mathbf{p})$

(b) $\mathbf{m} \cdot (\mathbf{n} \cdot \mathbf{p}) = (\mathbf{m} \cdot \mathbf{n}) \cdot \mathbf{p}$

(c) $\mathbf{m} \cdot \mathbf{n} = \mathbf{n} \cdot \mathbf{m}$

(d) $\mathbf{m}^{\mathbf{n}+\mathbf{p}} = \mathbf{m}^{\mathbf{n}} \cdot \mathbf{m}^{\mathbf{p}}$

(e) $(\mathbf{m}^{\mathbf{n}})^{\mathbf{p}} = \mathbf{m}^{\mathbf{n} \cdot \mathbf{p}}$

(f) $(\mathbf{m} \cdot \mathbf{n})^{\mathbf{p}} = \mathbf{m}^{\mathbf{p}} \cdot \mathbf{n}^{\mathbf{p}}$

There are also results connecting the arithmetic on $\mathbb{N}$ with its order. Recall that in Section 3.2 we defined an order on $\mathbb{N}$ by $\mathbf{m} < \mathbf{n}$ if $\mathbf{m} \in \mathbf{n}$ and that we derived various results about this order.

Theorem 3.13

The following hold for all $\mathbf{a}, \mathbf{m}, \mathbf{n} \in \mathbb{N}$.

1. If $\mathbf{m} < \mathbf{n}$ then $\mathbf{a} + \mathbf{m} < \mathbf{a} + \mathbf{n}$.
2. If $\mathbf{a} > \mathbf{0}$ and $\mathbf{m} < \mathbf{n}$ then $\mathbf{a} \cdot \mathbf{m} < \mathbf{a} \cdot \mathbf{n}$.
3. If $\mathbf{a} > \mathbf{1}$ and $\mathbf{m} < \mathbf{n}$ then $\mathbf{a}^{\mathbf{m}} < \mathbf{a}^{\mathbf{n}}$.

Proof

We shall prove the second of these results and leave the rest as an exercise for you.

Suppose that $\mathbf{a} > \mathbf{0}$. We shall prove the result by induction on $\mathbf{n}$ for fixed $\mathbf{a}$ and $\mathbf{m}$. As multiplication is defined in terms of addition, we should not be surprised if need some prior result about addition and order, and indeed we shall assume that the first result in this theorem has already been proved. (There should be no circularity in this argument, as the proof of the result for addition is unlikely to exploit multiplication.)

Note that the result holds vacuously for all $\mathbb{N}$ such that $\mathbf{n} \leq \mathbf{m}$, on the logical principle that a statement of the form 'if P then Q' is true when P is false. So that the smallest $\mathbf{n}$ for which there is anything significant to prove, namely because $\mathbf{m} < \mathbf{n}$ is true, is $\mathbf{m}^+$. We have

By Theorem 3.10 in the previous section $\mathbf{m}^+$ is the least number greater than $\mathbf{m}$.

$$\begin{aligned} \mathbf{a} \cdot \mathbf{m}^+ &= (\mathbf{a} \cdot \mathbf{m}) + \mathbf{a} \quad \text{(by definition of multiplication)} \\ &> (\mathbf{a} \cdot \mathbf{m}) + \mathbf{0} \quad \text{(by the theorem's result for } +, \text{ as } \mathbf{a} > \mathbf{0}) \\ &= \mathbf{a} \cdot \mathbf{m} \quad \text{(by definition of multiplication),} \end{aligned}$$

so that the result holds for $\mathbf{n} = \mathbf{m}^+$.

For the inductive step, suppose that the result holds for $\mathbf{n}$ where $\mathbf{m} < \mathbf{n}$. (As remarked earlier, we do not have to do anything for $\mathbf{n}$ with $\mathbf{n} \leq \mathbf{m}$.) Then

$$\begin{aligned} \mathbf{a} \cdot \mathbf{n}^+ &= (\mathbf{a} \cdot \mathbf{n}) + \mathbf{a} \quad \text{(by definition of multiplication)} \\ &> (\mathbf{a} \cdot \mathbf{n}) + \mathbf{0} \quad \text{(by the first result of the theorem for } + \text{ as } \mathbf{a} > \mathbf{0}) \\ &= \mathbf{a} \cdot \mathbf{n} \quad \text{(by definition of multiplication)} \\ &> \mathbf{a} \cdot \mathbf{m} \quad \text{(by the inductive hypothesis)}, \end{aligned}$$

so that the result holds for $\mathbf{n}^+$. The result follows by induction. ■

Exercise 3.29

Prove the rest of Theorem 3.13, i.e. for all $\mathbf{a}, \mathbf{m}, \mathbf{n} \in \mathbb{N}$:

(a) if $\mathbf{m} < \mathbf{n}$ then $\mathbf{a} + \mathbf{m} < \mathbf{a} + \mathbf{n}$;

(b) if $\mathbf{a} > 1$ and $\mathbf{m} < \mathbf{n}$ then $\mathbf{a}^{\mathbf{m}} < \mathbf{a}^{\mathbf{n}}$.

Note incidentally the order in which we have written the terms in these results, $\mathbf{a} + \mathbf{m}$ rather than $\mathbf{m} + \mathbf{a}$, etc. As ever, this is because it ties in immediately with the way in which addition etc. is defined.

Exercise 3.30

Show the following for all $\mathbf{m}, \mathbf{n}, \mathbf{a} \in \mathbb{N}$:

(a) if $\mathbf{a} + \mathbf{m} = \mathbf{a} + \mathbf{n}$ then $\mathbf{m} = \mathbf{n}$;

(b) for $\mathbf{a} > \mathbf{0}$, if $\mathbf{a} \cdot \mathbf{m} = \mathbf{a} \cdot \mathbf{n}$ then $\mathbf{m} = \mathbf{n}$;

(c) for $\mathbf{a} > \mathbf{1}$, if $\mathbf{a}^{\mathbf{m}} = \mathbf{a}^{\mathbf{n}}$ then $\mathbf{m} = \mathbf{n}$.

One can go on to develop even more of the standard theory of the natural numbers in terms of their representations as sets. Virtually none of this development requires looking at the detail of this representation – one builds on established results and makes considerable use of induction! The details of the representation are, however, very useful to us within set theory as giving a way of describing what we mean by a finite set, and this is what we shall look at in the next section.

Further exercises

Exercise 3.31

Let $\mathbf{m}, \mathbf{n} \in \mathbb{N}$.

(a) Show that $\mathbf{m} + \mathbf{n} = \mathbf{0}$ if and only if $\mathbf{m} = \mathbf{n} = \mathbf{0}$.

(b) Show that $\mathbf{m} \cdot \mathbf{n} = \mathbf{0}$ if and only if $\mathbf{m} = \mathbf{0}$ or $\mathbf{n} = \mathbf{0}$.

Exercise 3.32

Show that for all $\mathbf{m}, \mathbf{n} \in \mathbb{N}$, if there is some $\mathbf{k} \in \mathbb{N}$ such that $\mathbf{m} + \mathbf{k} = \mathbf{n}$ then $\mathbf{m} \leq \mathbf{n}$.

Exercise 3.33

Show that for all $\mathbf{m}, \mathbf{n} \in \mathbb{N}$, if $\mathbf{m} \leq \mathbf{n}$ there is a unique $\mathbf{k} \in \mathbb{N}$ such that $\mathbf{m} + \mathbf{k} = \mathbf{n}$. [This can be done without using the subtraction operation which is covered in the next exercise!]

Exercise 3.34

An operation $\dot{-}$ can be defined by recursion as follows, for all $\mathbf{m}, \mathbf{n} \in \mathbb{N}$ with $\mathbf{n} \geq \mathbf{m}$:

For $\mathbf{n} \geq \mathbf{m}$, $\mathbf{n} \dot{-} \mathbf{m}$ obviously represents the usual $n - m$ within ZF. One can define $\mathbf{n} \dot{-} \mathbf{m}$ to equal $\mathbf{0}$ when $\mathbf{n} < \mathbf{m}$; but for this exercise we are only interested in the case when $\mathbf{n} \geq \mathbf{m}$.

$$\mathbf{m} \dot{-} \mathbf{m} = \mathbf{0},$$
$$\mathbf{n}^+ \dot{-} \mathbf{m} = (\mathbf{n} \dot{-} \mathbf{m})^+, \quad \text{for } \mathbf{n} \geq \mathbf{m}.$$

(a) Show that for all $\mathbf{m}, \mathbf{n} \in \mathbb{N}$ with $\mathbf{n} \geq \mathbf{m}$

$$\mathbf{m} + (\mathbf{n} \dot{-} \mathbf{m}) = \mathbf{n}.$$

(b) Show that for all $\mathbf{m}, \mathbf{k} \in \mathbb{N}$

$$(\mathbf{m} + \mathbf{k}) \dot{-} \mathbf{m} = \mathbf{k}.$$

(c) Show that for all $\mathbf{m}, \mathbf{n}, \mathbf{k} \in \mathbb{N}$ with $\mathbf{n} \geq \mathbf{m}$

$$(\mathbf{n} \dot{-} \mathbf{m}) + \mathbf{k} = (\mathbf{n} + \mathbf{k}) \dot{-} \mathbf{m}.$$

(d) Show that for all $\mathbf{m}, \mathbf{n}, \mathbf{k} \in \mathbb{N}$ with $\mathbf{n} \geq \mathbf{m}$

$$(\mathbf{n} \dot{-} \mathbf{m}) \cdot \mathbf{k} = (\mathbf{n} \cdot \mathbf{k}) \dot{-} (\mathbf{m} \cdot \mathbf{k}).$$

Exercise 3.35

Prove the *quotient–remainder theorem*, i.e. if $\mathbf{a}, \mathbf{b} \in \mathbb{N}$ with $\mathbf{b} > \mathbf{0}$, then there are $\mathbf{q}, \mathbf{r} \in \mathbb{N}$ such that $\mathbf{a} = (\mathbf{b} \cdot \mathbf{q}) + \mathbf{r}$, where $\mathbf{r} < \mathbf{b}$. [Hint: use induction on $\mathbf{a}$.]

The number $\mathbf{q}$ is the *quotient* and $\mathbf{r}$ the *remainder*.

Show further that $\mathbf{q}, \mathbf{r}$ are unique.

3.4 Finite sets

The main point of interest in this book is the theory of infinite sets. What makes a set infinite? Clearly this question complements the question of when a set is finite: a set is infinite when it is not the case that it is finite. And there are several different ways of answering these questions. In this section we shall define 'finite', so that 'infinite' will be defined as 'not finite'. And our definition will make crucial reference to our representation of natural numbers by sets.

We shall meet an alternative definition in Exercise 6.38 in Section 6.4.

Intuitively, by a finite set we mean one whose elements can be counted off by natural numbers up to a particular number, e.g. as

1st element, 2nd element, 3rd element, ..., 25th (and final) element.

This counting off can be described in terms of a bijection between the set and some suitable set of natural numbers. The way we have defined natural numbers means that $\mathbf{0}$ $(= \varnothing)$ contains 0 elements, $\mathbf{1}$ $(= \{\mathbf{0}\})$ contains 1 element, $\mathbf{2}$ $(= \{\mathbf{0}, \mathbf{1}\})$ contains 2 elements,..., $\mathbf{53}$ contains 53 elements, and so on. This suggests that suitable sets of natural numbers to which we can refer finite sets are the natural numbers themselves, and that the set $\mathbf{n}$ is somehow itself an n-element set. We thus have the following definitions.

> ***Definitions***
>
> A set X is *finite* if there is a bijection $f\colon \mathbf{n} \longrightarrow X$ for some $\mathbf{n} \in \mathbb{N}$. If there is no such bijection for any $\mathbf{n} \in \mathbb{N}$, X is *infinite*.
>
> If there is a bijection $f\colon \mathbb{N} \longrightarrow X$, then X is *countably infinite*. A *countable* set is one which is either finite or countably infinite.

We shall look at infinite sets in more detail in Chapter 6.

Clearly this definition of 'finite' depends crucially on the way we have chosen to represent natural numbers by sets. As an example of using the definition of finite, the set {blue, green, red} of colours is finite because there is a bijection between it and the natural number $\mathbf{3}$ $(= \{\mathbf{0}, \mathbf{1}, \mathbf{2}\})$ defined by

We could give a technical definition of 'X has $\mathbf{n}$ elements' as 'there is a bijection from $\mathbf{n}$ to X'. So this set of colours has $\mathbf{3}$ elements.

$$
\begin{aligned}
f\colon \mathbf{3} &\longrightarrow \{\text{blue, green, red}\}\\
\mathbf{0} &\longmapsto \text{blue},\\
\mathbf{1} &\longmapsto \text{green},\\
\mathbf{2} &\longmapsto \text{red}.
\end{aligned}
$$

This definition means that results about finiteness are essentially results about the elements of $\mathbb{N}$. Take for instance the *pigeon-hole principle*, which says that

> for any finite set A, any function from A into a proper subset of itself must map at least two elements of A to the same image.

This can be rephrased as

Have a think about it!

> for any finite set A, any function from A to itself which is not onto cannot be one–one.

And this is logically equivalent to saying

> for any finite set A, any one–one function from A to itself must be onto.

An alternative definition of A being 'finite', due to Dedekind, is indeed that any one–one function from A to itself must be onto. The equivalence of this to our definition is non-trivial – see Exercise 6.38 in Chapter 6.

If we can prove this last principle holds for all natural numbers $\mathbf{n} \in \mathbb{N}$, i.e. if a function $f\colon \mathbf{n} \longrightarrow \mathbf{n}$ is one–one then it is onto, then the principle holds for all finite sets A.

Exercise 3.36

Justify this last remark!

Solution

Suppose that A is a finite set and that $g\colon A \longrightarrow A$ is one–one. As A is finite there is a bijection $h\colon \mathbf{n} \longrightarrow A$, for some $\mathbf{n} \in \mathbb{N}$. Then the function f defined by $f = h^{-1} \circ g \circ h$,

$$h^{-1} \circ g \circ h\colon \mathbf{n} \xrightarrow{h} A \xrightarrow{g} A \xrightarrow{h^{-1}} \mathbf{n},$$

is the composition of one–one functions and is thus one–one (from $\mathbf{n}$ into $\mathbf{n}$). Supposing that we have proved the pigeon-hole principle for $\mathbf{n}$, this means that f is onto. But this forces the original function g on A to be onto.

If f is onto then as $g = h \circ f \circ h^{-1}$ is the composition of onto maps, g is also onto.

So we shall now prove the pigeon-hole principle for natural numbers.

> ***Theorem 3.14 Pigeon-hole principle***
>
> For any $\mathbf{n} \in \mathbb{N}$, if $f: \mathbf{n} \longrightarrow \mathbf{n}$ is a one–one function, then f is onto.

Proof

It should be no surprise that our method of proof is induction on $\mathbf{n}$. Our inductive hypothesis for $\mathbf{n}$ will be that the result holds for all one–one functions from $\mathbf{n}$ into itself.

The result is vacuously true for $\mathbf{n} = \mathbf{0}$ $(= \varnothing)$, on the grounds that one cannot find a one–one function from the empty set into itself which is not onto. If you are not happy with basing the induction on this case, you might be happier basing it on the case $\mathbf{n} = \mathbf{1}$! As $\mathbf{1} = \{\mathbf{0}\}$, the only possible one–one function f from $\mathbf{1}$ into itself is defined by $f(\mathbf{0}) = \mathbf{0}$ and is thus also onto.

Of course, this f is the only function from $\mathbf{1}$ to $\mathbf{1}$, let alone the only one–one function!

For the inductive step we assume that the result holds for $\mathbf{n}$ and shall show that it holds for $\mathbf{n}^+$. So we shall consider a one–one function $f: \mathbf{n}^+ \longrightarrow \mathbf{n}^+$. Our strategy is to look at the restriction $f|_\mathbf{n}$ of f to the subset $\mathbf{n}$ of its domain, adapt this to obtain a one–one function g from $\mathbf{n}$ into $\mathbf{n}$, use the inductive hypothesis to obtain that g is onto, and then use this information to infer that the original function f on $\mathbf{n}^+$ must also be onto. Much use will be made of the fact that $\mathbf{n}^+ = \mathbf{n} \cup \{\mathbf{n}\}$, so that $\mathbf{n}^+$ consists of the elements of $\mathbf{n}$ along with the one extra element, the set $\mathbf{n}$ itself. The argument needs some care!

The function $f|_\mathbf{n}$ has domain $\mathbf{n}$ and rule $f|_\mathbf{n}(\mathbf{i}) = f(\mathbf{i})$ for all $\mathbf{i} \in \mathbf{n}$.

By Theorem 3.6 we have $\mathbf{n} \notin \mathbf{n}$, so that the element $\mathbf{n}$ of $\mathbf{n}^+$ is genuinely an extra element.

We shall consider two cases, depending on whether the range of $f|_\mathbf{n}$, which we write as $\mathrm{Range}(f|_\mathbf{n})$, is a subset of $\mathbf{n}$ or not. The easy case is when $\mathrm{Range}(f|_\mathbf{n}) \subseteq \mathbf{n}$. We can then regard the restriction $f|_\mathbf{n}$ as a function from $\mathbf{n}$ into itself.

For any function h we write $\mathrm{Range}(h)$ for the *range* of h, namely the set of images of h.

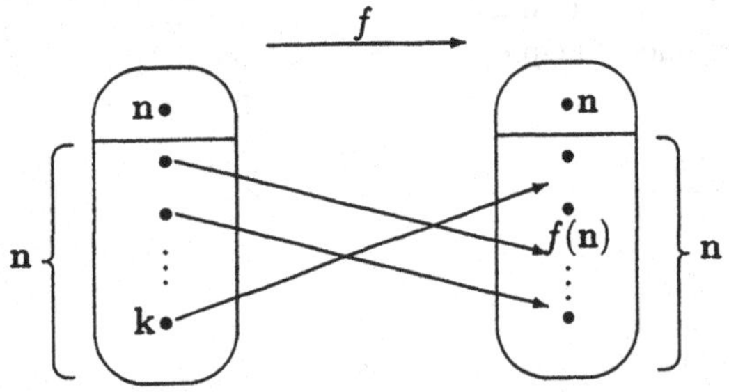

As f is one–one, $f|_\mathbf{n}$ is also one–one (a very easy exercise for you) so that by the inductive hypothesis $f|_\mathbf{n}$ is onto. Turning to the question of whether the original function f is onto, i.e. of whether each element of the codomain $(\mathbf{n}^+)$ is the image under f of some element of the domain (also $\mathbf{n}^+$), this means that every element of the codomain other than $\mathbf{n}$ itself is accounted for as the image of an element of the subset $\mathbf{n}$ of the domain. As f is one–one this forces the image $f(\mathbf{n})$ to be an element of $\mathbf{n}^+$ not in $\mathbf{n}$, so that $f(\mathbf{n})$ can only equal $\mathbf{n}$. Thus every element of $\mathbf{n}^+ (= \mathbf{n} \cup \{\mathbf{n}\})$ is an image under f, so that f is onto as required.

The more complicated case is when Range($f|_{\mathbf{n}}$) is not a subset of **n**. This means that **n**, the only element of $\mathbf{n}^+$ not in **n**, is the image $f(\mathbf{k})$ of some element **k** of **n**. And as f is one–one, it follows that $f(\mathbf{n})$ cannot also equal **n**, so that $f(\mathbf{n})$ is some element of **n** (as any element of $\mathbf{n}^+ = \mathbf{n} \cup \{\mathbf{n}\}$ not equal to **n** is an element of **n**).

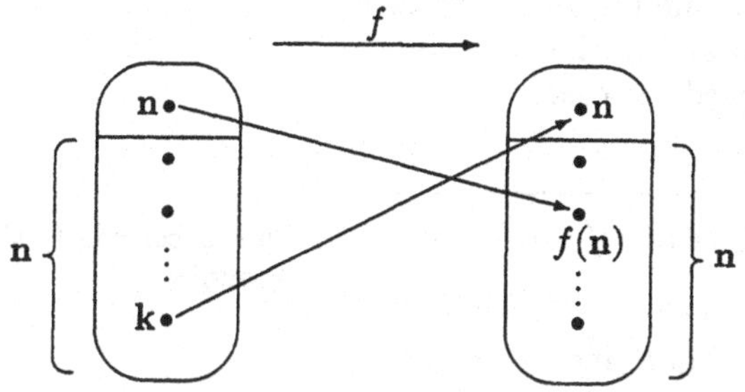

To be able to use the inductive hypothesis we shall adapt the restriction $f|_{\mathbf{n}}$ to create a one–one function g from **n** to **n**, as follows. Define g by

$$g\colon \mathbf{n} \longrightarrow \mathbf{n}$$
$$g(\mathbf{i}) = \begin{cases} f(\mathbf{n}), & \text{if } \mathbf{i} = \mathbf{k}, \\ f(\mathbf{i}), & \text{otherwise.} \end{cases}$$

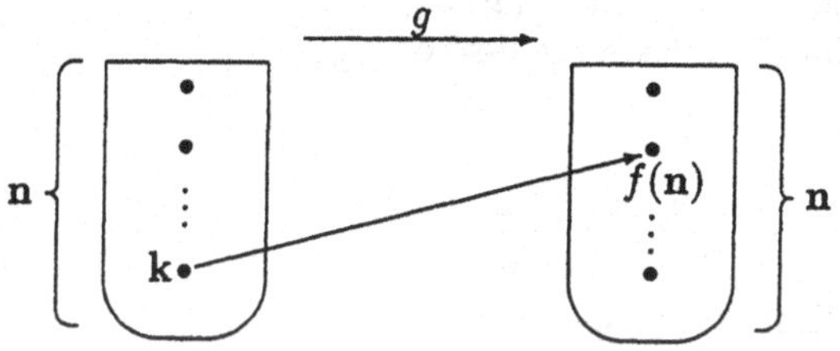

We need to check that the images of g do actually lie in **n** and that g is one–one. First of all, $g(\mathbf{k})$ is in **n** as $f(\mathbf{n}) \in \mathbf{n}$. And for $\mathbf{i} \neq \mathbf{k}$ we have $g(\mathbf{i})$ equal to $f(\mathbf{i})$ which cannot equal **n**, as the image **n** was used up as $f(\mathbf{k})$ and f is one–one, so must also be in **n**. Thus g is into **n**.

We shall leave you to check that g is one–one as an exercise. The inductive hypothesis then gives that g is onto and we can then argue that the original f is onto, as follows. We must show that every element of the codomain $\mathbf{n}^+$ of f is the image of something in its domain. We already know that the element **n** of $\mathbf{n}^+$ is $f(\mathbf{k})$. How about the elements of **n**, i.e. the remaining elements of $\mathbf{n}^+$? As g is onto, every element of **n** is $g(\mathbf{i})$ for some $\mathbf{i} \in \mathbf{n}$; and, with the exception of the one element which is $g(\mathbf{k})$, each $g(\mathbf{i})$ is by definition the same as $f(\mathbf{i})$. The one exception for which $g(\mathbf{i}) \neq f(\mathbf{i})$, namely $g(\mathbf{k})$, was originally $f(\mathbf{n})$, so is an image of f. Thus f is indeed onto. Thus the theorem follows by induction. ■

Exercise 3.37

Suppose that $h\colon A \longrightarrow B$ is a one–one function and that $C \subseteq A$. Show that the restriction $h|_C$ of h to C is also one–one.

Exercise 3.38

Show that the function g used in the proof above is indeed one–one.

For our definitions of finite and infinite set to correspond to our firm intuitions, we would hope to be able to show that $\mathbb{N}$ is infinite and that there is no bijection between different natural numbers $\mathbf{m}$ and $\mathbf{n}$. The pigeon-hole principle is a vital aid in showing that these are indeed the case.

Exercise 3.39

Let $\mathbf{n}$ be a natural number. Show that there is no bijection $f\colon \mathbf{n} \longrightarrow \mathbb{N}$.

Thus according to the definition, $\mathbb{N}$ is infinite.

Solution

Our best bet is to assume that there is such a bijection and try to derive a contradiction. So assume that there is a bijection $f\colon \mathbf{n} \longrightarrow \mathbb{N}$. Then its inverse function $f^{-1}\colon \mathbb{N} \longrightarrow \mathbf{n}$ is also a bijection. Consider the restriction of this map to the subset $\mathbf{n}$ of $\mathbb{N}$,

$$\begin{aligned} f^{-1}|_{\mathbf{n}}\colon \mathbf{n} &\longrightarrow \mathbf{n} \\ \mathbf{i} &\longmapsto f^{-1}(\mathbf{i}). \end{aligned}$$

As f^{-1} is one–one, so is $f^{-1}|_{\mathbf{n}}$. But then, by the pigeon-hole principle, $f^{-1}|_{\mathbf{n}}$ is also onto. Thus wherever f^{-1} maps elements of the complement $\mathbb{N} \setminus \mathbf{n}$, e.g. $\mathbf{n}$ or $\mathbf{n}^+$, to images in $\mathbf{n}$, f^{-1} can no longer be one–one. We have the desired contradiction, so that $\mathbb{N}$ is indeed infinite.

Exercise 3.40

Suppose that $\mathbf{m}, \mathbf{n} \in \mathbb{N}$ with $\mathbf{m} \neq \mathbf{n}$, so without loss of generality $\mathbf{m} \in \mathbf{n}$. Show that there is no bijection between $\mathbf{n}$ and $\mathbf{m}$. [Hint: suppose that there is a bijection $f\colon \mathbf{n} \longrightarrow \mathbf{m}$, and consider $f|_{\mathbf{m}}$.]

Exercise 3.41

Does the pigeon-hole principle hold for $\mathbb{N}$ itself, i.e. is it the case that every one–one function from $\mathbb{N}$ to itself is onto?

Defining finite sets in terms of our representations of natural numbers within set theory has other advantages. For instance the theory of finite linearly ordered sets is in some sense just the theory of $<$ on the sets $\mathbf{n} \in \mathbb{N}$, as we shall see in Chapter 7.

See e.g. Exercise 7.15.

To end this chapter, let us investigate the relationship between the arithmetic on $\mathbb{N}$ and sizes of sets, in the following sense. Take, for instance, our definition of $+$. Once we have built it up within a proper set theory, it will, for each set $\mathbf{m}$ and $\mathbf{n}$ in $\mathbb{N}$, produce a set which we write as $\mathbf{m} + \mathbf{n}$. The provable behaviour of $\mathbf{m} + \mathbf{n}$ has so far conformed to our expectations of addition of our everyday natural numbers, e.g. commutativity, $\mathbf{0}$ is an additive identity and so on. We now have a way of measuring the size of a finite set, and can observe that, for instance, the set $\mathbf{2} + \mathbf{3}$, which can be computed from the definition to equal the set $\mathbf{5}$, has the same number of elements, 5 in the real world, as we would expect from joining a 2-element set and a (disjoint) 3-element set. Indeed, for any manageably small particular values of $\mathbf{m}$ and

$\mathbf{n}$ we would expect to be able to show that the set $\mathbf{m}+\mathbf{n}$ has '$\mathbf{m}$ plus $\mathbf{n}$' elements in the same way. This, and similar results for $\mathbf{m}\cdot\mathbf{n}$ and $\mathbf{m}^{\mathbf{n}}$, can indeed be shown to be true for all $\mathbf{m}$ and $\mathbf{n}$. Of course, as they are results for all natural numbers, the likely method of proof will be induction.

Once $\mathbf{m}$ and $\mathbf{n}$ become large, say both equal to $\mathbf{10^{10}}$, directly working out $\mathbf{m}+\mathbf{n}$ is not very practicable! That's one reason why we shall seek a general proof.

For the addition result, we need a way of achieving a set with '$\mathbf{m}$ plus $\mathbf{n}$' elements – it won't do to take $\mathbf{m}\cup\mathbf{n}$, as our construction gives that one of $\mathbf{m},\mathbf{n}$ must be a subset of the other, so that $\mathbf{m}\cup\mathbf{n}$ just won't be the right size. We shall adopt a useful trick to create *disjoint* sets with $\mathbf{m}$ and $\mathbf{n}$ elements using ordered pairs, namely

This trick will be used extensively in this book.

$$\mathbf{m}\times\{\mathbf{0}\}=\{(\mathbf{i},\mathbf{0}):\mathbf{i}\in\mathbf{m}\}$$

So we shall need ordered pairs within our set theory.

and

$$\mathbf{n}\times\{\mathbf{1}\}=\{(\mathbf{j},\mathbf{1}):\mathbf{j}\in\mathbf{n}\}.$$

Exercise 3.42

(a) Show that there is a bijection between $\mathbf{m}$ and $\mathbf{m}\times\{\mathbf{0}\}$ and similarly one between $\mathbf{n}$ and $\mathbf{n}\times\{\mathbf{1}\}$.

(b) Explain why $\mathbf{m}\times\{\mathbf{0}\}$ and $\mathbf{n}\times\{\mathbf{1}\}$ are disjoint.

Solution

(a) Clearly f given by

$$\begin{aligned} f:\mathbf{m} &\longrightarrow \mathbf{m}\times\{\mathbf{0}\}\\ \mathbf{i} &\longmapsto (\mathbf{i},\mathbf{0}) \end{aligned}$$

is a bijection. The $\mathbf{n}$ case is similar.

(b) As the second coordinate of any $(\mathbf{i},\mathbf{0})$ in $\mathbf{m}\times\{\mathbf{0}\}$, namely $\mathbf{0}$, is different from the $\mathbf{1}$ which is the second coordinate of any $(\mathbf{j},\mathbf{1})$ in $\mathbf{n}\times\{\mathbf{1}\}$, we cannot have any $(\mathbf{i},\mathbf{0})$ equal to any $(\mathbf{j},\mathbf{1})$.

We shall have to ensure that ordered pairs in set theory have this property, that if the coordinates of two ordered pairs don't match exactly, then the ordered pairs aren't equal.

Theorem 3.15

For any $\mathbf{m},\mathbf{n}\in\mathbb{N}$ there is a bijection between the set $\mathbf{m}+\mathbf{n}$ and $(\mathbf{m}\times\{\mathbf{0}\})\cup(\mathbf{n}\times\{\mathbf{1}\})$.

Proof

We shall prove the result for any fixed $\mathbf{m}\in\mathbb{N}$ and all $\mathbf{n}\in\mathbb{N}$ by induction on $\mathbf{n}$.

For $\mathbf{n}=\mathbf{0}$ we have

$$\mathbf{m}+\mathbf{0}=\mathbf{m}\quad\text{(by definition of +)},$$

while

$$\begin{aligned}(\mathbf{m}\times\{\mathbf{0}\})\cup(\mathbf{0}\times\{\mathbf{1}\}) &= (\mathbf{m}\times\{\mathbf{0}\})\cup(\varnothing\times\{\mathbf{1}\})\quad(\text{as } \mathbf{0}=\varnothing)\\ &= (\mathbf{m}\times\{\mathbf{0}\})\cup\varnothing\\ &= (\mathbf{m}\times\{\mathbf{0}\}).\end{aligned}$$

As we saw in Exercise 3.42, there is an obvious bijection between $\mathbf{m}$ and $\mathbf{m} \times \{\mathbf{0}\}$, so that our result holds for $\mathbf{n} = \mathbf{0}$.

Now suppose that the result holds for $\mathbf{n}$, so that there is a bijection $f\colon \mathbf{m}+\mathbf{n} \longrightarrow (\mathbf{m} \times \{\mathbf{0}\}) \cup (\mathbf{n} \times \{\mathbf{1}\})$. We shall show that the result holds for $\mathbf{n}^+$, i.e. there is a bijection between $\mathbf{m}+\mathbf{n}^+$ and $(\mathbf{m} \times \{\mathbf{0}\}) \cup (\mathbf{n}^+ \times \{\mathbf{1}\})$. For $\mathbf{m}+\mathbf{n}^+$ we have

$$\begin{aligned} \mathbf{m}+\mathbf{n}^+ &= (\mathbf{m}+\mathbf{n})^+ \quad \text{(by definition of +)} \\ &= (\mathbf{m}+\mathbf{n}) \cup \{\mathbf{m}+\mathbf{n}\}. \end{aligned}$$

As $\mathbf{k} \notin \mathbf{k}$ for all $\mathbf{k} \in \mathbb{N}$, this is a union of *disjoint* sets – the set $\mathbf{m}+\mathbf{n}^+$ genuinely has one more element than $\mathbf{m}+\mathbf{n}$.

On the other hand we have

$$\begin{aligned} (\mathbf{m} \times \{\mathbf{0}\}) \cup (\mathbf{n}^+ \times \{\mathbf{1}\}) &= (\mathbf{m} \times \{\mathbf{0}\}) \cup ((\mathbf{n} \cup \{\mathbf{n}\}) \times \{\mathbf{1}\}) \\ &= (\mathbf{m} \times \{\mathbf{0}\}) \cup (\mathbf{n} \times \{\mathbf{1}\}) \cup (\{\mathbf{n}\} \times \{\mathbf{1}\}) \\ &= (\mathbf{m} \times \{\mathbf{0}\}) \cup (\mathbf{n} \times \{\mathbf{1}\}) \cup \{(\mathbf{n}, \mathbf{1})\}. \end{aligned}$$

This is also a union of disjoint sets. The ordered pair $(\mathbf{n}, \mathbf{1})$ cannot be in $(\mathbf{m} \times \{\mathbf{0}\})$ as its second coordinate is $\mathbf{1}$ and cannot be in $\mathbf{n} \times \{\mathbf{1}\}$ as its first coordinate is not an element of $\mathbf{n}$: so it is a genuinely extra element. We can thus extend our bijection f to a bijection g defined as follows:

It should be clear that g is indeed a bijection.

$$g\colon \mathbf{m}+\mathbf{n}^+ \longrightarrow (\mathbf{m} \times \{\mathbf{0}\}) \cup (\mathbf{n} \times \{\mathbf{1}\})$$
$$g(\mathbf{i}) = \begin{cases} f(\mathbf{i}), & \text{if } \mathbf{i} \in \mathbf{m}+\mathbf{n}, \\ (\mathbf{n}, \mathbf{1}), & \text{if } \mathbf{i} = \mathbf{m}+\mathbf{n}. \end{cases}$$

Thus the result holds for $\mathbf{n}^+$ as required. The theorem follows by induction. ■

Exercise 3.43

Show that for all $\mathbf{m}, \mathbf{n} \in \mathbb{N}$ with $\mathbf{m}, \mathbf{n} \geq \mathbf{1}$ there are bijections between the following:

(a) the natural number $\mathbf{m} \cdot \mathbf{n}$ and the Cartesian product $\mathbf{m} \times \mathbf{n}$;

(b) the natural number $\mathbf{m}^{\mathbf{n}}$ and the set of all functions from the set $\mathbf{n}$ to the set $\mathbf{m}$, i.e. domain $\mathbf{n}$ and codomain $\mathbf{m}$.

In everyday maths, there are m^n functions from an n-element set to an m-element set.

Solution

(a) This one is left for you!

(b) Let us introduce a temporary piece of notation, *Functions*$(\mathbf{n}, \mathbf{m})$, for the set of functions from $\mathbf{n}$ to $\mathbf{m}$. We shall use induction on $\mathbf{n} \geq \mathbf{1}$ for fixed $\mathbf{m}$. (Induction on $\mathbf{m}$ for fixed $\mathbf{n}$ is not recommended!)

In the next chapter we shall introduce the notation X^Y for the set of all functions from Y to X. With this notation *Functions*$(\mathbf{n}, \mathbf{m})$ gets written as $\mathbf{m}^{\mathbf{n}}$, which would make this question very confusing! You will, however, see in Chapter 6 how creative this confusion can become.

For $\mathbf{n} = \mathbf{1}$ we have $\mathbf{m}^{\mathbf{1}} = \mathbf{m}$ (using the result of Exercise 3.26). What functions are there from $\mathbf{1}$ to $\mathbf{m}$? As $\mathbf{1} = \{\mathbf{0}\}$ each such function f is determined by which element of $\mathbf{m}$ is chosen as $f(\mathbf{0})$. We can thus define a straightforward bijection $\theta\colon \mathbf{m}^{\mathbf{1}} \longrightarrow$ *Functions*$(\mathbf{1}, \mathbf{m})$ by

$$\theta(\mathbf{i}) = \text{ the function } f \text{ such that } f(\mathbf{0}) = \mathbf{i}.$$

Suppose that the result holds for $\mathbf{n} \geq \mathbf{1}$, so that there is a bijection $\theta\colon \mathbf{m}^{\mathbf{n}} \longrightarrow$ *Functions*$(\mathbf{n}, \mathbf{m})$. We shall exploit this θ to construct a bijection from $\mathbf{m}^{\mathbf{n}^+}$ to *Functions*$(\mathbf{n}^+, \mathbf{m})$.

First note that by definition of exponentiation $\mathbf{m}^{\mathbf{n}^+} = \mathbf{m}^{\mathbf{n}} \cdot \mathbf{m}$. By the result of the first part of this exercise there is a bijection, ψ say, between $\mathbf{m}^{\mathbf{n}^+}$ and $\mathbf{m}^{\mathbf{n}} \times \mathbf{m}$.

As the definition of exponentiation exploits that of multiplication, it is no surprise that proving results about the first operation requires lemmas about the second.

Now think about how $\mathit{Functions}(\mathbf{n}^+, \mathbf{m})$ is related to $\mathit{Functions}(\mathbf{n}, \mathbf{m})$. We have $\mathbf{n}^+ = \mathbf{n} \cup \{\mathbf{n}\}$ and the element $\mathbf{n}$ of $\mathbf{n}^+$ is an extra element not already in $\mathbf{n}$ (as $\mathbf{n} \notin \mathbf{n}$). Thus any function f from $\mathbf{n}^+$ to $\mathbf{m}$ is determined by what it does to the *subset* $\mathbf{n}$ of $\mathbf{n}^+$, or equivalently the restriction function $f|_{\mathbf{n}}$, and by where it sends the *element* $\mathbf{n}$ of $\mathbf{n}^+$, i.e. the element in the codomain $\mathbf{m}$ which is $f(\mathbf{n})$.

By our assumption, for each $\mathbf{i} \in \mathbf{m}^{\mathbf{n}}$ there is a corresponding function $\theta(\mathbf{i})$ in $\mathit{Functions}(\mathbf{n}, \mathbf{m})$. So we can make each pair $(\mathbf{i}, \mathbf{j}) \in \mathbf{m}^{\mathbf{n}} \times \mathbf{m}$ correspond to the function f in $\mathit{Functions}(\mathbf{n}^+, \mathbf{m})$ such that $f|_{\mathbf{n}} = \theta(\mathbf{i})$ and $f(\mathbf{n}) = \mathbf{j}$. By composing the bijection ψ (from $\mathbf{m}^{\mathbf{n}^+}$ to $\mathbf{m}^{\mathbf{n}} \times \mathbf{m}$) with this correspondence, we can obtain the desired bijection from $\mathbf{m}^{\mathbf{n}^+}$ to $\mathit{Functions}(\mathbf{n}^+, \mathbf{m})$. Thus our result also holds for $\mathbf{n}^+$, so that it holds for all $\mathbf{n} \geq \mathbf{1}$ by induction.

Further exercises

Exercise 3.44

Show that for all $\mathbf{m}, \mathbf{n} \in \mathbb{N}$ there is a bijection between $\mathbf{m}$ and $(\mathbf{m} + \mathbf{n}) \setminus \mathbf{n}$, i.e. the complement of $\mathbf{n}$ in $(\mathbf{m} + \mathbf{n})$.

Exercise 3.45

Show that for all $\mathbf{m}, \mathbf{n} \in \mathbb{N}$ with $\mathbf{n} \geq \mathbf{m}$, there is a bijection between $\mathbf{n} \dot{-} \mathbf{m}$ (as defined in Exercise 3.34 in the previous section) and $\mathbf{n} \setminus \mathbf{m}$, the complement of $\mathbf{n}$ in $\mathbf{m}$.

Exercise 3.46

(a) Suppose that X is a proper subset of $\mathbf{n}$ for $\mathbf{n} \in \mathbb{N}$. Show that there is a bijection between X and some natural number $\mathbf{m} \in \mathbf{n}$. [Hint: use induction on $\mathbf{n}$.]

(b) Show that any subset of a finite set is finite.

Exercise 3.47

(a) Let A and B be finite sets. Show that their union $A \cup B$ is finite.

(b) Show that the union of finitely many finite sets is finite.

4 THE ZERMELO–FRAENKEL AXIOMS

4.1 Introduction

So far in this book we have given the impression that sets are needed to help explain the important number systems on which so much of mathematics (and the science that exploits mathematics) is based. Dedekind's construction of the real numbers, along with the associated axioms for the reals, completes the process of putting the calculus (and much more) on a rigorous footing. We have to a large extent followed an important strand of the historical development of the subject. But we have leapt ahead in at least one important respect, by giving a construction of the natural numbers which was done well after the deaths of Dedekind and Cantor, the key developers of the subject. This construction was produced because in the meantime the theory of sets themselves, rather than the numbers which they were being used to explain, had thrown up alarming and deep problems. The root of these problems is, perhaps, the issue of infinity, with which we shall deal later, but which, with the work of Cantor, provides the exciting and revolutionary mathematics that set theory primarily seeks to underpin. To some extent, the axioms for set theory which we shall present in this chapter are designed to avoid these problems.

It is important to realize that there are schools of mathematics that would reject 'standard' real analysis and, along with it, Dedekind's work.

The struggle to resolve these problems pervading the discipline of mathematical logic (and more) has resulted in some of the most important work in mathematics and philosophy of the 20th-century.

Let us look at where one the major problems arose. We have already mentioned (in Section 3.1) Frege's definition of the natural numbers, in which he used a logical formula to describe the property that a set contains exactly one element and then exploited this to define $\mathbb{N}$ by an inductive process. It is not the purpose of this book to explain all of Frege's work, but it is to be understood that it was seminal in the development of modern logic – it certainly allows us to tie down some of the problems caused by the imprecision in Cantor's work. Frege identified two complementary ways of describing sets, by extension and intension. Defining a set S by *extension* means explicitly giving its elements, while defining it by *intension* means giving a property which all its elements, and only its elements, possess. For instance,

You might find it helpful to think of the 'extent' of a set as an overview of everything in the set, as opposed to its 'intent', what is intended to be in it.

$$\{2, 4, 6, 8, 10, 12, 14\}$$

and

$$\{x \in N : 0 < x < 15 \text{ and } x \text{ is even}\}$$

both define the same set: the first definition is by extension and the second by intension. The general form of a definition of a set S by intension is

$$\{x : x \text{ has property } P\},$$

where P is a shorthand for some property which objects like x might, or might not, possess. When dealing with infinite sets, we frequently appear to define sets by extension, e.g.

$$S = \{\ldots, -4, -2, 0, 2, 4, \ldots\},$$

Built into this is the idea that given any property P and object x, either x has property P or it doesn't. This alone generates philosophical debate, especially when linked to infinite sets and the issue of *decidability*.

but when challenged to explain, with some sort of finite description, what is meant by the dots ... ('and so on'!), we fall back on a definition by intension, like

$$\{x \in \mathbb{Z} : x \text{ is even}\}.$$

A key feature of the set theory and mathematics in this book is that every mathematical object that we discuss can indeed be described in a finite way. As the most exciting part of set theory deals with infinite sets, and as such a set can never be finitely described by extension, definition by intension is crucial. Furthermore, given any finite set described by extension, e.g.

$$\{1, 7, 43\}$$

whose elements don't seem to have any elegant mathematical description, we can always give an alternative (albeit cumbersome) finite description of it by intension, like

$$\{x : x = 1 \text{ or } x = 7 \text{ or } x = 43\}.$$

Frege's hope was to be able to define the natural numbers in terms of pure logic and sets, and the sets themselves from pure logic. He created a formal system for reasoning in which sound arguments would be represented by symbolic formulas standing for mathematical assertions being manipulated by agreed, mechanical, rules. Within such a system, one could represent the property, of a set C, that it contains exactly one element, by a formal version of

If you have encountered propositional or predicate calculus, you will have encountered the fruits of Frege's work.

$$(\text{there exists } x)\big(x \in C \text{ and } (\text{for all } y)(\text{if } y \in C \text{ then } y = x)\big).$$

This defines 'C has one element'.

Frege also tried to explain what sets are and what properties they have. Crucial in his analysis was that if P is a property of sets x, then there is a set, defined by intension, consisting of precisely those sets x with property P, i.e.

$$\{x : x \text{ is a set with property } P\}$$

is a set. Frege's definition of the natural number 1 was then as

This principle, of creating sets from other sets, is so important as to merit a special name, the *comprehension principle*.

$$\{x : x \text{ is a set and } x \text{ has one element}\},$$

i.e. the set of all one-element sets. There is no obvious circularity in this definition, thanks to the logical formula by which Frege defined 'has one element', so that the definition seemed very attractive.

Unfortunately the comprehension principle, on which Frege's definition rests, has fatal flaws. First of all, it gives rise to some uncomfortable circularities, as follows. Let us use $[\![1]\!]$ to stand for the set of all one-element sets. Then, surely, we can form the one-element set $\{[\![1]\!]\}$, whose only element is $[\![1]\!]$, and which thus belongs to the set of all one-elements sets, namely $[\![1]\!]$ itself: i.e.

This looks like the notation for an equivalence class. Indeed Frege's definition of 1 can be regarded as an equivalence class – what's the equivalence relation?

$$[\![1]\!] \in \{[\![1]\!]\} \in [\![1]\!].$$

Does this fit our intuition of sets, that we have sets x, y such that $x \in y \in x$? We can do better than this, again using the comprehension principle, to construct the set $\mathscr{V}$ of all sets,

We shall call $\mathscr{V}$ the *universe* of sets. In what follows we shall ensure that $[\![1]\!]$ and $\mathscr{V}$ are not treated as normal sets.

$$\mathscr{V} = \{x : x \text{ is a set}\}.$$

Then as $\mathscr{V}$ is a set, we have

$$\mathscr{V} \in \mathscr{V}.$$

Exercise 4.1

On the basis of what we've just done, give an example of a set x for which there are sets y, z with $x \in y \in z \in x$.

Solution

Just take $x = y = z = \mathscr{V}$!

Such circularities are not necessarily wrong. The idea of a set belonging to itself is at this stage merely worrying. Surely it is more likely that a set won't belong to itself. Alas, it is precisely this idea which, along with the comprehension principle, delivers something fairly devastating, as follows. Use the comprehension principle to define the set $\mathscr{R}$ of all sets which do not belong to themselves, i.e.

$$\mathscr{R} = \{x : x \text{ is a set and } x \notin x\}.$$

This leads to something worse than a circularity. Either $\mathscr{R}$ belongs to itself, or it doesn't do so. If $\mathscr{R}$ does belong to itself, then it must be one of the sets x such that $x \notin x$, so that it does *not* belong to itself, i.e.

$$\text{if } \mathscr{R} \in \mathscr{R} \text{ then } \mathscr{R} \notin \mathscr{R}.$$

And if $\mathscr{R}$ does not belong to itself, so that it is an x such that $x \notin x$, it must belong to $\mathscr{R}$, so that it *does* belong to itself, i.e.

$$\text{if } \mathscr{R} \notin \mathscr{R} \text{ then } \mathscr{R} \in \mathscr{R}.$$

This gives a contradiction known as *Russell's paradox*, which was communicated in 1902 by the British mathematician and philosopher Bertrand Russell (1872–1970) to Frege, just as Frege was about to publish his major work on the subject. The consequence of this argument was highly alarming, that set theory as an attractive basis for explaining key mathematical systems like $\mathbb{R}$ and $\mathbb{N}$ was contradictory. The damage was far greater in that set theory was also the framework for Cantor's revolutionary theory of infinity (of which we shall see a great deal in this book), not to mention an increasing amount of other mathematics of the day. The paradox came on top of a number of other apparent paradoxes arising from the theory of infinite sets, so some sort of resolution was urgently required.

The famous correspondence between Russell and Frege is reproduced in [11].

We call a mathematical system from which we can prove a *contradiction*, namely some statement and its negation, an *inconsistent* system. Classical logic then shows that for such a system, *any* statement is a theorem. So Russell's paradox is drastic!

With hindsight, it is perhaps not surprising either that ideas about sets should be in need of some sorting out or that even then they remain difficult and problematic, any more than with the ideas behind calculus or geometry. Nor is it surprising that the crisis should arise after there had been great mathematical discoveries through these ideas. The point of much of this book is to show both these discoveries and how, partly as a consequence of the crisis caused by paradoxes, these have been put onto a more rigorous foundation. So, how might the paradoxes be avoided?

First of all, we must face up to our inability to say exactly what a set is. Sets, and the key relationship between them, namely that of one set belonging to another, remain undefined. But we shall have to be more careful about what sets there are, and how new sets can be constructed from known sets. The comprehension principle, which allows us to collect sets together on the basis of sharing a common property, seems fundamental and thus needs to be rescued in some way, given that Russell's paradox stems from it. Russell's solution was to regard sets as living in a hierarchy of *types*, essentially levels, whereby collecting together some of the sets in one level gives a set at the next highest level. In this way, defining the Russell set using sets x at one level for which $x \notin x$ gives a set $\mathscr{R}$ at the next highest level, not at the same level; and the point of the levels is that there is no question of a set at a higher level being a member of one at a lower level. In this approach, Frege's definition of the number 1 can be used, taking sets at the lowest level to give a set at the next level. And the sets at this next level which have a single element can be collected to form a new version of the number 1 at the next highest level; and so on. This gives an annoying proliferation of number 1s! A more popular framework is to regard there as just being the one 'level' of sets and to restrict the ways in which we construct new sets from old in an attempt to avoid known pitfalls, like Russell's paradox. This is the route we shall take, but we shall borrow something from a third framework, which is to have two types of 'set', namely *sets* and *classes*. All sets will also be classes, but not vice versa! Collecting together sets sometimes still gives one a set, but sometimes gives an object 'too big' to be regarded as a set, yet still meaningful as a class, like the universe $\mathscr{V}$ of all sets mentioned earlier.

This approach, developed by the Hungarian and later American mathematician John von Neumann (1903–57) is well described in the standard textbook by E. Mendelson[16].

Our definition below of *class* is motivated by the central importance of the comprehension principle in defining sets. As you have seen, it would be catastrophic for the theory to regard every object described by use of this principle as a set. But precisely because such objects have descriptions, it is very useful to have a name for them, as in the following definition.

Definition

A *class* is any object of the form

$$\{x : x \text{ is a set with property } P\}.$$

We shall avoid the (important philosophical) issue of what sort of object a class is. One important feature of classes is that some, like $\mathscr{R}$, cannot be regarded as sets (assuming that it is desirable that mathematics is consistent!), while many others, like $\{x : x \text{ is a natural number}\}$, can safely be called sets. Another important feature is that classes provide a framework for discussing objects like the universe $\mathscr{V}$ which arise naturally within the theory.

As for what sets there are, and how we shall allow new sets to be constructed from known ones, this shall be done in terms of axioms, expressed in a formal language for logic. The axioms are based on those proposed by the German mathematician Ernst Zermelo (1871–1953) in 1905. It is important to realize, however, that Zermelo's motivation for producing axioms was not so much the avoidance of paradoxes, but primarily to give a framework for a controversial result about sets, of much interest to mathematicians of the day, and still of great significance. This result concerned a problem of Cantor's, whether it was possible to define a well-ordering on the set of real numbers, and Zermelo's argument, exploiting a principle called the axiom of choice, had previously been rejected by other mathematicians. By enunciating the

Cantor himself, the architect of the theory of infinite sets and many of the questions this then posed, appears not to have been surprised or worried by the paradoxes, regarding them as inevitable when feeling one's way with sets, especially infinite sets. See e.g. Cantor's 1899 article in [11].

assumptions about sets underlying his argument, Zermelo hoped to make it harder for others to reject it.

In the author's view, it is important to bear in mind the spirit in which Zermelo provided his axioms, and in which others later modified them. Such axioms might in part be motivated by representing in a formal way what is intuitively true about sets, from the ways in which we are accustomed to think about them and use them. And the axioms attempt to steer us clear of contradictions. But these axioms are also intended to underpin important results, like the original work of Cantor, as well as Zermelo's own arguments, and sometimes go beyond intuition (but not necessarily against it).

Our intuition of sets might, in any case, be hazy, so there is scope for axioms to sharpen up intuition in a variety of directions.

For a further analysis of the status of the axioms, see Maddy [18].

In the rest of this chapter, we shall give axioms for set theory and the logical framework within which they sit. We have a shopping list of requirements for the theory, e.g. that it permits the representation of natural numbers as sets as in the previous chapter, and likewise the representation of the reals in terms of the rationals. And we shall prove that some of these requirements are indeed met. Much effort will go into showing that certain classes, like $\{x : x$ belongs to both the sets a and $b\}$, are sets, without appearing to create any contradictions in the theory. But we cannot actually prove that we are avoiding paradoxes. This turns out to be impossible, for reasons beyond the scope of this book – Gödel's incompleteness theorems – but inspired in part by the considerations stemming from the development of set theory.

If a class cannot be treated as a set without entailing a contradiction, we shall call it a *proper class.*

Paradoxes: some further remarks

This book is primarily intended to present the foundations of modern set theory, a subject which has in many ways left Russell's and other paradoxes far behind. The Zermelo–Fraenkel axioms in this chapter and the remarks in the introduction above further this intention. But the paradoxes are of very major interest, not just historically, but also because of their consequences for 20th century mathematics, philosophy and computer science, so we shall say a little bit more about them here as an appendix.

There is a long history of important and tricky paradoxes involving infinity, and perhaps it is thus unsurprising that Cantor's work should generate so many new ones. The first of the latter was found by Burali-Forti in 1897 and involved ordinals, one of Cantor's two sorts of infinite number. The essence of the paradox is that, putting $\mathcal{O}$ as the set of all ordinals, one can then show that $\mathcal{O}$ is itself an ordinal, so that $\mathcal{O} \in \mathcal{O}$: this in turn contradicts an essential feature of ordinals.

A good discussion can be found in Adrian Moore [8].

Ordinals will be dealt with in Chapter 8 and this paradox discussed in Exercise 8.7.

Cantor's reaction to the Burali-Forti paradox was that some infinite collections were too 'big' to be treated as sets, that they were 'inconsistent'. Cantor reacted in the same way to a similar paradox involving his other sort of infinite numbers, namely cardinals, should one form the 'set' of all cardinals. But Russell's paradox appeared to undermine even more fundamental ideas about what constituted a set.

See Cantor's paper of (1899) in [11].

This paradox is the subject of Exercise 9.6 in Section 9.3.

It is important to realize that even before the discovery of these paradoxes, Cantor's work had met with some very hostile reaction on grounds other than that it involved logical contradictions: results like those about the size of $\mathbb{R}$

compared to that of ℕ were very controversial. It is something of a matter of wonder that his work survived the crisis!

Russell's paradox relies in large part on our ability to make statements which refer to themselves. Such statements can then be paradoxical in the sense that judging them as either true or false results in a contradiction. Russell and then others have produced a multitude of such paradoxical statements in everyday language, but even the limited language for set theory which we shall look at in the next section is rich enough to be able to express Russell's paradox and ultimately support Gödel's incompleteness theorems, of which one major consequence is that we cannot prove that the Zermelo–Fraenkel axiom system is free from contradictions.

For Russell's original example, the one of such serious set-theoretic import, the relevant statement can be taken as $\mathscr{R} \in \mathscr{R}$. If this statement is true, then it's false; and if it's false, then it's true!

And Gödel's theorems themselves exploit the capacity of the language to make statements referring to themselves.

4.2 A formal language

Before we give axioms for set theory, we must specify the framework within which these axioms sit. That is the aim of this section.

We shall write the axioms using a very limited language, one that fits a formal logical treatment using the predicate calculus. For the purposes of this book, it is important to be able to use and interpret the formal language, but not to construct formal proofs using it. It is, however, important to realize that such formal proofs can in principle be given.

The simplest statements in this language will be of the form $x \in y$ and $x = y$, for sets x, y. More complicated statements will be built up from these by combining or prefixing them by a variety of *logical connectives*, symbols with intended meanings as follows:

symbol	*meaning*
$\wedge$	and
$\vee$	or
$\neg$	not
$\rightarrow$	implies
$\leftrightarrow$	if and only if
$\forall$	for all sets
$\exists$	there exists a set

Our formal language uses these logical connectives, brackets, the symbols $\in$ and $=$, and letters (subscripted should we desire) like

$$x, y, z, \ldots, x_0, x_1, \ldots, y_5, \ldots, X, Y, Z, \ldots$$

as *variables* to represent individual sets. The statements, or equivalently *formulas*, of the language are all expressions of the forms

$$x \in y \quad \text{and} \quad x = y,$$

The $\in$ symbol was introduced by Peano.

where x, y are any variables, and any expression built up recursively from these by finitely many applications of the following formation rules:

The finite use of these rules means that all statements involve only finitely many symbols.

if ϕ, ψ are formulas, then so are

$\neg\phi$,

$(\phi \wedge \psi)$,

$(\phi \vee \psi)$,

$(\phi \rightarrow \psi)$,

$(\phi \leftrightarrow \psi)$,

$\forall v\phi$, for any variable v,

$\exists v\phi$, for any variable v.

The construction rules use brackets where necessary to ensure that there is just the one way of reading a statement.

The important thing for now is to explore the expressive power of this language. For instance

$$\neg\exists x\forall y\ y \in x$$

says that 'there is no set x such that all sets y are elements of x', which can be less awkwardly rephrased as 'there is no set to which all sets belong'. And

$$\forall x\forall y(x \in y \ \vee\ (x = y \ \vee\ y \in x))$$

says that 'for any two sets, one belongs to the other or they are equal'. Notice that this is surely a false statement about sets. Writing down a statement about sets doesn't entail that it has to be true!

The symbol $\vee$ is intended to mean the *inclusive* 'or', i.e. $\phi \vee \psi$ means 'ϕ or ψ or both'.

We shall often use the formal language to express that a set x (or y or whatever) has a particular property. For instance

$$\forall y \neg\ y \in x$$

says that x is empty. The variable x occurs in this formula in a way which is usually described as *free*. In general, the variable v occurring in a formula ϕ is described as a *free variable* of ϕ if no stage of the building up of ϕ has involved putting $\forall v$ or $\exists v$ in front of a statement in which v appears. (Actually, in any standard treatment of predicate calculus, the term 'free variable' is used in a slightly less restricted way, so that e.g. the underlined occurrence of y in

$$(\exists y\ y \in z\ \wedge\ \neg\ \underline{y} = z)$$

is a free variable of the formula – our definition gives only that the z is free. We shall avoid any problem in this book by only writing formulas in which variables free in one part of the formula are not used elsewhere in a non-free way.)

Note that all the objects we are describing with our formal language are sets. Thus if x has no sets y as elements, it has no elements at all – there can't be some element (in x) which is not a set.

Exercise 4.2

What do each of the following formulas say?

(a) $\forall x\exists y \neg\ x \in y$

(b) $\forall x\forall y((x \in y\ \wedge\ y \in z)\ \rightarrow\ x \in z)$

(c) $\forall x\exists y(\phi(x,y)\ \wedge\ \forall y'(\phi(x,y')\ \rightarrow\ y' = y))$, where $\phi(x,y)$ is a formula with free variables x and y.

We shall eventually lapse into use of the standard abbreviation $x \notin y$ for the formula $\neg\ x \in y$. Likewise we shall use $x \neq y$ for $\neg\ x = y$. For this section we shall stick to our limited language without using abbreviations.

Solution

(a) For every set x, there is a set y of which x is not an element. This might be expressed in a less stilted way by saying that for any set there is a set to which it doesn't belong.

(b) All elements of any element of z are themselves elements of z, or equivalently, any element of z is a subset of z – think about it!

(c) This says that for any x there is exactly one y for which $\phi(x,y)$ holds. Thus $\phi(x,y)$ can be used to define a function f, where, for each x, $f(x)$ equals *the* y for which $\phi(x,y)$ holds.

It is often quite difficult to translate statements about sets in an unstilted way!

Examples of sets with this property are the natural numbers **n** defined in Chapter 3, following from the transitivity of $\in$ on $\mathbb{N}$, as in Theorem 3.6.

Exercise 4.3

Express the each of the following by a formula in the formal language.

(a) The set x contains at least one element.

(b) There is a set with at least one element.

(c) The set x contains exactly one element.

Solution

(a) The formula $\exists y\ y \in x$ will do.

(b) We merely need to say that there is some x with the property in the previous part of the exercise. So we can take the formula as $\exists x \exists y\ y \in x$.

(c) It would be nice to write something like $\exists y\ x = \{y\}$; but this uses symbols, { and }, not in our limited language. We can capture the idea, however, by saying that $y \in x$ and that any y' in x is in fact this y, as follows:

$$\exists y(y \in x\ \wedge\ \forall y'(y' \in x\ \rightarrow\ y' = y)).$$

Let us move closer towards the main way in which we shall use the formal language to state the Zermelo–Fraenkel axioms. Most of these axioms will state that certain sets exist. In Section 4.1 we discussed the two natural ways of describing a set, by intension and extension, and how in practice we need only use intension. So some of the axioms will be of the form

For example, the set $\{1, 7, 43\}$ defined by extension could be defined by intension as

$$\{x : (x = 1\ \vee\ (x = 7\ \vee\ x = 43))\}.$$

there is a set x of all sets y which have property$P$.

We shall confine ourselves to those properties P of sets y which can be expressed by a formula $\phi(y)$ within our formal language with a free variable y. We want to capture the idea that x is the set of precisely those y with the property $\phi(y)$ by a formula of our language; but as we have already remarked, this language doesn't contain symbols for the set brackets { and }. The trick is to exploit the symbol $\leftrightarrow$ for 'if and only if' to say exactly when y is an element of x, by

As you will see, confining ourselves to properties expressible in our formal language is not a great hardship! But it is a restriction of a sort – see Exercise 6.43 in Section 6.4.

$$(y \in x\ \leftrightarrow\ \phi(y)).$$

Then the existence of a set x consisting of all these ys is represented by the formula

$$\exists x \forall y(y \in x\ \leftrightarrow\ \phi(y)).$$

For instance, suppose that we wish to state within the language that for given sets a, b there is an intersection set x, where $x = a \cap b$. The property that characterizes elements y of $a \cap b$ is of course that $y \in a$ and $y \in b$. So the existence of such an x can be given by

$$\exists x \forall y (y \in x \leftrightarrow (y \in a \wedge y \in b)).$$

Notice that we are extending our use of variable symbols by permitting such symbols, in this case a and b, to stand as *names* for given sets. In general, if some of the free variables in a formula ϕ are replaced by names, we shall describe the formula as *referring to named sets* and only count the remaining variables as its free variables.

Exercise 4.4

Express each of the following by a formula within the formal language.

(a) The union of a and b is a set (where a, b are given sets).

(b) For any sets, their union is a set.

Solution

(a) $\exists x \forall y (y \in x \leftrightarrow (y \in a \vee y \in b))$

(b) $\forall a \forall b \exists x \forall y (y \in x \leftrightarrow (y \in a \vee y \in b))$

Exercise 4.5

Express each of the following by a formula within the formal language.

(a) z is subset of x.

(b) For any set, there is a set consisting of all its subsets.

This set of all subsets is called a *power set*.

Solution

(a) $\forall w (w \in z \rightarrow w \in x)$

(b) If we use the shorthand $z \subseteq x$ for the formula $\forall w (w \in z \rightarrow w \in x)$, then the following formula would do:

$$\forall x \exists y \forall z (z \in y \leftrightarrow z \subseteq x).$$

The set y in this formula would be the set of all subsets of x. Eliminating this piece of shorthand gives the following formula of our language:

$$\forall x \exists y \forall z (z \in y \leftrightarrow \forall w (w \in z \rightarrow w \in x)).$$

Let us take this as the formal definition of the symbol $\subseteq$ and of 'z is a subset of x'.

We shall soon move towards using shorthand notations like $z \subseteq x$ to make our formulas easier to decipher.

Of course, just because we can write down a formula that says a certain set exists, doesn't mean it does exist! For instance, we can construct a formula which says that the Russell 'set' exists. Try this as the next exercise.

Exercise 4.6

Express by a formula within the formal language that there is a set consisting of all sets which are not elements of themselves.

Solution

$\exists x \forall y (y \in x \leftrightarrow \neg\, y \in y)$

Now that we have specified our formal language, we can formulate within it some axioms for set theory, which we do in the next section. And one of our hopes is that these axioms will not allow Russell's paradox to occur.

4.3 Axioms 1 to 3

In this section, we shall give the first three axioms of set theory as proposed by Zermelo and modified by others, including Fraenkel. The theory is described as Zermelo–Fraenkel set theory – *ZF* for short. And we shall show how to construct some of the basic building blocks of the theory on the basis of these axioms, for instance how to represent ordered pairs and functions as sets. We shall give the remaining axioms, and prove some of their consequences, in Sections 4.4 and 4.5.

Abraham Fraenkel (1891–1965) was a German and then Israeli mathematician. Accounts of the process of modifying the axioms and of alternative systems can be found in e.g. Fraenkel, Bar-Hillel and Levy [19] or Kneebone [20].

Recall that Zermelo's main aim in producing axioms was to underpin an argument involving complicated set concepts, like well-ordering $\mathbb{R}$. His axioms were, to some extent, simple, straightforward and non-controversial statements about sets; but they were also designed to serve the needs of his more complicated argument, and are thus not immune from a degree of scepticism! They do, however, underpin the replication of large amounts of standard mathematics within a framework of sets, and in particular (with some refinements introduced by other mathematicians) place Cantor's revolutionary work on infinite sets on a more rigorous basis. They rapidly won acceptance from the mathematical community as being the basis for most of the mainstream work on sets and the foundations of mathematics.

A further feature of the Zermelo–Fraenkel axioms, in common with many other mathematical axiom systems, is that they are designed with a form of economy in mind. Rather than give a large number of axioms, for instance asserting the existence of intersections of sets, *ZF* has relatively few axioms and requires ingenuity in deriving such a thing from the axioms. We shall not give you all the axioms in one go, but derive one or two initial results from the first few axioms before going on to the next batch.

Writing down axioms for sets in a formal language is also a way of dodging the issue of what sets really are, just as the standard presentation of the theory of the real numbers using axioms avoids committing oneself to saying what the reals really are. The symbols we use for variables, x, y, etc., and the symbols $\in$ and $=$ have intended interpretations, respectively as sets, 'is an element of' and 'equals'. But we do not have to say what the words 'set' , 'element', etc., mean. The theory of sets which follows in this book is, strictly speaking, what the axioms and deductions from them tell us about sets. As one aim of the axioms is to permit us to perform familiar, everyday (and uncontroversial) manipulations with sets, most of the axioms will be easy to read and understand in terms of our informal notions about sets. This means that we can continue to talk about sets fairly informally. But we shall have to be on our guard to ensure that what we claim can indeed be backed up within the formal theory, from the axioms.

As we saw in Chapter 2, real numbers can be defined in terms of something 'more basic', although it might be argued that any such definition doesn't match any intuition of a real number as a number, rather than as e.g. a set of other objects. Axioms for the reals avoid the need to say what sort of object a real number is. For sets it is not clear in terms of which 'more basic' objects one might define 'set'.

There are nine axioms of *ZF* set theory, which we shall label as *ZF1* to *ZF9*, but which we shall often refer to by their names. For each axiom we shall try to give an informal description of what it says, as well as write it formally. If you don't always understand the significance of an axiom just from reading, don't worry, we shall discuss each of them in more detail later on! Here now are the first three axioms.

Axioms of ZF

ZF1: Axiom of extensionality

$$\forall x \forall y (x = y \leftrightarrow \forall z (z \in x \leftrightarrow z \in y))$$

Two sets are equal if and only if they contain the same elements.

ZF2: Empty set axiom

$$\exists x \forall y \; y \notin x$$

There is a set with no elements.

ZF3: Axiom of pairs

$$\forall x \forall y \exists z \forall w (w \in z \leftrightarrow (w = x \lor w = y))$$

For any two sets, there is a set whose elements are precisely these sets.

From now on, we will write $y \notin x$ for $\neg\, y \in x$.

Zermelo bundled our *ZF2* and *ZF3* into a single axiom of elementary sets, along with an axiom saying that if x is a set then there is a set with x as its only element. This latter axiom is in fact a consequence of *ZF2* and *ZF3*.

The axiom of extensionality is a statement of the very standard connection between set equality and set membership. You will probably have exploited this connection in everyday mathematics, and we have used it implicitly many times so far in this book. (You may wonder why we have both $\in$ and $=$ as basic symbols of our formal language, given that we could use this axiom to define $=$ in terms of $\in$. The reason is that standard formal proof systems for predicate calculus treat $=$ as a basic symbol and include rules which permit replacing a set mentioned in a statement by an equal set.)

The empty set axiom is important in two ways. It says that there is a set with no elements – the idea of such a set often strikes students, when encountering it for the first time in mathematics, as exciting or dangerous, rather like the number 0! But it is also one of the only axioms of *ZF* which asserts that there are any sets. Most of the other axioms have the character of the axiom of pairs, which is of the form 'if ... are sets, then so is ...'. (Note that the empty set axiom doesn't say that there is exactly one empty set. We shall show that this is a consequence of the *ZF* axioms.)

The axiom of pairs, which gives an unexceptional way of constructing new sets from known sets, is one of those whose importance is really in its role in constructing interesting sets, as we shall soon see.

Although the axiom of extensionality is unexceptional, you might be interested in the consequence of it arising in the following exercise.

Exercise 4.7

If one was trying to speculate on the nature of the universe of sets, it might be tempting to think sets are built up from 'atomic' sets, where by an atomic set we mean one that cannot be reduced in any way to other sets. In particular, an atomic set would be one which had no elements. How many atomic sets could there be in *ZF* set theory?

Set theories which allow more than one atomic set have been fruitfully explored.

Solution

Suppose that x and y are both atomic sets, so both have no elements. Then by the axiom of extensionality (and the standard way we use 'if and only if'), x and y must be equal. So there is at most one atomic set. By the empty set axiom there is at least one set with no elements. Thus there is exactly one atomic set.

The discussion in the solution shows that there is exactly one set in *ZF* which has no elements, so we are entitled to refer to it as *the* empty set. The axiom of pairs is similar to the empty set axiom in that it says that given two sets x and y, there is a set z whose elements are these, but it doesn't say that this is a unique set. It can be shown that it is indeed unique.

Exercise 4.8

Show that given sets x and y there is a *unique* set z whose elements are x and y.

Solution

By the axiom of pairs there is at least one set z whose elements are x and y. If z' is another such set, then any element of z', namely x or y, is also an element of z, and vice versa. By the axiom of extensionality we have $z = z'$, so that the set z is indeed unique.

We could describe the set z above as the *pair set* of x and y. And we shall introduce the familiar curly bracket notation for sets by writing $\{x, y\}$ for z. We shall be careful in this section about justifying extending the use of this notation beyond pair sets: for instance, we shall have to prove that for given sets x, y and z, there is a unique set with these three sets as its elements. We would of course want to write this set as $\{x, y, z\}$.

Exercise 4.9

What can we say about $\{x, y\}$ if x and y are the same set?

Solution

By the axiom of extensionality, $\{x, x\}$ must equal the set whose only element is x. The normal name for such a set is a *singleton*.

This last exercise, along with the *ZF* axioms introduced so far, yields the following theorem.

Theorem 4.1

For any set x there is a set whose only element is x.

Proof

As x is a set, the axiom of pairs gives that there is a set whose elements are x and x (i.e. $\{x, x\}$). By the result of Exercise 4.9, this equals the set whose only element is x (i.e. $\{x\}$), which shows that this is indeed a set. ■

The results so far, some of which are about the uniqueness of objects as well as their existence, justify the following definitions.

Definitions

The *empty set*, written as $\varnothing$, is the set with no elements.

Given sets x and y, we write $\{x, y\}$ for the set whose elements are x and y. This set is called a *pair* and also an *unordered pair*.

Given a set x, we write $\{x\}$ for the set whose only element is x. This set is called a *singleton*.

As our definition of the pair $\{x, y\}$ does not require that $x \neq y$, it could be the case that a pair might actually be a singleton! This sounds potentially confusing, but in practice it won't be.

You may have been surprised by our slipping in the term 'unordered pair' above as a way of describing the set $\{x, y\}$. We did this because we now wish to discuss how to represent *ordered* pairs within set theory. We have already seen how ordered pairs can be used to play a part in defining one sort of number in terms of more basic numbers: for example, a rational number was defined in terms of ordered pairs of integers. So we need to represent ordered pairs as sets for this reason alone. But in fact ordered pairs and sets of ordered pairs will be a vital tool for representing vastly more mathematics within *ZF*, for instance functions and orderings. What we require, for each pair of sets x and y, is a set representing the ordered pair (x, y) that allows us to distinguish between the 'first' coordinate of the pair, x, and the 'second' coordinate, y. Using $\{x, y\}$ to represent this ordered pair will not do. By the axiom of extensionality, the set $\{x, y\}$ is the same as the set $\{y, x\}$. Even though we have written the x before the y in $\{x, y\}$, someone else, to whom we hand the pair set (as a single object, not with this particular written description), might not be able to reconstruct the order in which we had written the two elements down – that's why we call $\{x, y\}$ an unordered pair. So we have to find some other way to represent an ordered pair.

Our requirements are as follows.

1. For any sets x and y, we want a unique set to stand for (x, y).
2. We want to be able to distinguish (x, y) from (y, x), or indeed from any other (a, b) where $a \neq x$ or $b \neq y$.

The definition which follows can be shown to meet these requirements just using *ZF1*, *ZF2* and *ZF3*. To distinguish between our informal notion of the ordered pair (x, y) (which we shall continue to use where appropriate) and the particular set representing it in *ZF*, we shall write the latter as $\langle x, y\rangle$. (We shall use the notation $\langle x, y\rangle$ where we think it is important to realize that we are using this particular representation of the ordered pair as a set.)

Definitions

Given sets x and y, we write $\langle x, y\rangle$ for the set

$$\{\{x\},\{x,y\}\},$$

which is called an *ordered pair* . The x and y are called the *first* and *second coordinates*, respectively, of the ordered pair.

Exercise 4.10

Show that for sets x, y, there is a unique set $\langle x, y\rangle$. [Hint: this is a straightforward use of *ZF1* and *ZF3* along the lines on earlier work in this section.]

Thus $\langle x, y\rangle$ satisfies requirement 1 above. Requirement 2 requires much more work and we shall treat it as a theorem.

Theorem 4.2 The ordered pair property

For any sets x, y, u, v, if $\langle x, y\rangle = \langle u, v\rangle$, then $x = u$ and $y = v$.

We give this result a grand name as it is the critical property of ordered pairs!

Proof

Our proof will make extensive use of the axiom of extensionality (*ZF1*) and the standard properties of = (like $x = y$ implies $\{x\} = \{y\}$). Suppose that $\langle x, y\rangle = \langle u, v\rangle$, i.e.

$$\{\{x\},\{x,y\}\} = \{\{u\},\{u,v\}\}.$$

By extensionality every element of the set on the left-hand side of this equation (LHS for short) is an element of the right-hand side (RHS for short), and vice versa – we shall make repeated use of this principle.

First take the element $\{x\}$ of the LHS. This must equal one of the elements $\{u\}$ and $\{u, v\}$ of the RHS. We'll deal with these two possibilities as separate cases.

It might conceivably equal both of them, meaning, of course, that $\{u\}$ and $\{u, v\}$ are themselves equal.

Case $\{x\} = \{u\}$

By *ZF1* we have $x = u$. Now consider the element $\{x, y\}$ of the LHS. This must equal one of $\{u\}$ and $\{u, v\}$, giving us a couple of sub-cases. Just for the sake of variety, we shall investigate these in terms of whether $x = y$ or $x \neq y$.

If $x = y$ then by *ZF1* we have $\{x, y\} = \{x\}$, so that by *ZF1*

$$\begin{aligned}\{\{x\},\{x,y\}\} &= \{\{x\},\{x\}\}\\ &= \{\{x\}\}.\end{aligned}$$

As the $\{u, v\}$ on our original RHS must be an element of the LHS, this forces, again by *ZF1*, $\{u, v\} = \{x\}$, which in turn forces $u = x$ and $v = x$, by *ZF1*. This gives $x = y = u = v$, so that $x = u$ and $y = v$ as required.

If $x \neq y$ then we cannot have $\{x, y\} = \{u\}$, as *ZF1* would force $x = u$ and $y = u$, giving $x = y$, a contradiction! Thus we must have $\{x, y\} = \{u, v\}$. As

$u = x$, this gives $\{x, y\} = \{x, v\}$. By *ZF1* the element y of $\{x, y\}$ equals one of the elements of $\{x, v\}$, i.e. x or v. As $y \neq x$, we can only have $y = v$. Thus $x = u$ and $y = v$.

Hence if $\{x\} = \{u\}$ we must have $x = u$ and $y = v$.

Case $\{x\} = \{u, v\}$

This case is left as the next exercise for you. ■

Exercise 4.11

Complete the proof of Theorem 4.2 above by dealing with the case when $\{x\} = \{u, v\}$.

Solution

By *ZF1* $u = x$ and $v = x$, so that $u = v$. Thus by *ZF1* $\{u, v\} = \{u\}$, so that

$$\begin{aligned}\{\{u\}, \{u, v\}\} &= \{\{u\}, \{u\}\}\\ &= \{\{u\}\}\end{aligned}$$

and our original equation becomes

$$\{\{x\}, \{x, y\}\} = \{\{u\}\}.$$

The elements $\{x\}$ and $\{x, y\}$ of the LHS must both be elements of the RHS; but the RHS only has one element, namely $\{u\}$. Thus by *ZF1*

$$\{x\} = \{x, y\} = \{u\}.$$

From this, a similar argument using *ZF1* gives $x = y = u$. As $x = v$, this gives the required result that $x = u$ and $y = v$.

Our requirements for satisfactory representation of an ordered pair, in particular that we have defined a unique set and that this set can be 'decoded' in only one way, are typical of many set constructions.

We can exploit our definition of an ordered pair to define an ordered triple, an ordered quadruple and so on.

> ***Definition***
>
> Suppose that $x_1, x_2, x_3, \ldots, x_n$ are sets, where $n \in \mathbb{N}$, $n \geq 3$. Then the *ordered n-tuple* $\langle x_1, x_2, x_3, \ldots, x_n\rangle$ is defined recursively for $n \geq 3$ by
>
> $$\langle x_1, x_2, x_3, \ldots, x_n\rangle = \langle x_1, \langle x_2, x_3, \ldots, x_n\rangle\rangle,$$
>
> i.e. the ordered pair with first coordinate x_1 and second coordinate $\langle x_2, x_3, \ldots, x_n\rangle$.

A 3-tuple means the same as a triple.

For $n = 3$ this gives the ordered triple $\langle x, y, z\rangle$ as $\langle x, \langle y, z\rangle\rangle$. For any specific value of n, this definition gives a finite construction, coded by the n, of an n-tuple. You might wish to investigate the case $n = 3$ in the next exercise.

Exercise 4.12

Let x, y, z be sets.

(a) Show that, with the definition above, $\langle x, y, z\rangle$ is a unique set.

(b) Show that if u, v, w are sets with $\langle x, y, z\rangle = \langle u, v, w\rangle$, then $x = u$, $y = v$ and $z = w$.

This is the analogue of the ordered pair property for ordered triples.

(c) An alternative candidate for the definition of an ordered triple is $\langle\langle x, y\rangle, z\rangle$. Show that this also satisfies the properties in the preceding parts (so could have been used as as the definition). Does it represent the ordered triple by the same set as $\langle x, y, z\rangle$?

It is perhaps curious that, given three sets x, y, z, we can now define the ordered triple $\langle x, y, z\rangle$ as a set, whereas we cannot yet justify that there is a set $\{x, y, z\}$ with precisely these elements. Less surprising is that although we can construct an ordered pair as a set, we cannot yet construct interesting *sets* of ordered pairs, like Cartesian products $A \times B$. The batch of axioms in the next section will rectify this situation and enable us to do much more.

Further exercises

Exercise 4.13

Show that $\{x, \{y\}\}$ is not suitable as a definition of the ordered pair (x, y), because it does not have the ordered pair property (as in Theorem 4.2).

Exercise 4.14

Which, if any, of the following constructions give a suitable definition of an ordered triple $\langle x, y, z\rangle$? (The key point is whether the definition has the property dealt with in Exercise 4.12(b) above.)

In parts (a), (c) and (d), you may assume that it is possible to form sets with three elements!

(a) $\{\{x\}, \{x, y\}, \{x, y, z\}\}$

(b) $\langle\{x, y\}, \{y, z\}\rangle$

(c) $\{\langle x, y\rangle, \langle y, z\rangle, \langle z, x\rangle\}$

(d) $\{\langle x, y\rangle, \langle x, z\rangle, \langle y, z\rangle\}$

(e) $\{\langle x, y\rangle, \langle y, z\rangle\}$

(f) $\{\langle x, \langle x, y\rangle\rangle, \langle\langle y, z\rangle, z\rangle\}$

4.4 Axioms 4 to 6

In the previous section we used axioms *ZF1* to *ZF3* to define ordered pairs as sets. In this section we shall look at the next three axioms of *ZF* which will, amongst other things, allow us to construct interesting sets of ordered pairs like Cartesian products $A \times B$. Using Cartesian products, we shall then show how to represent functions by sets within *ZF*.

Axioms of ZF

ZF4: Axiom of separation

$\forall x \exists y \forall z (z \in y \leftrightarrow (z \in x \wedge \phi(z)))$, where $\phi(z)$ is any statement of the formal language (possibly referring to named sets) with free variable z.

For any set x there is a set consisting of all z in x for which $\phi(z)$ holds.

ZF5: Power set axiom

$\forall x \exists y \forall z (z \in y \leftrightarrow z \subseteq x)$

For any set x there is a set consisting of all subsets of x.

ZF6: Union axiom

$\forall x \exists y \forall z (z \in y \leftrightarrow \exists w (z \in w \wedge w \in x))$

For any set x there is a set which is the union of all the elements of x.

In the previous section, we showed how to define $z \subseteq x$ in terms of the basic formal language. We can therefore use this abbreviation here.

Before discussing these axioms, let us introduce some important and familiar notation associated with them. Each axiom has the form 'for all sets x there is a set y such that ...', and by an argument using the axiom of extensionality along the lines of Exercise 4.8 in the previous section, we can show that for any given x the y is unique. So we can introduce some notation for this y.

Definitions

Let x be a set.

For each formula $\phi(z)$ (possibly referring to named sets) we write $\{z \in x : \phi(z)\}$ for the set y in *ZF4* such that

$$\forall z (z \in y \leftrightarrow (z \in x \wedge \phi(z))).$$

We write $\mathscr{P}(x)$, called the *power* set of x, for the set y in *ZF5* such that

$$\forall z (z \in y \leftrightarrow z \subseteq x).$$

We write $\bigcup x$, called the *union* of x, for the set y in *ZF6* such that

$$\forall z (z \in y \leftrightarrow \exists w (z \in w \wedge w \in x)).$$

This now extends our legitimate use of the curly bracket notation for sets within *ZF* in a big way!

Of course we can also call this the set of all subsets of x!

Exercise 4.15

Let x be a set. Show that there is a *unique* set y whose elements are all the subsets of x.

This new notation makes it easier to see that the axiom of separation is a version of the comprehension principle; but it is a very restricted version. First of all, it limits the properties which we can use to define sets by the comprehension principle to those expressible by a formula $\phi(z)$ of our formal language (where $\phi(z)$ might involve references to named sets): this isn't enough of a limitation to bar Russell's paradox, as you may recall from Exercise 4.6 in Section 4.2. The real restriction is that, rather than allowing us to create the set of all zs with the property $\phi(z)$, it only allows us to 'separate out' the elements z of a *given set* x with this property. The axiom is thus saying that, given a set x, there are certain sets definable as subsets of x.

Does axiom *ZF4* avoid Russell's paradox? See Exercise 4.32 below.

By way of contrast, the power set axiom says that there is a set $\mathscr{P}(x)$ of all subsets of a set x without telling us anything about any of these subsets. This axiom may seem reasonable, given our experience with finite sets x, even though such a set with n elements has rather a lot of subsets, namely 2^n. Whether it represents an intuitive statement about infinite sets is debatable. But as it leads to some of the most exciting parts of Cantor's original work on infinite numbers, there is not much point in arguing about its inclusion!

The union axiom and the notation $\bigcup x$ might take a bit of time to absorb. We are all comfortable with the idea of the union $a \cup b$ of two set a and b. Imagine the set x as having as elements the sets $a, b, c, d, \ldots$: then $\bigcup x$ is their union $a \cup b \cup c \cup d \cup \ldots$. Or perhaps you have seen the symbol $\bigcup$ used in everyday mathematics in something like $\bigcup\{A_i : i \in I\}$, in which case think of $\bigcup x$ as the same as $\bigcup\{y : y \in x\}$. Why do we have an axiom about unions in this form? Why don't we have an axiom guaranteeing us that $a \cup b$ is a set for any sets a, b? Indeed, can we show that $a \cup b$ is a set?

$z \in \bigcup x$ if and only if $z \in y$ for some $y \in x$.

We shall occasionally backslide on 'legitimate' use of the curly bracket notation by writing $x = \{y : y \in x\}$, instead of $x = \{y \in x : y \in x\}$ – the latter is legitimate by *ZF4*.

Exercise 4.16

What are the elements of $\bigcup\{a, b\}$?

Solution

For any set z, $z \in \bigcup\{a, b\}$ if and only if $z \in y$ for some $y \in \{a, b\}$. The only ys in $\{a, b\}$ are a and b. Thus $z \in \bigcup\{a, b\}$ if and only if $z \in a$ or $z \in b$, i.e z is an element of what we customarily call $a \cup b$.

As for any sets a and b there is a set $\{a, b\}$ (by *ZF3*), so that $\bigcup\{a, b\}$ is a set (by *ZF6*), we can indeed define $a \cup b$ within *ZF*.

Definition

The *union* of two sets a and b, written as $a \cup b$, is the set $\bigcup\{a, b\}$.

Exercise 4.17

Why do you think we need the union axiom in its given form, rather than simply have an axiom guaranteeing that $a \cup b$ is a set?

Solution

If we want to take the union of infinitely many sets, say $A_0, A_1, A_2, \ldots, A_n, \ldots$, we cannot do this using the $\cup$ operator, as this would require an infinitely long formula

$$A_0 \cup A_1 \cup A_2 \cup \ldots \cup A_n \cup \ldots.$$

Providing that there is a set $\{A_n : n \in \mathbb{N}\}$, the $\bigcup$ operator will allow us to represent the union by a finitely long formula, namely

$$\bigcup\{A_n : n \in \mathbb{N}\}.$$

The union axiom, like the power set axiom, gives a way of showing that there are bigger sets around than those currently in our list. For instance, we can now show (with a spot of ingenuity!) that if we have three sets x, y, z then there is a set with these as its elements. First of all, we can form the sets $\{x, y\}$ and $\{z\}$ (essentially using *ZF3*). We then take the union of these two sets $\{x, y\} \cup \{z\}$. This is of course the desired set with elements x, y, z and we are entitled to extend the use of our curly bracket notation within *ZF* and write this as $\{x, y, z\}$.

Exercise 4.18

Let $x_1, x_2, \ldots, x_n$ be sets, for $n \in \mathbb{N}$ with $n \geq 1$.

(a) Show that there is a set with $x_1, x_2, \ldots, x_n$ as its elements.

(b) Show that $x_1 \cup x_2 \cup \ldots \cup x_n$ is a set.

The result of this allows us to extend our curly bracket notation and write $\{x_1, x_2, \ldots, x_n\}$ for the set with $x_1, x_2, \ldots, x_n$ as its elements.

Exercise 4.19

Show that $\bigcup\{x\} = x$.

Exercise 4.20

Show that if $y \in x$ then $y \subseteq \bigcup x$.

Exercise 4.21

Show that if x is a set, then so is x^+, where $x^+ = x \cup \{x\}$.

This is an item on our shopping list for sets arising from our construction of $\mathbb{N}$ using sets.

We have shown that for sets a and b their union $a \cup b$ is a set. How about their intersection $a \cap b$? It is tempting to say that this is a set because

$$a \cap b = \{z : (z \in a \ \wedge \ z \in b)\};$$

but this use of the comprehension principle hasn't been authorized by our axioms. The property that the zs must have is fine: it is indeed given by a formula of the language. But we also need the zs to be 'separated' out of some set x. We can find a suitable set x by taking it to be $a \cup b$, which we already know to be a set. And now we can correctly use the axiom of separation to justify that $a \cap b$ is a set, justifying the definition below.

> ***Definition***
>
> The *intersection* of two sets a and b, written as $a \cap b$, is the set $\{z \in a \cup b : (z \in a \ \wedge \ z \in b)\}$.

The method we have just used for showing that $a \cap b$ is a set is very powerful. We can frequently define a 'set' using the comprehension principle, $\{z : z \text{ has property } P\}$, where the property P is expressible by a formula $\phi(z)$ in our formal language. Our definition almost looks like the format for the axiom of separation, except that we haven't got a set x from which to 'separate out' the zs. We can often construct a suitable 'big' set x using one or both of *ZF5* and *ZF6*; and then use *ZF4* to obtain the set we really wanted.

Exercise 4.22

Let x be a non-empty set. Show that the *intersection* of x, written as $\bigcap x$, and defined informally by

$$\bigcap x = \{z : z \in y \text{ for all } y \in x\},$$

is indeed a set in *ZF*. [Hints: show that $\bigcap x \subseteq \bigcup x$, and then use the axiom of separation.]

So if $x = \{a, b, c, \ldots\}$ then $\bigcap x = a \cap b \cap c \cap \ldots$

Exercise 4.23

Let x be a set. Show that $\{\{y\} : y \in x\}$ is a set.

Now that we have dealt with union and intersection, let us deal with the other familiar operation on sets, namely *complement*.

Exercise 4.24

One might sloppily attempt to describe the complement of a set x as the set of all objects not in x, namely $\{z : z \notin x\}$. Is this a set?

Solution

If x is a set and $\{z : z \notin x\}$ is also a set, then their union would be a set. But this union equals $\mathscr{V}$, the universe of sets, so that $\mathscr{V}$ is a set. We could then use the axiom of separation to show that

$$\{z \in \mathscr{V} : z \notin z\}$$

is a set – the $\mathscr{V}$ takes the place of the set x in the standard format for this axiom. As $z \in \mathscr{V}$ says no more than that z is a set, we have really shown that

$$\mathscr{R} = \{z : z \notin z\}$$

is a set. But this means that Russell's paradox can be derived in *ZF*.

Thus if the complement $\{z : z \notin x\}$ is a set, *ZF* is inconsistent. For reasons which we shall discuss later, this does not prove that the complement is not a set: it only shows us that it is highly undesirable that this should be the case!

Note that this last solution also shows that assuming $\mathscr{V}$ is a set leads to Russell's paradox.

In practice in mathematics, we never use the complement of x in the way above. We always use the complement of x in another set, y say, often written as $y \setminus x$. That this is a set is easily shown using separation, as follows immediately from the definition below. (The class $\{z : z \notin x\}$ is often called the *absolute complement* of x.)

> ***Definition***
>
> Let x and y be sets. Then the *complement* of x in y, written as $y \setminus x$, is the set $\{z \in y : z \notin x\}$.

We have already defined ordered pairs. We shall now show that Cartesian products consisting of sets of them exist. Given sets X and Y, their *Cartesian product* $X \times Y$ is the set $\{\langle x, y\rangle : x \in X, y \in Y\}$. This is of course the normal mathematical definition, but it doesn't match the axiom of separation. We can get closer to the correct format of this axiom by writing the set as

$$\{z : \exists x \exists y ((x \in X \ \wedge \ y \in Y) \ \wedge \ z = \langle x, y\rangle)\}.$$

As $\langle x, y\rangle$ can be defined within our formal language, the defining property is also expressible within it. All we want is some known 'big' set in which the zs sit.

Cartesian products play a key role in set theory. They provide a way of representing by sets important objects like functions and order relations, which don't immediately strike one as being sets. Thus a theory only of sets can deal simultaneously with objects like numbers and sets of numbers and also any mathematical relationships between them.

A typical z is of the form $\{\{x\}, \{x, y\}\}$, where $x \in X$ and $y \in Y$. Now x and y are both elements of $X \cup Y$, which is known to be a set. The key observation is to note that the element $\{x, y\}$ of z is a *subset* of $X \cup Y$, so that

$$\{x, y\} \in \mathscr{P}(X \cup Y).$$

Likewise the other element $\{x\}$ is also an element of $\mathscr{P}(X \cup Y)$, so that, repeating the observation about subsets,

$$z = \{\{x\}, \{x, y\}\} \subseteq \mathscr{P}(X \cup Y),$$

giving

$$z \in \mathscr{P}(\mathscr{P}(X \cup Y)).$$

As X and Y are sets, so is $\mathscr{P}(\mathscr{P}(X \cup Y))$. Thus we can now define $X \times Y$ in *ZF*, because we can use *ZF4* to separate out those elements of $\mathscr{P}(\mathscr{P}(X \cup Y))$ which are ordered pairs, as follows.

> ***Definition***
>
> For sets X and Y the *Cartesian product*, written as $X \times Y$, is the set
>
> $$\{z \in \mathscr{P}(\mathscr{P}(X \cup Y)) : \exists x \exists y ((x \in X \ \wedge \ y \in Y) \ \wedge \ z = \langle x, y\rangle)\}.$$

Exercise 4.25

Let $X_1, X_2, \ldots, X_n$ be sets, where $n \in \mathbb{N}$ with $n \geq 2$. Show that the Cartesian product $X_1 \times X_2 \times X_3 \times \ldots \times X_n$ is a set, where this is defined recursively by

$$X_1 \times X_2 \times X_3 \times \ldots \times X_n = X_1 \times (X_2 \times X_3 \times \ldots \times X_n).$$

This definition matches that of ordered n-tuple, so that $X_1 \times X_2 \times X_3 \times \ldots \times X_n$ is a set of ordered n-tuples. There are alternative definitions. In what follows we shall use Cartesian products of more than two sets without being too fussy about their precise definition in *ZF*.

Now that we can represent Cartesian products by sets within *ZF*, we can also represent important objects like relations (e.g. orders) and functions by sets. For instance the order $<_{\mathbb{N}}$ on $\mathbb{N}$ can be represented as a set by

$$\{\langle \mathbf{m}, \mathbf{n}\rangle \in \mathbb{N} \times \mathbb{N} : \mathbf{m} \in \mathbf{n}\},$$

which will be a set by the axiom of separation, once we have shown that $\mathbb{N}$ is a set. For the rest of this section, we shall concentrate on how Cartesian products are used to represent functions as sets. The key observation from everyday mathematics is that virtually all the information about a function $f\colon A \longrightarrow B$ is contained in the set of ordered pairs $\{(a, f(a)) : a \in A\}$. This set is often called the *graph* of the function f.

For the benefit of purists, the only information which gets lost is that the codomain of the function is B. Given the graph of f, one can recover the image set, or range, of f – all one knows about the original codomain is that it contains the range as a subset.

Exercise 4.26

Given a subset F of $A \times B$, how might we judge whether it is the graph of some function f with domain A and codomain B?

Solution

For f to be a function with domain A, we require that for each $a \in A$ there is a unique element to which a maps under f. In terms of ordered pairs in the graph of f, this means that there is a unique $b \in B$, namely $b = f(a)$, such that (a, b) is in the graph. So given a subset F of $A \times B$, F is the graph of some function with domain A if for each $a \in A$ there is a unique $b \in B$ such that $(a, b) \in F$.

In terms of our formal language, in which we would of course use our representation of ordered pairs by sets $\langle a, b\rangle$, this condition on F becomes

$$\forall a(a \in A \rightarrow \exists b(b \in B \wedge \langle a, b\rangle \in F \wedge \forall b'(\langle a, b'\rangle \in F \rightarrow b' = b))).$$

What we have just done becomes the first of a couple of definitions of functions represented by sets within *ZF*.

For the sake of ease of reading, we shall drop the brackets where several statements are joined by the connective $\wedge$: e.g. we shall write $(\phi \wedge \psi \wedge \theta)$ rather than $(\phi \wedge (\psi \wedge \theta))$ or $((\phi \wedge \psi) \wedge \theta)$ which are logically equivalent. We shall similarly drop brackets when we are joining several formulas together with $\vee$.

Definition

Let A, B and F be sets with $F \subseteq A \times B$. We say that F is a *function with domain A and codomain B* if

$$\forall a(a \in A \rightarrow \exists b(b \in B \wedge \langle a, b\rangle \in F \wedge \forall b'(\langle a, b'\rangle \in F \rightarrow b' = b))).$$

More colloquially, we say that F is a function *from A to B*.

As with other mathematical objects represented by sets within ZF, *we shall continue to use the familiar mathematical notation, here $f\colon A \longrightarrow B$, except where our particular representation of the object as a set is important.*

Once we are representing functions by the sets of pairs in their graphs, it will become useful to be able to talk about a set F of pairs as a function without necessarily specifying in advance what its domain is. The key property which we shall require is that for any a there is *at most* one b such that $\langle a, b\rangle \in F$. This can done by a simple formula. We shall then define the domain of F to be the set of those as for which there is some b with $\langle a, b\rangle \in F$. We would like to write this set as $\{a : \exists b\ \langle a, b\rangle \in F\}$; but we have the usual problem with the axiom of separation that we need to specify a set in which the as lie, from which we can then separate out the required set.

Exercise 4.27

Given a set F of ordered pairs $\langle a, b\rangle$, give a set in *ZF* in which all the as and bs lie.

Solution

A typical element $\langle a, b\rangle$ of F is a set $\{\{a\}, \{a, b\}\}$. Thus $\bigcup F$ is the set of all the $\{a\}$s and $\{a, b\}$s for $\langle a, b\rangle \in F$, so that $\bigcup(\bigcup F)$ is the set of all as and bs for which $\langle a, b\rangle \in F$. As F is a set, $\bigcup(\bigcup F)$ is also a set, by the union axiom.

The result of this last exercise will allow us to define not only the domain of a function using separation, but also its range (or image set).

Definitions

Let F be a set of ordered pairs. F is said to be a *function* if

$$\forall a \forall b \forall b' ((\langle a, b\rangle \in F \wedge \langle a, b'\rangle \in F) \rightarrow b = b').$$

The *domain* of a function F, written as $\mathrm{Dom}(F)$, is the set defined by

$$\mathrm{Dom}(F) = \{a \in \bigcup(\bigcup F) : \exists b\ \langle a, b\rangle \in F\}.$$

$\mathrm{Dom}(F)$ is a set by *ZF4* and *ZF6*.

There are some further important sets associated with functions, as in the following definitions.

Definitions

Suppose that the set F is a function.

The *range* of a function F, written as $\mathrm{Range}(F)$, is the set defined by

$$\mathrm{Range}(F) = \{b \in \bigcup(\bigcup(F)) : \exists a\ \langle a, b\rangle \in F\}.$$

$\mathrm{Range}(F)$ is a set by *ZF4* and *ZF6*. It is also called the *image set* of the function.

Suppose that X is a subset of $\mathrm{Dom}(F)$. Then the *restriction* of the function F to X, written as $F|_X$, is the set

$$\{\langle a, b\rangle \in F : a \in X\}.$$

$F|_X$ is a set by *ZF4*. It is a function with domain X – see Exercise 4.30.

In general, if G is a function with $G \subseteq F$, then G is said to be a *restriction* of F and F is said to be an *extension* of G.

Note that with this definition of a function as a set of pairs with a special property, two functions are equal exactly when they are equal as sets.

In many mathematics books, two functions are equal only when they have the same codomain, as well as having the same graph.

Exercise 4.28

Let F be the function $\{\langle \mathbf{0},\mathbf{1}\rangle, \langle \mathbf{1},\mathbf{2}\rangle, \langle \mathbf{3},\mathbf{1}\rangle\}$. Write down the elements of each of $\mathrm{Dom}(F)$, $\mathrm{Range}(F)$, and $F|_{\mathbf{2}}$.

Recall that $\mathbf{2} = \{\mathbf{0},\mathbf{1}\}$.

Solution

$$\mathrm{Dom}(F) = \{\mathbf{0},\mathbf{1},\mathbf{3}\};$$
$$\mathrm{Range}(F) = \{\mathbf{1},\mathbf{2}\};$$
$$F|_{\mathbf{2}} = \{\langle \mathbf{0},\mathbf{1}\rangle, \langle \mathbf{1},\mathbf{2}\rangle\}.$$

Exercise 4.29

Is the empty set $\varnothing$ a function?

Solution

We have not insisted that there are any ordered pairs in the set F representing a function. And the property

$$\forall a \forall b \forall b' ((\langle a,b\rangle \in F \ \wedge \ \langle a,b'\rangle \in F) \ \to \ b = b')$$

is vacuously true when F is empty. Thus $\varnothing$ is a function.

The argument is essentially that $\varnothing$ is a function because it is not the case that it is not a function!

Exercise 4.30

Let F be a function and X a subset of $\mathrm{Dom}(F)$. Show that $F|_X$ is indeed a function with domain X.

Exercise 4.31

Let $\mathscr{C}$ be a set of functions which has the following property:

for any $f,g \in \mathscr{C}$, $f \subseteq g$ or $g \subseteq f$.

Show that $\bigcup \mathscr{C}$ is a function and describe its domain in terms of the domains of the functions f in $\mathscr{C}$.

$\mathscr{C}$ is said to be a *chain* under ordering by $\subseteq$, or a $\subseteq$-*chain* for short.

An important set of functions is the set X^Y of all functions from a set Y to a set X. Every such function is a subset of $Y \times X$, so is an element of $\mathscr{P}(Y \times X)$. This allows us to define X^Y as a set using the axiom of separation as follows.

> ***Definition***
>
> Let X, Y be sets. Then the set of all functions from Y to X, written as X^Y, is defined by
>
> $$X^Y = \{F \in \mathscr{P}(Y \times X) : F \text{ is a function}\}.$$

Is the notation meant to look like exponentiation? Yes, it is! The reason will become clearer in Chapter 6 when we look at cardinal arithmetic. For the moment, reflect on the fact that the number of functions from a y-element set Y to an x-element set X, where $x, y \in \mathbb{N}$, is x^y, which matches the notation X^Y for the set of all these functions.

For each of the y elements of Y, there are x independent choices of its image in X under a function. This gives $\underbrace{x \cdot x \cdot \ldots \cdot x}_{y}$ different functions.

We might occasionally need a formal representation of sequences by sets. A sequence $\langle x_n\rangle$ can be regarded as a set of ordered pairs $\{\langle n, x_n\rangle : n \in \mathbb{N}\}$: this is as good a way as any of associating an object with each $n \in N$. Formally, we have the following.

> ***Definitions***
>
> A set S is said to be a *sequence of elements of the set* C if S is a function with domain $\mathbb{N}$ and codomain C. We can write $\langle S(\mathbf{n})\rangle$ for this sequence. We shall periodically follow standard mathematical practice by referring to the values of the function S as *elements of the sequence.*

We would normally write a sequence as $\langle x_n \rangle$, using subscript notation, and being a bit casual in using n instead of $\mathbf{n}$. And strictly speaking x_n shouldn't be called an element of the sequence, as within ZF it is in fact an element of the range of the function representing the sequence.

There's one minor problem with this definition and this is that we don't yet know that our system so far guarantees that $\mathbb{N}$, as we constructed it in Chapter 3, is a set. In fact it can be shown that the system so far does not guarantee the existence of *any* infinite sets, so that a further axiom is required, which we shall look at in the next section along with two further axioms of ZF.

Zermelo's main objective in producing axioms was to solve an important problem arising from Cantor's work. And a major objective of this book is to describe Dedekind's and Cantor's work within the framework of these axioms. But a subsidiary objective has to be ensuring that the theory avoids contradictions like Russell's paradox.

Exercise 4.32

Let x be a set. By the axiom of separation, there is a set y defined by

$$y = \{z \in x : z \notin z\},$$

which is similar to the construction involved in Russell's paradox. Is there anything paradoxical about this set y?

Solution

To match Russell's paradox, we would have to have that if $y \in y$ then $y \notin y$, and if $y \notin y$ then $y \in y$.

First, what happens if $y \in y$? Then indeed y is a $z \in x$ for which $z \notin z$, so that $y \notin y$, giving a contradiction. We can thus conclude that $y \notin y$.

But $y \notin y$ no longer automatically leads to the contradiction, as with Russell's paradox, that $y \in y$. This is because there is no reason to suppose that y is an element of x. Were it the case that $y \in x$, as well as $y \notin y$, so that y is a $z \in x$ for which $z \notin z$, we could indeed conclude that $y \in y$, giving us a contradiction. All this last argument does is to force us to conclude that in fact y is *not* an element of x, assuming that we are fighting to the last to avoid contradictions in ZF!

Note that we have not definitively shown that $y \notin x$. It might be the case that there is some set x for which the y is an element of x, from which it would follow that ZF is inconsistent. All we have really shown is that Russell's short and snappy argument doesn't automatically work for our set y.

The work of Kurt Gödel in the 1930s has the consequence that the consistency of ZF cannot be proved within ZF. So we cannot be sure that the axioms of ZF avoid some ultimate contradiction. The same applies to a very broad class of alternative axiom systems, so ZF is not at any particular fault. The important feature of the axioms is that they provide a framework for much of mathematics, especially the theories of the natural numbers, of the reals, and of infinite sets, and that so far no one has derived a contradiction from them.

A system is *consistent* if one cannot derive a contradiction within it.

For convenience we shall assume from now on that ZF *is consistent*. This will enable us to distinguish between proper classes and sets: a class is a proper class and not a set when assuming that it is a set leads to a contradiction. Thus we shall say that $\{x : x \notin x\}$ is not a set and is a proper class; and likewise the absolute complement of a set x, namely $\{z : z \notin x\}$, is not a set (by Exercise 4.24).

This is a key assumption. It essentially relieves us from making statements of the form 'if X is a set then we can derive Russell's paradox within *ZF*'. We can instead say 'X is not a set'.

Further exercises

Exercise 4.33

Which of the following statements are true for all sets A and B and which are false? In each case either prove the statement or give a counterexample as appropriate.

(a) $\bigcup(A \cup B) = \bigcup A \cup \bigcup B$
(b) $\bigcap(A \cap B) = \bigcap A \cap \bigcap B$
(c) $\bigcup(A \cap B) = \bigcup A \cap \bigcup B$
(d) $\bigcap(A \cup B) = \bigcap A \cap \bigcap B$
(e) $\bigcap(A \cap B) = \bigcap A \cup \bigcap B$
(f) $\bigcup \mathscr{P}(A) = A$
(g) $\mathscr{P}(\bigcup A) = \bigcup \mathscr{P}(A)$
(h) $\mathscr{P}(A \times B) = \mathscr{P}(A) \times \mathscr{P}(B)$

Exercise 4.34

Let M be a non-empty set. Explain why

$$\{A \times \{A\} : A \subseteq M, A \neq \varnothing\}$$

is a set.

Exercise 4.35

In Chapter 2 we defined real numbers in terms of rationals, rationals in terms of integers, and integers in terms of natural numbers. Assuming that $\mathbb{N}$ is a set within *ZF*, from which sets will we be able to separate out the sets $\mathbb{Z}$, $\mathbb{Q}$ and $\mathbb{R}$? Recall that an integer k in $\mathbb{Z}$ was defined as an equivalence class of pairs of natural numbers, so that k can be represented as a particular subset of $\mathbb{N} \times \mathbb{N}$. This means that $k \in \mathscr{P}(\mathbb{N} \times \mathbb{N})$, so that $\mathbb{Z}$ can be separated out of the set $\mathscr{P}(\mathbb{N} \times \mathbb{N})$.

(a) Treating real numbers as being given by Dedekind cuts of rationals (and the rationals as being given by equivalence classes of pairs of integers, etc.), show that $\mathbb{R}$ is a subset of

$$\mathscr{P}(\mathscr{P}(\mathscr{P}(\mathbb{N} \times \mathbb{N}) \times \mathscr{P}(\mathbb{N} \times \mathbb{N}))).$$

(b) Find a similar expression in terms of $\mathbb{N}$ for the real numbers treated as equivalence classes of Cauchy sequences of rationals.

Exercise 4.36

Use appropriate axioms out of *ZF1* to *ZF6* to show that neither of the following classes is a set.

(a) $\{x : x \text{ is a singleton}\}$

(b) $\{x : x \text{ is an ordered pair}\}$

Remember that this requires showing that if either is a set, then Russell's paradox or some other contradiction can be derived in *ZF*.

Exercise 4.37

In Section 3.4 of Chapter 3 we defined X to be finite when there is a bijection $f\colon \mathbf{n} \longrightarrow X$ for some $\mathbf{n} \in \mathbb{N}$. Show that '$X$ is finite' is representable by a formula within our formal language.

4.5 Axioms 7 to 9

We now come to the last three axioms of *ZF*. These have very different characters. One will ensure both that $\mathbb{N}$ is a set and that there are infinite sets (the latter being the main point of the book). The other two are considerably more technical, but one of them will have huge benefits, including underpinning defining functions and sets recursively within *ZF*, while the other will give an agreeable structure to the universe $\mathscr{V}$ of sets. We shall state the axioms and explain them in detail later on.

Axioms of ZF

ZF7: Axiom of infinity

$\exists x(\varnothing \in x \wedge \forall y(y \in x \rightarrow y \cup \{y\} \in x))$

There is an inductive set.

ZF8: Axiom of replacement

$\forall x \exists y \forall y'(y' \in y \leftrightarrow \exists x'(x' \in x \wedge \phi(x', y')))$, where $\phi(s,t)$ is a formula (possibly referring to named sets) such that $\forall s \exists t(\phi(s,t) \wedge \forall t'(\phi(s,t') \rightarrow t' = t))$.

If $\phi(s,t)$ is a class function, then when its domain is restricted to x, the resulting images form a set y.

ZF9: Axiom of foundation

$\forall x \exists y(y \in x \wedge x \cap y = \varnothing)$

Every set is *well-founded*, i.e. contains an $\in$-minimal element.

What does the axiom of infinity have to do with infinity? Axiom *ZF7* guarantees that there is an inductive set, so that we can define $\mathbb{N}$ as in Chapter 3 and show that $\mathbb{N}$ is a set within *ZF*. We can then define both 'finite' and 'infinite' within *ZF* in terms of this set $\mathbb{N}$, as done in Section 3.4. And we can then show that the set $\mathbb{N}$ itself is infinite, as in Exercise 3.39 of Section 3.4 so

Zermelo's original axiom of infinity was in a slightly different form. See Exercise 4.52. The other two axioms were added later (by Fraenkel and others) to enrich the formal theory.

that there is an infinite set. It can be shown that the remaining axioms of *ZF* do not entail that there is an infinite set: there is an interpretation of these axioms in which every set is finite. Given that the excitement of Cantor's work lies in the theory of infinite sets, it is thus crucial that we have an axiom like *ZF7* guaranteeing the existence of at least one such set. It is perhaps remarkable that from an axiom asserting the existence of just one infinite set, it follows that there are many more, as we shall see in later chapters.

Exercise 4.38

In Chapter 3 we defined $\mathbb{N}$ as the intersection of all inductive subsets of any inductive set y. Explain why $\mathbb{N}$ is a set within *ZF*.

Solution

By the axiom of infinity there is an inductive set x. Then $\mathscr{P}(x)$ is also a set, so that by the axiom of separation

$$\{z \in \mathscr{P}(x) : z \text{ is inductive}\}$$

'z is inductive' is easy to express in our formal language.

is a set. Then by Exercise 4.22 in the previous section

$$\bigcap\{z \in \mathscr{P}(x) : z \text{ is inductive}\}$$

is a set; and by definition, this is $\mathbb{N}$, so that $\mathbb{N}$ is a set.

The axiom of replacement, *ZF8*, was not one of Zermelo's original axioms. Fraenkel and others noticed that these axioms did not imply the existence of a set

$$\{\mathbb{N}, \mathscr{P}(\mathbb{N}), \mathscr{P}(\mathscr{P}(\mathbb{N})), \mathscr{P}(\mathscr{P}(\mathscr{P}(\mathbb{N}))), \ldots\},$$

We shall show that this is a set within *ZF* at the end of this section.

which has significance in Cantor's theory of infinite sets. The axiom of replacement makes sure that this is not only a set, but also has a host of desirable consequences. Not the least of the latter is that we can define functions on $\mathbb{N}$ by recursion, as we shall show later in this section; and perhaps even more importantly, we can do the same for functions on ordinals, as we shall see in Chapter 8.

Let us look at what the axiom of replacement says. The key ingredient is a formula $\phi(s,t)$ with the property that for every set s there is a unique set t such that $\phi(s,t)$ holds. So $\phi(s,t)$ is rather like a function.

Exercise 4.39

Why might $\phi(s,t)$ not actually describe a function in *ZF*?

Solution

Corresponding to $\phi(s,t)$ is the class of ordered pairs

$$\{\langle s,t\rangle : \phi(s,t) \text{ holds, } s \text{ any set}\}.$$

If this class is a set, it would indeed be a function within *ZF*. But if it is a set, so is the class of first coordinates of the ordered pairs it contains, namely the universe $\mathscr{V}$. But $\mathscr{V}$ is not a set, so neither is the class corresponding to $\phi(s,t)$.

Because the pairs $\langle s,t\rangle$ for which $\phi(s,t)$ holds form a proper class, we can describe $\phi(s,t)$ as being a *class function*. Now suppose that we restrict this class function to a set x as domain, to obtain

$$\{\langle s,t\rangle : s \in x \text{ and } \phi(s,t) \text{ holds}\}.$$

This looks very like a function in ZF. Its domain is a set and for each element in the domain there is a unique image to which it maps. Indeed we shall say that $\phi(s,t)$ *behaves like a function with domain* x. There is just one problem! We don't know to which set the images belong. What the axiom of replacement tells us is that there is a set y consisting of these images.

It seems so 'obvious' that $\phi(s,t)$ ought to define a function (as a set) that $ZF8$ is an uncontroversial axiom, unlike some of the others!

Exercise 4.40

Use the axiom of replacement to show that if $\phi(s,t)$ behaves like a function with domain x then it does indeed define a function in ZF.

Solution

By the axiom of replacement there is a set y consisting of all ts such that $\phi(s,t)$ holds for some $s \in x$. Thus all the corresponding ordered pairs $\langle s,t\rangle$ are elements of the set $x \times y$, so that by the axiom of separation $\{\langle s,t\rangle \in x \times y : \phi(s,t)\}$ is a set. Because $\phi(s,t)$ behaves like a function, this set is indeed a function within ZF.

Note that y is the image set, or range, of this function.

As an application of this axiom, we shall provide an alternative argument to that we hoped you used in solving Exercise 4.23, which asked you to show that if x is a set, so is $\{\{y\} : y \in x\}$. Let $\phi(s,t)$ be the formula

The intended solution for this exercise was one using $ZF4$ and $ZF5$.

$$t = \{s\}.$$

This behaves like a function when its domain is restricted to any set x (as for each $s \in x$ there is a unique t equal to $\{s\}$). So by the axiom of replacement there is a set consisting of all the corresponding images, i.e. $\{\{y\} : y \in x\}$ is a set.

We shall look at a major application of the axiom of replacement later in this section, where it plays a crucial role in defining functions within ZF by recursion. Meanwhile there are a couple of useful technical dodges in its use which we shall mention now. First of all, the statement of the axiom, writing $\phi(s,t)$ just involving two variables s and t, makes it look as though it can only be used to define functions of one variable. But for instance a function of two variables s_1, s_2 could be represented by a formula $\phi(s,t)$ where s is the ordered pair $\langle s_1, s_2\rangle$. For simplicity we would usually write the formula as $\phi(s_1, s_2, t)$ rather than $\phi(\langle s_1, s_2\rangle, t)$. The other dodge is used when we have a unique set t associated by a formula $\phi(s,t)$ with each s in a particular set x which interests us – our aim being to show that there is a set $\{t : s \in x \wedge \phi(s,t)\}$ using replacement – but we don't have any set t to associate with sets not in this x. All we do is bolt on an extra clause with 'and', associating some fixed set with each of these extra sets s, as in

$\{\langle s,t\rangle : \phi(s,t)\}$ would be the graph of a function of one variable mapping s to the t such that $\phi(s,t)$.

$$(s \in x \rightarrow \phi(s,t)) \wedge (s \notin x \rightarrow t = \varnothing):$$

The $\varnothing$ could be replaced by any fixed set.

this formula is now a class function, and it behaves like a function with domain x, so that the axiom of replacement gives that $\{t : s \in x \wedge \phi(s,t)\}$ is a set.

Let us now have a look at *ZF9*, the axiom of foundation. This says that each set x contains an element y such that $x \cap y = \varnothing$. This y is said to be an $\in$-*minimal element* of x and in general a set might have more than one such element: e.g. $\{\{1\}, \{1,3\}, \{\{1\},2,\{1,3\}\}\}$ contains two $\in$-minimal elements, $\{1\}$ and $\{1,3\}$, while the other element $\{\{1\},2,\{1,3\}\}$ is not $\in$-minimal, as it has elements which are also elements of the original set. As you can see, this is a somewhat more obscure axiom than some of the others – it may not be immediately obvious what its significance is! Try the next exercise for yourself, to see one of its desirable consequences.

Exercise 4.41

(a) Let z be any set. By using the axiom of foundation with $x = \{z\}$ show that $z \notin z$.

(b) Use the first part to deduce that the universe $\mathscr{V}$ of all sets is not itself a set.

Solution

(a) As z is a set, so is $\{z\}$. By the axiom of foundation there is an element y of $\{z\}$ such that $\{z\} \cap y = \varnothing$. As the only element of $\{z\}$ is z, this y can only be z, so that

$$\{z\} \cap z = \varnothing.$$

This means that the element z of the $\{z\}$ is not in the other set involved in this intersection, i.e. $z \notin z$, as required.

(b) If $\mathscr{V}$ was a set z, then by definition we would have $z \in z$, contradicting the result of the first part above.

The result of this last exercise can be regarded as a special case of the following theorem.

> ***Theorem 4.3 No infinite descending $\in$-chains***
>
> Suppose that $\langle x_n \rangle$ is a sequence. Then this sequence cannot form what is called an *infinite descending $\in$-chain*, i.e. we cannot have
>
> $$\ldots \in x_{n+1} \in x_n \in \ldots \in x_2 \in x_1 \in x_0.$$

Recall that a sequence in *ZF* is a function with domain $\mathbb{N}$.

Proof

We shall suppose that the sequence does form an infinite descending $\in$-chain and derive a contradiction. We use the axiom of foundation, taking x to be the set $\{x_n : n \in \mathbb{N}\}$. By the axiom there is some $y \in x$ such that $x \cap y = \varnothing$. As $y \in x$ we must have $y = x_n$ for some n. But then x_{n+1} is an element both of x and of y (as $x_{n+1} \in x_n$), so that

Strictly speaking, if we regard a sequence as a function with domain $\mathbb{N}$, then x is really the range of this function.

$$x_{n+1} \in x \cap y,$$

contradicting that $x \cap y = \varnothing$. ■

Exercise 4.42

Use Theorem 4.3 to show the following.

(a) There is no set x such that $x \in x$. (This is an alternative argument to that given in Exercise 4.41.)

(b) There are no sets x, y such that $x \in y$ and $y \in x$.

Solution

(a) If $x \in x$, then there is an infinite descending $\in$-chain

$$\ldots \in x \in x \in \ldots \in x \in x \in x,$$

contradicting the result of Theorem 4.3.

(b) This one is left for you!

Taking each x_n as x in Theorem 4.3.

Take any set x_0. Imagine investigating its elements, then their elements, and so on. This will give rise to lots of descending $\in$-chains

$$\ldots \in x_n \in \ldots \in x_2 \in x_1 \in x_0.$$

As a consequence of Theorem 4.3 all such chains are finitely long. What set can the bottom member of such a chain be? It has to be a set with no elements, so it can only be the empty set $\varnothing$.

This can be expressed more formally using the idea of *transitive closure*. See Exercise 4.53 at the end of the section.

Exercise 4.43

For one of these descending $\in$-chains, say $\varnothing \in x_{n-1} \in \ldots \in x_2 \in x_1 \in x_0$, what sort of set x_{n-1} might come just above the bottom member of the chain?

Solution

It is tempting to say that x_{n-1} has to be $\{\varnothing\}$, but the situation could be much more complicated than this. In general x_{n-1} might have lots of other chains going down through it, not just the one we were given, e.g.

Recall that $\mathbf{0} = \varnothing$, $\mathbf{1} = \{\mathbf{0}\}$ and $\mathbf{2} = \{\mathbf{0}, \mathbf{1}\}$.

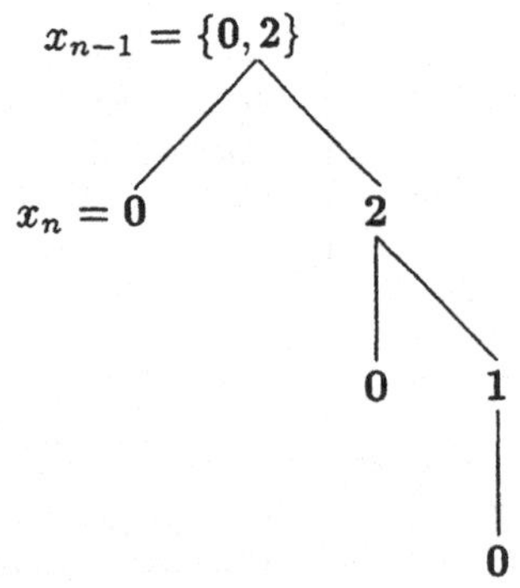

In spite of this complication, a consequence of the axiom of foundation, along with the other axioms of ZF, is that all sets are built up from the empty set. Indeed, thanks to this axiom one can show that the universe of sets $\mathscr{V}$ can be constructed by an iterative process starting from the empty set and repeatedly forming power sets and unions. This process, forming what is called the *cumulative hierarchy of sets*, begins as follows:

$$\mathscr{V}_0 = \varnothing,$$
$$\mathscr{V}_{\mathbf{n}^+} = \mathscr{P}(\mathscr{V}_{\mathbf{n}}), \text{ for } \mathbf{n} \in \mathbb{N},$$

Strictly speaking, these $\mathscr{V}_{\mathbf{n}}$s are defined by recursion, which we shall soon justify as Theorem 4.5.

and collects together all these $\mathscr{V}_{\mathbf{n}}$s by forming

$$\bigcup\{\mathscr{V}_{\mathbf{n}} : \mathbf{n} \in \mathbb{N}\}.$$

This latter class is a set, as a consequence of the axiom of replacement: let's call it X. As X is a set, we can carry on with the process, repeatedly taking power sets, giving

When we have defined the ordinal ω, we shall write this X as $\mathscr{V}_\omega$.

$$X,\ \mathscr{P}(X),\ \mathscr{P}(\mathscr{P}(X)),\ \mathscr{P}(\mathscr{P}(\mathscr{P}(X))),\ \ldots,$$

taking the union of these along with all the earlier $\mathscr{V}_{\mathbf{n}}$s

$$\bigcup\{\mathscr{V}_0,\ \mathscr{V}_1,\ \mathscr{V}_2,\ \ldots,\ X,\ \mathscr{P}(X),\ \mathscr{P}(\mathscr{P}(X)),\ \mathscr{P}(\mathscr{P}(\mathscr{P}(X))),\ \ldots\}$$

then taking power sets again, and so on. It can be shown that every set will eventually appear as an element of one of the sets in this hierarchy. It would clearly be very useful to have a way of labelling the sets like X and $\mathscr{P}(X)$ that follow on from the $\mathscr{V}_{\mathbf{n}}$s so that we can count where they lie in the hierarchy. We shall be able to do this using ordinals, which we shall discuss in Chapter 8.

We shall look at this again in Chapter 9.

Exercise 4.44

Give a sketch of the descending $\in$-chains (along the lines of our solution to Exercise 4.43) starting with the set $\mathbb{N}$ as x_0 and explain why none of these chains is infinite (without reference to Theorem 4.3).

To end this chapter, we shall now give a proof of a version of the recursion principle for $\mathbb{N}$ used in Chapter 3. We delayed giving a proof of this until we had established $\mathbb{N}$ as a genuine set within *ZF* (requiring, not least, the axiom of infinity) and had other important machinery in place, e.g. functions represented by sets within *ZF*. Much use will be made of induction – scarcely surprising, as this is the key feature of $\mathbb{N}$ – and one crucial step will exploit the axiom of replacement.

Theorem 4.4 The recursion principle

Let y_0 be any element of a set Y and $h\colon \mathbb{N} \times Y \longrightarrow Y$ a function on pairs $(x, y) \in \mathbb{N} \times Y$. Then there exists a unique function $f\colon \mathbb{N} \longrightarrow Y$ such that

$$f(\mathbf{0}) = y_0,$$
$$f(\mathbf{n}^+) = h(\mathbf{n}, f(\mathbf{n})), \text{ for all } \mathbf{n} \in \mathbb{N}.$$

Theorem 3.11 of Section 3.3, recursion on $\mathbb{N}$, is the special case when Y is the set $\mathbb{N}$.

Proof

It will be helpful, just for the purposes of this proof, to give a name to a function which approximates to the desired f in the following sense. We shall call g an **n**-*function*, for $n \in \mathbb{N}$, if the following hold:

(i) g is a function;

(ii) the domain of g is the set n^+;

(iii) $g(0) = y_0$;

(iv) for all $i \in n$, g satisfies the rule $g(i^+) = h(i, g(i))$.

We shall show that there is a unique **n**-function, let's say g_n, for each $n \in \mathbb{N}$ and that these can be arranged in a $\subseteq$-chain

$$g_0 \subseteq g_1 \subseteq g_2 \subseteq \ldots \subseteq g_n \subseteq g_{n^+} \subseteq \ldots .$$

We shall then define f as the union of this chain (this is the step that requires the axiom of replacement) and check that it has the required properties. First of all we need the following lemma.

> *Lemma:* Suppose that g is an **n**-function and g' an **n'**-function. Put **m** equal to $n \cap n'$. Then
>
> $$g|_{m^+} = g'|_{m^+} .$$
>
> *Proof of Lemma:* We shall use induction to show that $g(i) = g'(i)$ for all elements i of m^+. Both $g(0)$ and $g'(0)$ are defined to be y_0, so the result holds for $i = 0$. Suppose now that the result holds for $i \in m$, so that $g(i) = g'(i)$: we shall show that it also holds for i^+. We have
>
> $$\begin{aligned} g(i^+) &= h(i, g(i)) \quad \text{(as g is an **n**-function)} \\ &= h(i, g'(i)) \quad \text{(by our assumption that $g(i) = g'(i)$)} \\ &= g'(i^+) \quad \text{(as g' is an **n'**-function),} \end{aligned}$$
>
> which shows that the result holds for i^+, as required. Induction gives that the result holds for all $i \in m^+$, completing the proof of the lemma. $\square$

As a consequence of this lemma, if both g and g' are **n**-functions (so that $n = n'$ in the lemma), then $g = g'$. Thus if there is any **n**-function, it is unique.

We shall now show that there is an **n**-function, which we shall call g_n, for each $n \in \mathbb{N}$. We shall do this by induction on **n**.

For $n = 0$ we require a function g_0 with domain 0^+. As the only element of the domain is 0, all we need do is to set $g_0(0)$ equal to y_0 to guarantee that g_0 is a 0-function. So define g_0 as a set of ordered pairs by

$$g_0 = \{\langle 0, y_0 \rangle\}.$$

This is indeed a function with domain 0^+ satisfying requirements (iii) and (iv) (the latter vacuously) for a 0-function.

For the inductive step we assume that an **n**-function g_n exists for **n** and show that there is an n^+-function. All we need to do is bolt an extra ordered pair onto g_n to cope with the extra element n^+ in the domain, n^{++}, of an n^+-function; and the extra pair simply has to meet requirement (iv) for an n^+-function. So we define g_{n^+} by

$$g_{n^+} = g_n \cup \{\langle n^+, h(n, g_n(n)) \rangle\}.$$

Remember that functions are represented by sets of ordered pairs.

Between them (iii) and (iv) cover the values of $g(j)$ for all **j** in n^+, the domain of g.

Remember that each g_n is a set.

By Exercise 3.12 in Section 3.2 $n \cap n'$ equals $\min\{n, n'\}$, and is thus a natural number, one of **n** and **n'**.

As every element of m^+, other than 0, is of the form i^+ for some $i \in m$, our argument does cover all the elements of m^+.

$$\begin{aligned} 0^+ &= 0 \cup \{0\} \\ &= \varnothing \cup \{0\} \\ &= \{0\} \end{aligned}$$

Formally within ZF we show that $\{n \in \mathbb{N} : \exists g (g \text{ is an **n**-function})\}$ is inductive. And we can represent 'g_n is an **n**-function' by a formula along the lines of

$$\begin{aligned} \big((g_n \text{ is a function}) \\ \wedge \ \mathrm{Dom}(g_n) = n^+ \\ \wedge \ g_n(0) = y_0 \\ \wedge \ \forall i \big(i \in n \ \rightarrow \ g_n(i^+) = h(i, g_n(i))\big)\big) \end{aligned}$$

We shall leave it as a straightforward exercise for you to show that this $g_{\mathbf{n}^+}$ meets all the requirements to be an $\mathbf{n}^+$-function. The existence of an $\mathbf{n}$-function for each $\mathbf{n} \in \mathbb{N}$ then follows by induction.

We shall now construct the required function f. From our earlier lemma it follows that if $\mathbf{m} \in \mathbf{n}$ (so that $\mathbf{m} \cap \mathbf{n} = \mathbf{m}$) then

$$\begin{aligned} g_{\mathbf{m}} &= g_{\mathbf{m}}|_{\mathbf{m}^+} \quad (\mathbf{m}^+ \text{ is just the domain of } g_{\mathbf{m}}) \\ &= g_{\mathbf{n}}|_{\mathbf{m}^+} \quad (\text{by the lemma, as } \mathbf{m} = \mathbf{m} \cap \mathbf{n}) \\ &\subseteq g_{\mathbf{n}}. \end{aligned}$$

We can thus regard all the $g_{\mathbf{n}}$s as forming a $\subseteq$-chain

$$g_0 \subseteq g_1 \subseteq g_2 \subseteq \ldots \subseteq g_{\mathbf{n}} \subseteq g_{\mathbf{n}^+} \subseteq \ldots .$$

We define f to be the union of this chain, i.e.

$$f = \bigcup\{g_{\mathbf{n}} : \mathbf{n} \in \mathbb{N}\},$$

and shall show that this satisfies the requirements of the theorem. But we must first show that f is a set in *ZF*. This follows from the union axiom once we have shown that $\{g_{\mathbf{n}} : \mathbf{n} \in \mathbb{N}\}$ is a set. And this latter fact follows from the axiom of replacement taking the formula $\phi(s,t)$ to be

$$(s \in \mathbb{N} \rightarrow (t \text{ is an } s\text{-function})) \wedge (s \notin \mathbb{N} \rightarrow t = \varnothing).$$

The clause $(s \notin \mathbb{N} \rightarrow t = \varnothing)$ in $\phi(s,t)$ ensures that for all sets s, not just those in $\mathbb{N}$ which really interest us, there is a unique set t for which $\phi(s,t)$ holds.

We have shown that for each natural number $\mathbf{n}$ there is a unique $\mathbf{n}$-function, so that $\phi(s,t)$ is a class function. Taking x in the axiom of replacement to be the set $\mathbb{N}$, the resulting set of images is $\{g_{\mathbf{n}} : \mathbf{n} \in \mathbb{N}\}$, so that the latter is indeed a set.

From the result of Exercise 4.31 in the previous section, it follows that f is a function with domain equal to the union of the domains of the $g_{\mathbf{n}}$, namely $\mathbb{N}$. Whenever $\mathbf{m} \in \mathbf{n}$ we have $g_{\mathbf{m}} \subseteq g_{\mathbf{n}} \subseteq f$, so that $f(\mathbf{m}) = g_{\mathbf{n}}(\mathbf{m})$. Thus we have

$$\begin{aligned} f(\mathbf{0}) &= g_{\mathbf{n}}(\mathbf{0}) \text{ for any } \mathbf{n} \\ &= y_0; \end{aligned}$$

and for any $\mathbf{n} \in \mathbb{N}$ we have

$$\begin{aligned} f(\mathbf{n}^+) &= g_{\mathbf{n}^+}(\mathbf{n}^+) \\ &= h(\mathbf{n}, g_{\mathbf{n}^+}(\mathbf{n})) \quad (\text{as } g_{\mathbf{n}^+} \text{ is an } \mathbf{n}^+\text{-function}) \\ &= h(\mathbf{n}, f(\mathbf{n})). \end{aligned}$$

Thus f is the function required for the theorem.

There is one final detail of the theorem to be checked, namely that the function f is actually unique. This is left for you to check, as part (b) of the next exercise. ■

Exercise 4.45

Complete the proof above by filling in the following gaps.

(a) Show that if $g_{\mathbf{n}}$ is an $\mathbf{n}$-function, then $g_{\mathbf{n}^+}$ defined by

$$g_{\mathbf{n}^+} = g_{\mathbf{n}} \cup \{(\mathbf{n}^+, h(\mathbf{n}, g_{\mathbf{n}}(\mathbf{n})))\}$$

is an $\mathbf{n}^+$-function.

(b) Show that if two functions f and f' have the desired properties of Theorem 4.4, then $f = f'$.

Among the applications of this version of the recursion principle is in underpinning the proof of Theorem 3.3 of Section 3.1, which shows that two Peano systems are isomorphic.

To round off this chapter, we would like to introduce one further, even more powerful, version of the recursion principle. Rather than state it straightaway, let us look at a situation which it will resolve. We mentioned earlier that the axiom of replacement was added to Zermelo's axioms when it was found that the class

$$\{\mathbb{N}, \mathscr{P}(\mathbb{N}), \mathscr{P}(\mathscr{P}(\mathbb{N})), \mathscr{P}(\mathscr{P}(\mathscr{P}(\mathbb{N}))), \ldots\}$$

could not be shown to be a set. This class has particular significance within Cantor's theory of infinite sets, hence the interest in it being a set. We would like to argue as follows. Use Theorem 4.4 with h defined by $h(x, y) = \mathscr{P}(y)$ to define a function f with domain $\mathbb{N}$ by

This set corresponds to the cardinal number $\aleph_\omega$. We shall meet such numbers later in the book.

$$f(\mathbf{0}) = \mathbb{N},$$
$$f(\mathbf{n}^+) = \mathscr{P}(f(\mathbf{n})), \text{ for all } \mathbf{n} \in \mathbb{N}.$$

Then Range(f), which is precisely the class we want, is a set.

Exercise 4.46

There is a flaw in our argument defining the f above. What is it?

Solution

The statement of Theorem 4.4 requires a set Y for which $h\colon \mathbb{N} \times Y \longrightarrow Y$ is a function. In this case Y would have to include as a subset precisely the class $\{\mathbb{N}, \mathscr{P}(\mathbb{N}), \mathscr{P}(\mathscr{P}(\mathbb{N})), \mathscr{P}(\mathscr{P}(\mathscr{P}(\mathbb{N}))), \ldots\}$ which we are trying to show is a set!

One might easily think that this flaw means that what otherwise looks like a desirable and natural mathematical definition of f is doomed. But in fact the statement of Theorem 4.4 can be modified by replacing the function h by what we have described earlier as a class function $h\colon \mathbb{N} \times \mathscr{V} \longrightarrow \mathscr{V}$, corresponding to a formula $\phi(x, y, z)$ such that for each pair of sets x, y there is a unique set z such that $\phi(x, y, z)$ holds. Here we would take $\phi(x, y, z)$ to be the formula

$$z = \mathscr{P}(y),$$

which yields the class function h where $h(x, y) = \mathscr{P}(y)$ for all $x \in \mathbb{N}$ and all sets y. The modified form of the theorem is as follows.

Theorem 4.5 Recursion principle for $\mathbb{N}$, class form

Let $\phi(x, y, z)$ be a formula of set theory such that for each $x \in \mathbb{N}$ and set y there is a unique set z such that $\phi(x, y, z)$ holds. Thus $\phi(x, y, z)$ essentially defines a class function

$$h: \mathbb{N} \times \mathscr{V} \longrightarrow \mathscr{V},$$

where $\mathscr{V}$ is the universe of sets. And let y_0 be a given set. Then there is a unique function f with domain $\mathbb{N}$ such that

$$f(0) = y_0,$$
$$f(\mathbf{n}^+) = h(\mathbf{n}, f(\mathbf{n})), \text{ for all } \mathbf{n} \in \mathbb{N}.$$

Proof

The proof is exactly the same as that of Theorem 4.4! The only place where one has to be slightly more conscious of what's going on is the stage left as part of Exercise 4.45, in showing that $g_{\mathbf{n}^+}$ defined by

$$g_{\mathbf{n}^+} = g_{\mathbf{n}} \cup \{\langle \mathbf{n}^+, h(\mathbf{n}, g_{\mathbf{n}}(\mathbf{n}))\rangle\}$$

is actually a set (which is part of what is required for it to be a function). As the induction hypothesis at this stage of the proof was that $g_{\mathbf{n}}$ is a function, this means that there is a unique set z, namely $h(\mathbf{n}, g_{\mathbf{n}}(\mathbf{n}))$, such that $\phi(\mathbf{n}, g_{\mathbf{n}}(\mathbf{n}), z)$ holds. Then by the ordered pair construction we have that $\langle \mathbf{n}^+, h(\mathbf{n}, g_{\mathbf{n}}(\mathbf{n}))\rangle$ is a set, so that the singleton $\{\langle \mathbf{n}^+, h(\mathbf{n}, g_{\mathbf{n}}(\mathbf{n}))\rangle\}$ is a set, giving that the union $g_{\mathbf{n}} \cup \{\langle \mathbf{n}^+, h(\mathbf{n}, g_{\mathbf{n}}(\mathbf{n}))\rangle\}$ is a set. ■

We frequently disguise the use of this theorem by saying that we are defining a sequence of sets $\langle A_n \rangle$, rather than saying that we are constructing a function f with domain $\mathbb{N}$. An immediate consequence of the theorem is that the f has an image set, which we would write in terms of the A_ns as $\{A_n : n \in \mathbb{N}\}$. This sort of construction is so common that we shall honour it as the following loosely stated theorem, which is a simple consequence of Theorem 4.5.

Once f is known to be a function in ZF, it then has an image set.

Theorem 4.6

Suppose that $g: \mathscr{V} \longrightarrow \mathscr{V}$ is a class function and let y_0 be a given set. Then a sequence of sets $\langle A_n \rangle$ can be defined by

$$A_0 = y_0,$$
$$A_{n+1} = g(A_n), \text{ for all } \mathbf{n} \in \mathbb{N},$$

and $\{A_n : n \in \mathbb{N}\}$ is a set.

We have now seen all the axioms of Zermelo–Fraenkel set theory. These axioms provide the most widely used basis for both a formal framework within which mathematics can be done and the study of sets themselves. For the rest of the book we shall usually be very relaxed about the use of these axioms, rarely referring to them in any detailed way. Rest assured that the new ideas we shall be describing, primarily Cantor's work on infinite sets, can be rigorously described within the theory, and that theorems about them can be shown to follow from the axioms. But there will be occasions where we have to be very careful to explain that certain sets, or constructions involving them, can really be represented within the theory, not least so that we can be reasonably confident that we are avoiding contradictions. And in the next chapter we shall investigate an area where such caution is needed. The subject of the chapter is the only one of Zermelo's original axioms which was deliberately not included in the eventual list of axioms of *ZF*, because of its subtle and controversial nature: the axiom of choice.

Further exercises

Exercise 4.47

Show that $\bigcup \mathbb{N} = \mathbb{N}$.

Exercise 4.48

Show that the ordered pair $\langle x, y\rangle$ could be defined in *ZF* as $\{x, \{x, y\}\}$. [Hint: you will need to exploit *ZF9*.]

Exercise 4.49

Explain why there are sets $\mathscr{V}_\mathbf{n}$ for all $\mathbf{n} \in \mathbb{N}$ and a set $\{\mathscr{V}_\mathbf{n} : \mathbf{n} \in \mathbb{N}\}$ such that

$$\mathscr{V}_\mathbf{0} = \varnothing,$$
$$\mathscr{V}_{\mathbf{n}^+} = \mathscr{P}(\mathscr{V}_\mathbf{n}), \text{ for } \mathbf{n} \in \mathbb{N}.$$

Exercise 4.50

(a) Show that $\mathscr{V}_\mathbf{1} = \mathbf{1}$, $\mathscr{V}_\mathbf{2} = \mathbf{2}$ and $\mathscr{V}_\mathbf{3} = \mathbf{3} \cup \{\mathbf{1}\}$. Compute $\mathscr{V}_\mathbf{4}$ and $\mathscr{V}_\mathbf{5}$.

(b) Show that $\mathbf{n} \subseteq \mathscr{V}_\mathbf{n}$ for all $\mathbf{n} \in \mathbb{N}$ (so that $\mathbf{n} \in \mathscr{V}_{\mathbf{n}^+}$). Deduce that $\mathbb{N} \subseteq \bigcup\{\mathscr{V}_\mathbf{n} : \mathbf{n} \in \mathbb{N}\}$.

(c) How many elements does $\mathscr{V}_\mathbf{n}$ have for each $\mathbf{n} \in \mathbb{N}$?

Exercise 4.51

Show that if $\mathbf{m}, \mathbf{n} \in \mathbb{N}$ with $\mathbf{m} < \mathbf{n}$ then $\mathscr{V}_\mathbf{m} \subseteq \mathscr{V}_\mathbf{n}$. [Hint: use induction on $\mathbf{n}$ for fixed $\mathbf{m}$.]

Exercise 4.52

Zermelo's original infinity axiom was as follows:

$$\exists x\big(\varnothing \in x \land \forall y(y \in x \rightarrow \{y\} \in x)\big).$$

(Zermelo actually used

$$\varnothing,\ \{\varnothing\},\ \{\{\varnothing\}\},\ \{\{\{\varnothing\}\}\},\ \ldots$$

to represent the natural numbers.)

Show that the existence of such a set x follows from the axioms of *ZF*.

Exercise 4.53

Let Y be a set. A function f with domain $\mathbb{N}$ is defined by

$$f(\mathbf{0}) = Y,$$
$$f(\mathbf{n}^+) = \bigcup f(\mathbf{n}), \text{ for all } \mathbf{n} \in \mathbb{N}.$$

The *transitive closure* of Y, written as $T(Y)$, is defined by

$$T(Y) = \bigcup \text{Range}(f),$$

so that, informally,

$$T(Y) = Y \cup \bigcup Y \cup \bigcup\bigcup Y \cup \bigcup\bigcup\bigcup Y \ldots.$$

(a) Explain why $T(Y)$ is a set.

(b) Show that if $x \in y \in T(Y)$ then $x \in T(Y)$ (or equivalently if $y \in T(Y)$ then $y \subseteq T(Y)$).

$T(Y)$ is said to be $\in$-*transitive*.

(c) Show that if $X \in Y$ then $T(X) \subseteq T(Y)$.

(d) Let Z be a set with the properties that $Y \subseteq Z$ and that whenever $x \in z \in Z$ then $x \in Z$. Show that $T(Y) \subseteq Z$.

So that $T(Y)$ is the smallest $\in$-transitive set containing Y as a subset.

(e) By the axiom of foundation there is a $y \in T(Y)$ such that $y \cap T(Y) = \varnothing$. What set must y be? Prove your answer.

5 THE AXIOM OF CHOICE

5.1 Introduction

In this chapter we shall have our first look at the only one of Zermelo's original axioms not to be included among the *ZF* axioms, the axiom of choice. This principle, initially very controversial, turns out to play a pivotal role in the theory of infinite sets and in much 'ordinary' mathematics involving infinite sets.

One of the most interesting activities of abstract mathematics is finding new mathematical objects. This involves describing the object and convincing others that the object does exist, often by giving a construction. The issues of the existence of objects, what constitutes a description and what constitutes a construction are heavily interlinked, and full of complications. For instance $\{x : x \notin x\}$ appears to describe a set, but, as you know, such a set cannot exist without creating a contradiction in set theory. And there are arguments which justify the existence of an object, for instance by assuming its non-existence and obtaining a contradiction, without giving a way of constructing it.

Set theory is very careful about asserting the existence of set objects: some of the axioms assert the existence of sets, and others assert the existence of certain sets on the basis of constructions involving known sets; and the development of the subject includes showing that other, more sophisticated, constructions of sets are legitimate. The axiom of choice, which can be expressed in several provably equivalent ways, likewise asserts the existence of a set – usually in the form of a function. One of the reasons why the axiom is of interest is that for some people it sounds plausible that the existence of this function should already be deducible from the other standard axioms; whereas for others the axiom seems unacceptable because it doesn't give a construction of the function. Before we state the axiom, think about the following question:

> Suppose that $f: A \longrightarrow B$ is an *onto* function. Is there a *one–one* function $g: B \longrightarrow A$ such that $f(g(b)) = b$ for all $b \in B$?

g is called a *transversal.*

Do you feel that the answer to this question is yes, or no, or it depends? What would constitute an argument that there is such a function g? Although there might be more roundabout methods, a direct construction of g would surely be one's first choice.

Exercise 5.1

Let $f: \mathbb{N} \longrightarrow B$ be an *onto* function. Show that there is a *one–one* function $g: B \longrightarrow \mathbb{N}$ such that $f(g(b)) = b$ for all $b \in B$ by giving a construction of g.

Solution

We shall construct g by specifying $g(b)$ for each $b \in B$. The requirement that $f(g(b)) = b$ means that $g(b)$ has to be an element of the set $f^{-1}(\{b\})$. As f is onto, we are guaranteed that this latter set is non-empty, for each b. How

$f^{-1}(\{b\})$ means $\{a \in \mathbb{N} : f(a) = b\}$.

might we specify a member of this set? – after all, we don't know too much about f. Luckily we know lots about $\mathbb{N}$, in particular that each non-empty subset of $\mathbb{N}$ has a least element: so we shall specify $g(b)$ by defining it to be $\min f^{-1}(\{b\})$.

An advantage of the above *construction* of g over simply asserting that some g exists, is that whoever uses it with a given f will always end up with the same g as any other user. For one of us to assert that g exists without this sort of precise construction does not allow us to share g in quite the same way.

Exercise 5.2

Let $f\colon \mathbb{R} \longrightarrow \mathbb{N}$ be an onto function. Does there exist a $g\colon \mathbb{N} \longrightarrow \mathbb{R}$ such that $f(g(b)) = b$ for all $b \in N$? (Think about this, but not for too long!)

Solution

We cannot simply reuse the construction of Solution 5.1, as a non-empty subset of $\mathbb{R}$ does not necessarily contain a least element (e.g. $(-\infty, -3)$). And we are not allowed to describe g by giving a table listing its values, as such a table would be infinitely long, and we want objects like g to have *finite* descriptions. But surely we know enough about $\mathbb{R}$, e.g. its arithmetic and order properties, to be able to describe, finitely, a way of obtaining $g(b)$ for each $b \in \mathbb{N}$? Alas it can be shown that there is no such way of doing so which only uses the framework of the set theory axioms we have introduced so far.

So we have run into a problem in describing g because $\mathbb{R}$ turns out to lack some vital piece of structure. But this is for a somewhat general onto function f with domain $\mathbb{R}$. Although the domain and codomain of f are specified, the rule of f is essentially arbitrary. To a large extent we are bound to run into problems precisely because the rule of the original function f is so arbitrary. If we have more information about f, e.g. its rule, we might sometimes be able to describe g.

Exercise 5.3

Suppose that $f\colon \mathbb{R} \longrightarrow \mathbb{R}$ is onto, continuous and non-decreasing. Show that there is a $g\colon \mathbb{R} \longrightarrow \mathbb{R}$ such that $f(g(b)) = b$ for each $b \in \mathbb{R}$. [You need some knowledge of real analysis for this exercise.]

Solution

Take any $b \in \mathbb{R}$. It is easy to show that $f^{-1}(\{b\})$ is a (non-empty) interval using the fact that f is non-decreasing and the intermediate value theorem. If this interval consists of just one point, α_b say, then the only candidate for $g(b)$ is α_b. If the interval has more than one point and is bounded below, then it has a greatest lower bound, α_b say. As f is continuous at α_b and takes the constant value b just to the right of α_b, we have that $f(\alpha_b) = b$, so that α_b is in fact the *least* element of $f^{-1}(\{b\})$. Then we can set $g(b)$ equal to α_b.

$I \subseteq \mathbb{R}$ is an *interval* if whenever $x, y \in I$ with $x < y$, then $[x, y] \subseteq I$.

We can summarize our construction so far by saying that if $f^{-1}(\{b\})$ is bounded below, then $g(b) = \min f^{-1}(\{b\})$. Lastly, what if $f^{-1}(\{b\})$ is not bounded below? From the given information about f we can show that this cannot happen. As f is onto, for any b there is an x for which $f(x) = b - 1$: as f is non-decreasing, this x must be a lower bound for $f^{-1}(\{b\})$. Thus $f^{-1}(\{b\})$ is always bounded below.

Thus our definition of g is

$$g(b) = \min f^{-1}(\{b\}).$$

Exercise 5.4

Suppose that $f\colon \mathbb{R} \longrightarrow B$ is onto, where B is finite. Show that there is a $g\colon B \longrightarrow \mathbb{R}$ such that $f(g(b)) = b$ for each $b \in B$.

Solution

There seems to be much less solid information to grab hold of than in the previous exercise! But as the image set B is finite, you can work your way (in a finite length of time) through the elements of B choosing an element of $f^{-1}(\{b\})$ to be $g(b)$ for each b, and then describe your function g to other mathematicians by means of a perhaps long but, importantly, finite list of cases.

If this does not satisfy you, you can dress up the argument by using induction on the number of elements of B. A suitable inductive hypothesis would be that for any onto function $f\colon C \longrightarrow D$, where D has n elements and $C \subseteq \mathbb{R}$, there is a suitable $g\colon D \longrightarrow C$. Completion of the proof is left to you.

This last exercise reminds us of an important feature of what constitutes a construction of something which we claim to exist: we should be able to describe the construction in a finite way, as well as in a way which enables others to replicate our construction. A long finite list of values of $g(b)$ may be an unsatisfying or boring way of describing g (compared to e.g. a single general rule for $g(b)$), but it *does* describe it.

Let us summarize what we have gleaned so far about our original problem concerning functions of the form $f\colon A \longrightarrow B$ and the existence of a corresponding function g. If B is finite there is no problem in describing g. But if B is infinite we need further information, about A or f or both, to ensure we can describe g. And if we cannot describe g, other than by stating the property desired of such a g (i.e. that $f(g(b)) = b$ for each b), on what basis can we assert that g exists? It is to resolve this problem that we introduce a special axiom, the axiom of choice, which we do in the following section. And it should come as no surprise that this axiom is intimately connected with issues about infinite sets, as we shall see in later chapters of the book.

5.2 The axiom of choice

There are number of different ways of formulating the axiom of choice – Zermelo, who first introduced it, used several himself. In this section we shall give one formulation and prove its equivalence to various others.

> ***Axiom of choice***
>
> Suppose that $\mathscr{F}$ is a family of non-empty sets. Then there is a function $h\colon \mathscr{F} \longrightarrow \bigcup \mathscr{F}$ such that for each $A \in \mathscr{F}$, $h(A) \in A$.
> h is said to be a *choice function* for $\mathscr{F}$.

We use 'family' to mean 'set'.

Note that $A \subseteq \bigcup \mathscr{F}$, as $A \in \mathscr{F}$.

A choice function h 'chooses' an element, namely $h(A)$, for each member of the family $\mathscr{F}$. The axiom doesn't say *how* the choice is made, only that such a function exists.

'Family' here just means 'set'.

Bertrand Russell [21] gave an entertaining illustration of the use of the axiom of choice, as follows. Think of how one might describe how to choose a shoe from each of an infinite set of pairs of shoes. By 'description' we mean much more than 'Pick any old shoe at random': we mean some sort of description which can be passed to other people and used by them so that, for any particular pair of shoes, each person would take the same shoe – of course, we are really talking about a choice function on the set of pairs of shoes: the people are really applying this function, not making their own individual choice.

A pair of shoes means what what it does in everyday life. In 1996 this still means two matching shoes, one intended for the left foot and one for the right foot, of the same person!

Without making even more of a meal of the situation, we can easily specify such a choice function by means of a finite description, for instance

Always choose the left shoe from each pair.

Now try to do the same thing for an infinite set of pairs of socks. There is no comparable finite description of a choice function on these pairs of socks. The unsubtle approach of specifying a choice function, e.g. by attaching some sort of sticky label to one sock of each pair, involves an infinitely long description, and the real issue here, for the abstract world of set theory, is that the description should be finite. The axiom of choice simply says that there is a choice function, which it conjures up from thin air!

Again, in 1996 a pair of socks means two socks which cannot be distinguished from each other, in theory at least!

Had there been only finitely many pairs of socks, such an unsubtle list would have been adequate. The analogous position for true sets is the description of a choice function on a finite family of sets by means of the full, but nevertheless finite, table of its values.

Back to proper set theory! How does the axiom of choice relate to the question discussed in the previous section:

Suppose that $f\colon C \longrightarrow D$ is an *onto* function. Is there a *one-one* function $g\colon D \longrightarrow C$ such that $f(g(d)) = d$ for all $d \in D$?

Well, given the function f, we need to choose an element out of $f^{-1}(\{d\})$ for each $d \in D$ to be able to define $g(d)$. So define the $\mathscr{F}$ in the statement of the axiom to be the family of sets $\{f^{-1}(\{d\}) : d \in D\}$. Then $\bigcup \mathscr{F}$ is just C (as f is onto), and $A \in \mathscr{F}$ means A is an $f^{-1}(\{d\})$. And the function h guaranteed by the axiom, such that $h(A) \in A$ for all $A \in \mathscr{F}$, allows us to construct the required function g, by setting

$$g(d) = h(f^{-1}(\{d\})) \text{ for all } d \in D.$$

$\{f^{-1}(\{d\}) : d \in D\}$ is a set by the axiom of replacement.

As pointed out in the previous section, we don't always need to use the axiom of choice to choose an element of $f^{-1}(\{d\})$, for instance if D is finite, or if f

or C has useful structure, as in the case $C = \mathbb{N}$, where we exploited the fact that any non-empty subset of $\mathbb{N}$ contains a least element. Although use of the axiom seems to have gained widespread acceptance in mathematics (largely for reasons with which we shall deal in the next section), it is often regarded as preferable to avoid its use where possible.

The useful structure of $\mathbb{N}$ here is that it is *well-ordered*. Well-order is interestingly related to the axiom of choice.

Let us now look at some of the equivalent formulations of the axiom and prove that they are indeed equivalent to it.

Theorem 5.1

The following statements are equivalent to the axiom of choice:

Disjoint family form
Suppose that $\mathscr{F}$ is a disjoint family of non-empty sets, i.e. for any distinct $A, B \in \mathscr{F}$, $A \cap B = \varnothing$. Then there is a function $h\colon \mathscr{F} \longrightarrow \bigcup \mathscr{F}$ such that for each $A \in \mathscr{F}$, $h(A) \in A$.

Power set form
Suppose that M is a non-empty set. Then there is a function $h\colon \mathscr{P}(M) \setminus \{\varnothing\} \longrightarrow M$ such that for all non-empty subsets A of M, $h(A) \in A$.

In both cases above it still seems natural to call h a choice function (for $\mathscr{F}$ and $\mathscr{P}(M)$ respectively). The disjoint family form is very subtly different from the axiom of choice – you might like to think why. We shall prove the theorem by showing that

> axiom of choice $\Rightarrow$ disjoint family form $\Rightarrow$ power set form $\Rightarrow$ axiom of choice.

Bear in mind that each statement is really preceded by a universal quantifier, 'for all families $\mathscr{F}$' or 'for all non-empty sets M'; and that our proof should only use set constructions which are allowed by the other axioms of *ZF*.

Proof

Axiom of choice $\Rightarrow$ disjoint family form
This implication is trivial. (It's the reverse implication which requires ingenuity!)

Disjoint family form $\Rightarrow$ power set form
Let M be a non-empty set. Then $\mathscr{P}(M) \setminus \{\varnothing\}$ is a family of non-empty sets. Alas, we cannot directly apply the disjoint family form of the axiom to this family, as in general pairs of subsets of M won't be disjoint. So we use a trick to turn the problem into one involving disjoint sets, by defining a new family of sets

There is a more direct proof that axiom of choice $\Rightarrow$ power set form which doesn't run into this problem. This is left for you as part of Exercise 5.9.

$$\mathscr{F} = \{A \times \{A\} : A \subseteq M, A \neq \varnothing\}.$$

Then for $\varnothing \neq A, B \subseteq M$, we have $(x, y) \in (A \times \{A\}) \cap (B \times \{B\})$ implies $x \in A \cap B$ and $y \in \{A\} \cap \{B\}$, which forces $y = A = B$. So if A and B are different non-empty subsets of M, the corresponding members of $\mathscr{F}$ are disjoint. So we can now apply the disjoint family form to $\mathscr{F}$ to get a choice function $g\colon \mathscr{F} \longrightarrow \bigcup \mathscr{F}$ such that for each $A \times \{A\} \in \mathscr{F}$, we have

We quite often need this trick to turn possibly overlapping sets into disjoint sets. Technically $\mathscr{F}$ is a set because it's a subset of $\mathscr{P}(M \times \mathscr{P}(M))$.

$g(A \times \{A\}) \in A \times \{A\}$. This means that $g(A \times \{A\})$ is of the form (a, A) for some $a \in A$.

We can now define the required function $h\colon \mathscr{P}(M) \setminus \{\varnothing\} \longrightarrow M$ by

$$h(A) = \pi_1(g(A \times \{A\})) \text{ for each non-empty } A \subseteq M,$$

where π_1 is the function

$$\begin{aligned} \pi_1\colon M \times \mathscr{P}(M) &\longrightarrow M \\ (x, y) &\longmapsto x. \end{aligned}$$

Power set form ⇒ axiom of choice

Given a family $\mathscr{F}$ of non-empty sets, the trick is to define $M = \bigcup \mathscr{F}$. Note that if $A \in \mathscr{F}$, then $A \subseteq M$. Now let g be the choice function $g\colon \mathscr{P}(M) \setminus \{\varnothing\} \longrightarrow M$ as in the power set form of the axiom of choice. This g chooses an element out of each non-empty subset of M, including each subset which also happens to be an A in $\mathscr{F}$. So to define our choice function h for $\mathscr{F}$, we need only take h to be $g|_{\mathscr{F}}$, i.e. the restriction of g to $\mathscr{F}$,

$$\begin{aligned} h\colon \mathscr{F} &\longrightarrow \bigcup \mathscr{F} \ (= M) \\ A &\longmapsto g(A). \end{aligned}$$

■

Each of these three forms of the axiom has its uses, and as they are equivalent we shall from now on adopt the following convention:

> ***Convention***
>
> By the axiom of choice, abbreviated as AC, we mean either the axiom as first stated or either of the two equivalent forms in Theorem 5.1.

Let us now look at some more equivalents to AC.

Exercise 5.5

We have already shown that AC implies the statement

> Suppose that $f\colon C \longrightarrow D$ is an *onto* function. Then there is a *one-one* function $g\colon D \longrightarrow C$ such that $f(g(d)) = d$ for all $d \in D$.

Prove that AC is in fact equivalent to this statement.

Solution

We need to show that the statement implies AC. It turns out to be most convenient to take AC in the disjoint family form, so suppose that $\mathscr{F}$ is a disjoint family of non-empty sets. We'll need to involve some sort of onto function f, and the trick is to take

$$\begin{aligned} f\colon \bigcup \mathscr{F} &\longrightarrow \mathscr{F} \\ x &\longmapsto \text{the } A \in \mathscr{F} \text{ to which } x \text{ belongs.} \end{aligned}$$

This f is well-defined as a function, as not only does each x in $\bigcup \mathscr{F}$ belong to at least one member of $\mathscr{F}$, by definition of $\bigcup \mathscr{F}$, but also the condition that $\mathscr{F}$ is a disjoint family ensures that x belongs to at most one such member.

And f is clearly onto. Then the transversal function $g: \mathscr{F} \longrightarrow \bigcup \mathscr{F}$ such that $f(g(A)) = A$ for all $A \in \mathscr{F}$ is precisely a choice function on $\mathscr{F}$. (The only elements of $\bigcup \mathscr{F}$ which f maps to A are those already in A: so $g(A)$ has to be an element of A.)

For the next equivalent to AC we need to refine the idea of a Cartesian product, like $A_1 \times A_2$, to cover the Cartesian product of infinitely many sets, say an indexed family of sets $\{A_i : i \in I\}$. We cannot simply write $A_i \times A_{i'} \times A_{i''} \times \ldots$ as this would give an infinitely long expression. What we do is treat each member of this infinite product as a sequence $\langle a_i \rangle_{i \in I}$, where $a_i \in A_i$ for each $i \in I$, and define the infinite product as the set of all such sequences. This leads to the following definitions.

Formally, such an *indexed family* of sets should really be represented by a function F with domain I where $F(i) = A_i$ for each $i \in I$, rather than the set of its images $\{A_i : i \in I\}$. Note that the A_is need not be distinct.

> ***Definitions***
>
> Given $a_i \in A_i$ for each $i \in I$, the *sequence* $\langle a_i \rangle_{i \in I}$ is the function $f: I \longrightarrow \bigcup\{A_i : i \in I\}$ where $f(i) = a_i$ for each $i \in I$. The *Cartesian product of the indexed family of sets* $\{A_i : i \in I\}$ is the set of all such sequences, i.e. of all functions $f: I \longrightarrow \bigcup\{A_i : i \in I\}$ such that $f(i) \in A_i$ for each $i \in I$. We shall write this product as $\prod_{i \in I} A_i$.

So, for instance, if $I = \mathbb{Z}$ and, for each $i \in \mathbb{Z}$, A_i is the open interval $(i, i+1)$ of $\mathbb{R}$, then one element of $\prod_{i \in I} A_i$ is $\langle i + \frac{1}{4} \rangle_{i \in \mathbb{Z}}$, i.e. the function $f: \mathbb{Z} \longrightarrow \mathbb{R} \setminus \mathbb{Z}$ defined by $f(i) = i + \frac{1}{4}$.

For a finite indexed family of sets, e.g. $\{A_1, A_2\}$, this definition of product does not produce the same set as $A_1 \times A_2$. But there is a reasonable correspondence between them. If it isn't clear from the context which product of finitely many sets to take, then stick to the usual one in terms of ordered n-tuples. (For an infinite indexed family of sets there's only the one possibility.)

Exercise 5.6

(a) Explain why $A_1 \times A_2$ and $\prod_{i \in \{1,2\}} A_i$ are different sets in *ZF*.

(b) Show that there is a bijection between $A_1 \times A_2$ and $\prod_{i \in \{1,2\}} A_i$.

Solution

(a) A typical member of $A_1 \times A_2$ in *ZF* is an ordered pair $\langle a_1, a_2 \rangle$, i.e. $\{\{a_1\}, \{a_1, a_2\}\}$, where $a_1 \in A_1, a_2 \in A_2$. And $\{\{a_1\}, \{a_1, a_2\}\}$ is a subset of $\mathscr{P}(A_1 \cup A_2)$.
A typical member of $\prod_{i \in \{1,2\}} A_i$ is a function $f: \{1,2\} \longrightarrow A_1 \cup A_2$ where $f(i) \in A_i$ for each $i \in \{1,2\}$. Technically within set theory f is a subset of $\{1,2\} \times (A_1 \cup A_2)$. So the sets $A_1 \times A_2$ and $\prod_{i \in \{1,2\}} A_i$ contain different sorts of element, and cannot be equal.

$\bigcup\{A_i : i \in \{1,2\}\} = A_1 \cup A_2$

As a set, a function is represented by its graph.

(b) Define $\theta: \prod_{i \in \{1,2\}} A_i \longrightarrow A_1 \times A_2$ by

$$\theta(f) = \langle f(1), f(2) \rangle.$$

It is straightforward to show that θ is one–one and onto.

Exercise 5.7

Show that AC is equivalent to the following statement:

> The Cartesian product of an indexed family of non-empty sets is itself non-empty.

Of course AC is not needed in the case that the indexed family is finite, for reasons similar to those in the solution to Exercise 5.4. The familiarity of the result for the finite case might make the statement in Exercise 5.7 seem more plausible than some of the other forms of AC.

Exercise 5.8

Suppose that $A_i = A$ for each $i \in I$, where $A \neq \varnothing$. Is AC needed to show that $\prod_{i \in I} A_i \neq \varnothing$?

Our discussion so far of AC has been entirely in terms of its significance for the basic abstract objects of set theory, particularly sets and functions. In the next section we shall look at some of the ways in which AC is used in everyday mathematics.

Further exercises

Exercise 5.9

Try to find direct proofs (as opposed to those in our proof of Theorem 5.1) of the following:

(a) AC $\Rightarrow$ power set form;

(b) power set form $\Rightarrow$ disjoint family form;

(c) disjoint family form $\Rightarrow$ AC.

Exercise 5.10

Consider the following statement:

> Suppose that $f\colon A \longrightarrow B$ is a one–one function. Then there is an onto function $g\colon B \longrightarrow A$ such that $g(f(a)) = a$ for each $a \in A$.

Does the proof of this statement require AC? Is the statement equivalent to AC?

Exercise 5.11

R is said to be a *relation between sets A and B* if $R \subseteq A \times B$. The *domain* of R is the set $\{a \in A : (a, b) \in R \text{ for some } b \in B\}$. Recall that a function F is a set of ordered pairs such that for each a there is at most one b for which $(a, b) \in F$. Show that AC is equivalent to the statement:

> Suppose that R is a relation between non-empty sets A and B. Then there is a function F with the same domain as R such that $F \subseteq R$.

Exercise 5.12

For each of the following families $\mathcal{F}$, construct, where possible, a choice function h for $\mathcal{F}$, i.e. $h: \mathcal{F} \longrightarrow \bigcup \mathcal{F}$ such that for each $A \in \mathcal{F}$, $h(A) \in A$.

(a) $\mathcal{F} = \mathcal{P}(\mathbb{Z}) \setminus \{\varnothing\}$

(b) $\mathcal{F} = \mathcal{P}(\mathbb{Q}) \setminus \{\varnothing\}$

(c) $\mathcal{F}$ is $\mathcal{P}(\mathbb{R}[x])$, the set of all polynomials with variable x and real coefficients.

(d) $\mathcal{F} = \{A_f : f \in \mathbb{C}[z]\}$, where $\mathbb{C}[z]$ is the set of all polynomials in z with complex coefficients and $A_f = \{z \in \mathbb{C} : f(z) = 0\}$.

Exercise 5.13

Let $\{A_i : i \in I\}$ be a (possibly infinite) family of disjoint non-empty sets. What conditions need to be imposed on the A_i for there to exist a one–one function $f: \bigcup\{A_i : i \in I\} \longrightarrow \prod_{i \in I} A_i$? Prove that f exists under these conditions, using AC if you like.

5.3 The axiom of choice and mathematics

The axiom of choice arose not so much from any introspection on the nature of sets, but as a principle justifying other seemingly valuable pieces of mathematics. The particular piece of mathematics which prompted Zermelo to formulate AC was the well-ordering principle:

> ***The well-ordering principle***
>
> Every set can be well-ordered.

We shall abbreviate the well-ordering principle as WO. We shall discuss the relationship between AC and WO later in the book, when we have more information about the theory of orders.

Cantor introduced this principle to develop some of his results on the theory of infinite subsets of $\mathbb{R}$. The principle was very controversial. Some mathematicians, for instance the very influential Hilbert, thought that the special case that $\mathbb{R}$ can be well-ordered was well worth investigating. But by and large mathematicians doubted WO as a general principle. Zermelo then formulated AC specifically to prove WO. This proof was received with some suspicion as it related the use of arbitrary choices in some parts of mathematics, hitherto somewhat unnoticed and relatively uncontroversial, with a principle which was much more controversial. (In fact this suspicion of his proof was what led Zermelo to formulate axioms for set theory, to defend his work.) In an atmosphere of heightened awareness of use of contentious principles of reasoning, it was discovered that some previously accepted arguments within mathematics made use of some form of AC and that there were other useful principles of reasoning, such as WO, which didn't look as though they involved choice functions, but were in fact equivalent to AC. (All this is detailed in the excellent book *Zermelo's Axiom of Choice* by Greg Moore[22].) To try to settle the issue of whether to accept or reject AC, mathematicians hoped to show that AC was actually either provable or disprovable from the other axioms of ZF.

As Zermelo wished to prove WO, he included AC as one of his original axioms. The controversy over AC led those who subsequently developed his axioms not to include it as a basic axiom in ZF.

In 1940 Gödel showed that AC was consistent with the rest of ZF, so it was not disprovable within ZF. And in 1963 Cohen proved that AC was independent of ZF, so could not be proved from ZF. This means that acceptance of AC has become more overtly a matter of taste, where one is influenced by the sort of mathematics entailed by, or underpinned by, AC. In the following we look at some of this mathematics.

Gödel showed that if ZF was consistent so was ZF along with AC. The consistency of ZF, however, cannot be proved within ZF, by Gödel's earlier incompleteness theorems.

Exercise 5.14

A function $f \colon \mathbb{R} \longrightarrow \mathbb{R}$ is said to be *sequentially continuous* at a if for every sequence $\langle x_n \rangle$ with $\lim_{n\to\infty} x_n = a$, we have $\lim_{n\to\infty} f(x_n) = f(a)$.

Prove that f is continuous at a, i.e. for every $\varepsilon > 0$ there is a $\delta > 0$ such that

$$\text{if } |x-a| < \delta \text{ then } |f(x) - f(a)| < \varepsilon$$

if and only if f is sequentially continuous at a.

Where in your argument have you used some form or consequence of the axiom of choice?

Solution

Suppose that f is continuous at a and that $\lim_{n\to\infty} x_n = a$. We shall show that for any $\varepsilon > 0$ there is an N such that if $n > N$ then $|f(x_n) - f(a)| < \varepsilon$.

Given $\varepsilon > 0$ there is a $\delta > 0$ such that if $|x - a| < \delta$ then $|f(x) - f(a)| < \varepsilon$, as f is continuous at a. For this $\delta > 0$ there is an N such that if $n > N$ then $|x_n - a| < \delta$, so that $|f(x_n) - f(a)| < \varepsilon$ as required. So f is sequentially continuous at a.

Conversely, suppose that f is not continuous at a. We shall show that f is not sequentially continuous at a, i.e. there is some sequence $\langle x_n \rangle$ with $\lim_{n\to\infty} x_n = a$ for which it is not the case that $\lim_{n\to\infty} f(x_n) = f(a)$.

f not continuous at a means that there is some $\varepsilon > 0$ such that for any $\delta > 0$, however small, there is an x with $|x - a| < \delta$ for which $|f(x) - f(a)| \geq \varepsilon$. So for each positive integer n, taking $\delta = \frac{1}{n}$ gives an x_n for which $|x_n - a| < \frac{1}{n}$ and $|f(x_n) - f(a)| \geq \varepsilon$. But this means that for this sequence $\langle x_n \rangle$, $\lim_{n\to\infty} x_n = a$ and $\lim_{n\to\infty} f(x_n) \neq f(a)$. Thus f is not sequentially continuous at a.

AC is needed in the second half of the proof, when for each n we choose an x_n out of the set $A_n = \{x : |x - a| < \frac{1}{n} \text{ and } |f(x) - f(a)| \geq \varepsilon\}$. So we need a choice function on the family $\{A_n : n \in \mathbb{N}, n > 0\}$.

There is no choice of x_n as something straightforward like $\min A_n$. It is easy to show that $\min A_n$ doesn't exist.

The above is a standard result in textbooks on real analysis which was proved more than 30 years before Zermelo's work and was apparently accepted without any qualms!

A standard and fundamental result in the theory of Lebesgue integration is:

> ***Theorem 5.2***
>
> A countable union of null sets is null.

(A subset X of $\mathbb{R}$ is said to be *null* if for any given $\varepsilon > 0$ there is a sequence $\langle I_n \rangle$ of intervals covering X with total length less than ε; i.e. $X \subseteq \bigcup\{I_n : n \in \mathbb{N}\}$ and $\sum_{n\in\mathbb{N}} l(I_n) < \varepsilon$.)

If I_n is any one of the intervals $(a_n, b_n), (a_n, b_n], [a_n, b_n), [a_n, b_n]$, the *length* $l(I_n)$ of I_n is $b_n - a_n$.

Proof

Suppose that $X = \bigcup\{X_m : m \in \mathbb{N}\}$, where each X_m is null. Then given $\varepsilon > 0$, there is, for each m, a sequence $\langle I_{m,n} \rangle$ of intervals covering X_m with total length less than $\frac{\varepsilon}{2^{m+1}}$. We define a sequence $\langle Y_k \rangle$ of intervals by working our way through a rectangular array of the $I_{m,n}$ as in the following diagram:

This argument is based on that used by Cantor to show that $\mathbb{Q}$ is countable, which you will meet in Section 6.4.

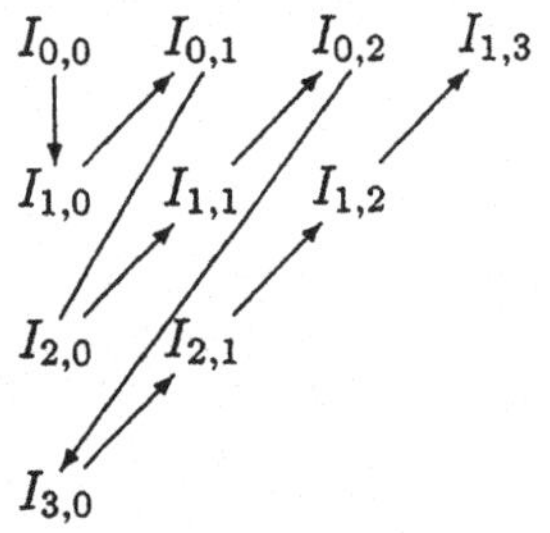

So $Y_0 = I_{0,0}$, $Y_1 = I_{1,0}$, $Y_2 = I_{0,1}$, $Y_3 = I_{2,0}$ and in general $I_{m,n}$ appears in the list as $Y_{\frac{1}{2}(m+n)(m+n+1)+n}$. It is easy to check that $X \subseteq \bigcup\{Y_k : k \in \mathbb{N}\}$ and that a bound for the total length of the Y_ks is given by

You might like to verify that $J(m,n) = \frac{1}{2}(m+n)(m+n+1)+n$ defines a bijection $J\colon \mathbb{N} \times \mathbb{N} \longrightarrow \mathbb{N}$.

$$\begin{aligned}\sum_{k\in\mathbb{N}} l(Y_k) &= \sum_{n\in\mathbb{N}} l(I_{0,n}) + \sum_{n\in\mathbb{N}} l(I_{1,n}) + \ldots + \sum_{n\in\mathbb{N}} l(I_{m,n}) + \ldots \\ &< \frac{\varepsilon}{2} + \frac{\varepsilon}{4} + \ldots + \frac{\varepsilon}{2^{m+1}} + \ldots \\ &= \varepsilon\end{aligned}$$

Thus given $\varepsilon > 0$ there is a suitable sequence of intervals covering X of length less than ε. Hence X is null. ■

Exercise 5.15

Where in the proof above have we used some form of AC?

Solution

There was a sneaky use of AC when we argued that

> ...given $\varepsilon > 0$, there is, for each m, a sequence $\langle I_{m,n} \rangle$ of intervals covering X_m with total length less than $\frac{\varepsilon}{2^{m+1}}$.

There will in general be many such sequences for each m, and we need to use some form of AC to choose one for each m in the infinite set $\mathbb{N}$.

Another consequence of AC for Lebesgue's work, in the related field of measure theory, is that there are non-measurable subsets of the spaces $\mathbb{R}^n$, as we shall explain. The idea of measure of a subset of $\mathbb{R}^n$ fleshes out intuitions of the length of an interval of $\mathbb{R}$ or a curve in $\mathbb{R}^2$, of the area of a shape in $\mathbb{R}^2$ and of the volume of a solid in $\mathbb{R}^3$. In the Lebesgue theory, a *measure* is a function μ from some set Y of subsets of $\mathbb{R}^n$ to $\mathbb{R} \cup \{\infty\}$ with the following properties:

(i) $\mu(X) \geq 0$, for any subset X in Y;

(ii) if X and Y are congruent subsets of $\mathbb{R}^n$, then $\mu(X) = \mu(Y)$;

(iii) μ is countably additive, i.e. if $X_0, X_1, X_2, \ldots, X_n, \ldots$ are countably many disjoint subsets of $\mathbb{R}^n$, then

$$\mu\left(\bigcup_{n\in\mathbb{N}} X_n\right) = \sum_{n=0}^{\infty} \mu(X_n).$$

X is said to be *measurable* with measure $\mu(X)$.

The need for countable additivity is seen in, for instance, probability theory, where one often wants the probability of some event occurring eventually, e.g. after some unspecified finite number of moves or throws. Modern probability theory makes considerable use of Lebesgue measure theory.

Usually one also imposes some sort of yardstick for a specific measure: for instance, for measuring lengths of subsets of the real line $\mathbb{R}$, one would usually define the measure of the unit closed interval, $\mu([0,1])$, as 1, and the measure of an interval $[a, b]$ to be $b - a$. Null sets, as in Theorem 5.2 above, are just the sets of measure 0. And it makes sense to regard the measure of $\mathbb{R}$ itself as ∞.

The question arises of whether there is a measure on all subsets of $\mathbb{R}^n$. Using AC, the answer is no. We shall show that there is some subset of the unit circle in $\mathbb{R}^2$, i.e. $\{(\cos\alpha, \sin\alpha) : 0 \leq \alpha < 2\pi\}$ which is not measurable. Our argument will be by assuming that the special subset is measurable, i.e. has measurable length, and obtaining a contradiction with the measure of the circumference of the whole circle being 2π.

> ***Theorem 5.3***
>
> There is a subset of the unit circle C with non-measurable length.

Proof

Let P_α be the point $(\cos\alpha, \sin\alpha)$ on C, so that $C = \{P_\alpha : 0 \leq \alpha < 2\pi\}$. Define a relation $\sim$ on C by

Note that P_α is the same point as $P_{\alpha+2k\pi}$ for any $k \in \mathbb{Z}$.

$$P_\alpha \sim P_\beta \text{ if and only if } \beta - \alpha = q\pi \text{ for some } q \in \mathbb{Q}.$$

It is easy to show that $\sim$ is an equivalence relation on the points of C.

Let r_θ be the rotation of the plane with centre the origin through an angle θ anti-clockwise. Then, given P_α, a typical point in C equivalent to it, which must be of the form $P_{\alpha +_{2\pi} q\pi}$ for some rational q in $[0, 2)$, can also be written as $r_{q\pi}(P_\alpha)$ for this q. Thus the set $\{r_{q\pi}(P_\alpha) : q \in [0, 2)\}$ is the equivalence class of P_α.

Define $+_{2\pi}$ on $[0, 2\pi)$ along the lines of addition modulo n, by

$$\alpha +_{2\pi} \beta = \begin{cases} \alpha + \beta, & \text{if } \alpha + \beta < 2\pi \\ \alpha + \beta - 2\pi, & \text{if } 2\pi \leq \alpha + \beta \end{cases}$$

Now use AC to choose a P_α from each of the equivalence classes and let X_0 be the set of all these P_αs, so that X_0 contains exactly one point from each class. Suppose that this set X_0 is measurable. For each rational $q \in [0, 2)$ let $X_{q\pi}$ be the result of rotating the set X_0 by $r_{q\pi}$, so that $X_{q\pi}$ is congruent to X_0 and consequently $\mu(X_{q\pi}) = \mu(X_0)$. The construction ensures that $X_{q\pi}$s for different values of q in $[0, 2)$ are disjoint, while we have also shown that the union of all the $X_{q\pi}$s must be C. Thus

$$\begin{aligned}\mu(C) &= \mu\left(\bigcup\{X_{q\pi} : q \in [0,2), q \text{ rational}\}\right)\\ &= \sum_{\substack{q\in[0,2)\\ q \text{ rational}}} \mu(X_{q\pi}) \quad (\mathbb{Q} \text{ is countable and } \mu \text{ is countably additive})\\ &= \sum_{\substack{q\in[0,2)\\ q \text{ rational}}} \mu(X_0) \quad (\text{as each } X_{q\pi} \text{ is congruent to } X_0)\\ &= \begin{cases} 0, & \text{if } \mu(X_0) = 0,\\ \infty, & \text{if } \mu(X_0) > 0.\end{cases}\\ &\neq 2\pi \quad (\text{which is the length of } C)\end{aligned}$$

We shall show that $\mathbb{Q}$ is countable in Chapter 6.

The contradiction arises at the stage where we suppose that X_0 is measurable, i.e. $\mu(X_0)$ exists. $\mu(X_0)$ does not exist, i.e. the subset X_0 has no measurable length. ■

Another way out of the contradiction is to reject AC! But the point of this chapter is to investigate the consequences of AC.

Note that lurking in the above proof is also a proof that the subset $\{\alpha : P(\alpha) \in X_0\}$ of $[0, 2\pi)$ is not measurable, and it wouldn't take much to turn the proof into one showing that there is a subset of the unit disc, $\{(r\cos\alpha, r\sin\alpha) : 0 \le r \le 1, \alpha \in [0,2)\}$, which is not a measurable subset of $\mathbb{R}^2$.

The existence of non-measurable subsets of $\mathbb{R}^n$, while of obvious importance to measure theorists, is nothing like as dramatic as the following theorem.

Theorem 5.4 Banach–Tarski theorem

Let S be a unit ball in $\mathbb{R}^3$, i.e. all points within a sphere of radius 1. Then S can be partitioned into finitely many subsets which can be moved, using translations and rotations, to produce *two* unit balls.

This theorem is often described as a paradox! The proof uses AC and is non-constructive. And most of the pieces of the ball are non-measurable sets, so very hard to visualize.

Should we regard the Banach–Tarski theorem as such an undesirable paradox that we should reject the axiom of choice? Or is the axiom of choice such a natural or useful principle that we should live with it and its consequences, even though some of the latter are a trifle surprising? In the next section we shall look at an equivalent principle to AC which helps us to see more of its mathematical consequences – of a desirable and non-paradoxical variety! And there are further equivalents (not just consequences!) of AC with importance in other parts of mathematics which you might encounter within undergraduate mathematics, for instance the following: *Tychonoff's theorem* in topology, which says that the (infinite) product of compact spaces is compact; and the *completeness theorem* for arbitrary infinite languages in first order predicate calculus, which says that a set of formulas is consistent if and only if it has a model. And later in the book we shall see more of how AC is intimately connected to the theory of infinite sets, and that it has other tempting mathematical equivalents. In the author's, and many other mathematicians', view,

AC and its equivalents have so many reasonable consequences that it seems desirable to accept it as a firm part of mathematics. From this point of view, results like the Banach–Tarski theorem and the existence of non-measurable sets serve to inform one's intuition, and lack of intuition, about complicated, infinite sets. But as this position is open to debate, we shall take a cautious position in the rest of this book, by seeking to avoid use of AC wherever possible, and highlighting any use of it.

Further exercise

Exercise 5.16

A *weak consequence* of AC, i.e. something which is a consequence of AC but which is not equivalent to AC, is the *principle of dependent choices*, as follows.

> Suppose that R is a relation between sets A and B, i.e. $R \subseteq A \times B$, such that the *range* of R, $\{b \in B : (a, b) \in R \text{ for some } a \in A\}$, is a subset of the domain of R, $\{a \in A : (a, b) \in R \text{ for some } b \in B\}$. Then there is a function $f : \mathbb{N} \longrightarrow A$ such that for all $\mathbf{n} \in \mathbb{N}$,
>
> $$f(\mathbf{n}) R f(\mathbf{n}^+).$$

(a) Show that AC implies the principle of dependent choices. [Hints: define f by recursion. How will AC get used?]

(b) Suppose that we were to drop the axiom of foundation, *ZF9*, from our axioms of set theory and that x was a set that was not well-founded, i.e. for this x there is no $y \in x$ such that $x \cap y = \varnothing$. Use the principle of dependent choices to show that there is an infinite descending $\in$-chain

$$\ldots \in x_{n+1} \in x_n \in \ldots \in x_2 \in x_1 \in x_0$$

where each x_n is an element of x.

(c) In Exercise 5.14 above, we used AC to show that sequential continuity implies continuity. Use the principle of dependent choices instead of AC to prove this result.

Many of the results of 19th-century real analysis which made unconscious use of AC can be proved from the weaker principle of dependent choices. Some mathematicians who reject the full version of AC are willing to take the latter principle as an axiom.

5.4 Zorn's lemma

Why have we singled out the axiom of choice from among the many assertions about, and axioms for, sets? One reason is that AC is intimately connected with what makes Cantor's original theory of infinite sets work, as we will discuss later. Another reason is that AC is equivalent to several other principles which do not look as though they have very much to do with choosing elements out of sets, but have interesting mathematical consequences. In this section we shall look at one such, called Zorn's lemma.

Max Zorn introduced the principle known as Zorn's lemma in 1925.

Zorn's lemma – ZL for short – concerns partially ordered sets. We shall look at ordered sets in some detail in Chapter 7, but only need a part of the terminology now.

Definitions

Let P be a set and R a subset of $P \times P$. (We shall often write xRy for $(x, y) \in R$.) R is a *weak partial order* on P, or P is *weakly partially ordered* by R, if it has the following properties:

reflexive for all x in P, xRx;

anti-symmetric for all x and y in P, if xRy and yRx then $x = y$;

transitive for all x, y and z in P, if xRy and yRz then xRz.

Associated with each weak partial order R on a set P there is a *strict* partial order S, where S is also a subset of $P \times P$ and, for all $x, y \in P$,

$$(x, y) \in S \text{ (or } xSy) \quad \text{exactly when} \quad xRy \text{ and } x \neq y.$$

The usual $\leq$ and $<$ on $\mathbb{N}$ are examples of a weak, and the corresponding strict, partial order. They have the additional property of being *linear*.

The most relevant example of a weak partial order for the applications in this section is to take P to be some set of subsets of a set Y, i.e. $P \subseteq \mathscr{P}(Y)$, and R to be $\subseteq$. For instance, P might be $\mathscr{P}(\mathbb{N})$, or the set of all infinite subsets of $\mathbb{N}$, or the set of all proper ideals of a ring S, or the set of all functions with domain some subset of A and codomain B, all of which are partially ordered by $\subseteq$.

This sort of example usually gives a partial order which is *not* linear.

We shall often just write 'partial order', instead of 'weak (or strict) partial order'. The context will usually make it clear whether a given order is weak or strict. Here, as $W \subseteq X$ includes the case when $W = X$, the partial order must be weak.

Exercise 5.17

Explain how the set of all functions with domain some subset of a set A and codomain the set B can be regarded as partially ordered by $\subseteq$.

Solution

Technically, a function f of this sort is represented in *ZF* by a set of pairs in $A \times B$, i.e. a subset of $A \times B$, where for each $a \in A$ there is at most one $b \in B$ for which $(a, b) \in f$. Thus it makes sense to compare two such functions, f and g, in terms of one being a subset of the other. We have

$$f \subseteq g$$

if and only if

$$\text{whenever } (a, b) \in f, \text{ then } (a, b) \in g$$

if and only if

$$f(a) = g(a), \text{ for all } a \in \mathrm{Dom}(f).$$

Recall the definition of the domain of f, $\mathrm{Dom}(f)$, as the subset $\{a \in A : (a, b) \in f \text{ for some } b \in B\}$ of A. Then for all $a \in \mathrm{Dom}(f)$, there is *exactly* one $b \in B$ with $(a, b) \in f$.

Exercise 5.18

Let Y be a set and suppose that $P \subseteq \mathscr{P}(Y)$. Show that P is partially ordered by $\subseteq$.

We hope that you find this exercise very routine indeed!

The statement of Zorn's lemma will refer to an 'upper bound' of a subset and a 'maximal' element within a partially ordered set. The definitions of these concepts rely on a convention of treating a weak partial order R as 'less than or equal to', rather than 'greater than or equal to'. Thus xRy is treated as though x was less than or equal to y, which makes the following definitions match our intuitions about the use of the words defined.

> ***Definitions***
>
> Let P be a set weakly ordered by R. If X is a non-empty subset of P, then an element y of P is an *upper bound* of X if xRy for all $x \in X$. An element x of P is a *maximal* element of P if there is no $y \in P$ such that $y \neq x$ and xRy.

Note that x being a maximal element just means there is no greater element y. It does not necessarily mean that x is the maximum element of P, in the sense of being greater than every other element of the set, as some of the examples below will illustrate.

Exercise 5.19

Let P be the set $\mathscr{P}(\mathbb{N}) \setminus \{\mathbb{N}\}$, i.e. all subsets of $\mathbb{N}$ other than $\mathbb{N}$ itself, partially ordered by $\subseteq$.

(a) Show that $\mathbb{N} \setminus \{0\}$ is a maximal element of P.

(b) Are there any other maximal elements of P? If so, what are they? And does P have a maximum element? If so, what is it?

(c) Find upper bounds, if they exist, for each of the following subsets of P.

(i) $X = \{\{2n, 2n+6\} : n \in \mathbb{N}\}$

(ii) $X = \{\{n, n+2\} : n \in \mathbb{N}\}$

Solution

(a) We need to show that there is no element of P which contains $\mathbb{N} \setminus \{0\}$ as a proper subset. The only candidate in $\mathscr{P}(\mathbb{N})$ for such an element is $\mathbb{N}$, but this is not an element of P. Thus $\mathbb{N} \setminus \{0\}$ is maximal.

(b) Similarly, $\mathbb{N} \setminus \{n\}$ is a maximal element of P for each $n \in \mathbb{N}$. The only subset of $\mathbb{N}$ which would be a candidate for a maximum element of P is $\mathbb{N}$ itself, but $\mathbb{N}$ isn't an element of P.

Of course $\mathbb{N}$ is the maximum element of $\mathscr{P}(\mathbb{N})$ when partially ordered by $\subseteq$. But the point of this exercise is that a partially ordered set need not contain a maximum element.

(c) (i) For a set P of subsets of Y partially ordered by $\subseteq$ and a subset X of P, the natural candidate for an upper bound of X is $\bigcup X$. This union of subsets of Y will also be a subset of Y, and if it is in the particular set P then it will be an upper bound for X in P, as

$$A \in X \text{ implies that } A \subseteq \bigcup X.$$

In addition, it will also be the *least* upper bound: if $B \in P$ is any upper bound for X, then $\bigcup X \subseteq B$.

For the subset X in this exercise,

$$\begin{aligned}\bigcup X &= \bigcup\{\{2n, 2n+6\} : n \in \mathbb{N}\}\\ &= \{2k : k \in \mathbb{N}\},\end{aligned}$$

which is an element of P. This is the least upper bound of X and any element of P containing it as a subset is also an upper bound of X.

(ii) In this case $\bigcup X$ equals $\mathbb{N}$, so that any upper bound for X in the ordered set P must contain $\mathbb{N}$ as a subset. But no element of P has $\mathbb{N}$ as a subset, so that X has no upper bound in P.

Exercise 5.20

Let P be the set of all finite subsets of $\mathbb{N}$. Does P have any maximal elements? If so, what are they?

Solution

P has no maximal elements. For any element A of P there is a larger finite subset of $\mathbb{N}$, hence also an element of P, containing A as a proper subset: thus A is not maximal.

Exercise 5.21

Let P be the set of all one–one functions with domain a subset of A and codomain B, partially ordered by $\subseteq$ (as in Exercise 5.17 above). Suppose that f is a maximal element of P. What special properties, if any, does this function f possess?

We are not claiming here that P necessarily has any maximal elements. The issue of whether a maximal element exists turns out to be of considerable importance, as you will see later.

Solution

Treat f as a subset of $A \times B$. As f is a maximal element of P, it must be impossible to add an extra pair (a, b) to f so that $f \cup \{(a, b)\}$ is also an element of P, i.e. f is also a one–one function. Clearly we could not add a new pair (a, b) for an a which is in $\mathrm{Dom}(f)$, as this would stop $f \cup \{(a, b)\}$ from being a function – it wouldn't be single-valued. This leaves us with two cases. Either $\mathrm{Dom}(f)$ has already used up all the as in A, i.e. $\mathrm{Dom}(f)$ equals A. Or there is still some $a \in A$ not in $\mathrm{Dom}(f)$, but $\mathrm{Range}(f)$ has used up all of B, so that adding an extra pair (a, b) to those in f creates a function $f \cup \{(a, b)\}$ which is not one–one, hence not in P.

In the case that $\mathrm{Dom}(f) = A$, we have a one–one function from A to B. In the case that $\mathrm{Range}(f) = B$, the inverse function f^{-1} is a one–one function from B to A.

To state Zorn's lemma, we need one further definition, as follows.

> ***Definition***
>
> Let P be a set weakly partially ordered by R, and let $\mathscr{C}$ be a non-empty subset of P. Then $\mathscr{C}$ is said to be a *chain* if it has the following additional property:
>
> **linear** for all $x, y \in \mathscr{C}$, xRy or yRx.
>
> In terms of the strict partial order S associated with R, this becomes
>
> for all $x, y \in \mathscr{C}$, xSy or $x = y$ or ySx.

Sometimes we shall use the description R-*chain* to emphasize that the chain is associated with the order R, rather than some other order also being discussed.

We shall look at linear orders in more detail in Chapter 7.

Examples of chains are

$$\{\{0, 1, 2, \ldots, n\} : n \in \mathbb{N}\} \text{ in } \mathscr{P}(\mathbb{N}) \text{ partially ordered by } \subseteq,$$

and

$$\{\text{open intervals } (\tfrac{1}{a}, a) : a \in \mathbb{R}, a > 1\} \text{ in } \mathscr{P}(\mathbb{R}) \text{ partially ordered by } \subseteq .$$

We can now state Zorn's lemma.

> ***Zorn's lemma***
>
> Let P be a non-empty set partially ordered by R with the property that every chain $\mathscr{C}$ in P has an upper bound in P. Then P contains at least one maximal element.

I don't know why it's called a lemma. According to Moore [22], Zorn himself regarded it as an axiom. Its status here in this book is as a principle equivalent to AC.

Before we look at the relationship between Zorn's lemma and AC, let's see how Zorn's lemma is used in some different parts of mathematics. An obvious application is to show that a set has a maximal subset with some sort of interesting structure, for instance the following:

a group contains a maximal Abelian subgroup,

and

a commutative ring with a multiplicative identity contains a maximal proper ideal.

It is worth noting here, even if you don't happen to know all these mathematical terms, that these results can all be proved without use of Zorn's lemma for *finite* groups and rings. It is when the latter are infinite that Zorn's lemma might be needed; and, as with AC, it is when they don't have 'enough other structure' that Zorn's lemma will be needed.

> ***Theorem 5.5***
>
> Let G be a group. Then, assuming Zorn's lemma, G has a maximal Abelian subgroup.

Proof

If G is Abelian then G itself is the required maximal subgroup. So we shall now deal with the case where G is not Abelian, so that any Abelian subgroup of G must be a *proper* subgroup.

To set up the use of Zorn's lemma, define P to be the set of all Abelian subgroups of G, and partially order P by $\subseteq$. Then a maximal element of P given by the lemma is just what we are looking for, namely a maximal Abelian subgroup of G. So we have to check that the conditions of the lemma are satisfied.

It is straightforward to show that P is non-empty – the trivial subgroup containing just the identity, e_G, of G is Abelian. The real work is in showing that if $\mathscr{C}$ is a chain in P, then $\mathscr{C}$ has an upper bound in P. In this context, $\mathscr{C}$ being a chain means that the elements of $\mathscr{C}$ are Abelian subgroups of G and that given any two of them, A and B, either $A \subseteq B$ or $B \subseteq A$. As suggested in the solution to Exercise 5.19(c)(ii), the sensible candidate for an upper bound of a $\subseteq$-chain, as here, is its union. Given that the partial order relation is $\subseteq$, we are guaranteed that $\bigcup \mathscr{C}$ is an upper bound of the chain $\mathscr{C}$. We have to check whether this upper bound is in P, i.e. whether $\bigcup \mathscr{C}$ is an Abelian subgroup of G.

Let's check whether $\bigcup\mathscr{C}$ is closed under the group operation, i.e. if $a, b \in \bigcup\mathscr{C}$, then $ab \in \bigcup\mathscr{C}$. So take any $a, b \in \bigcup\mathscr{C}$. This means that there are A, B in the chain $\mathscr{C}$ for which $a \in A$ and $b \in B$. As $\mathscr{C}$ is a chain, then either $A \subseteq B$ or $B \subseteq A$. Without loss of generality, we can suppose that $A \subseteq B$. Then both a and b are in B, so that as B is a subgroup of G (and hence closed under the group operation) the product ab is also in B. As $B \in \mathscr{C}$, this means that $ab \in \bigcup\mathscr{C}$, so that $\bigcup\mathscr{C}$ is closed, as required. Furthermore, as B is also Abelian and $a, b \in B$, we have that $ab = ba$, so that elements of $\bigcup\mathscr{C}$ commute.

This is where we make crucial use of the fact that $\mathscr{C}$ is a chain.

It is much more straightforward to show that $\bigcup\mathscr{C}$ satisfies the other properties for being a subgroup of G (e.g. containing the identity e_G of G and being closed under inverses).

Checking these properties doesn't require use of the fact that $\mathscr{C}$ is a chain, rather than any old set of Abelian subgroups of G.

Thus $\bigcup\mathscr{C}$ is an Abelian subgroup of P, so that the chain $\mathscr{C}$ has an upper bound in P. We can now conclude from Zorn's lemma that P has at least one maximal element, i.e. G has a maximal Abelian subgroup. ∎

Exercise 5.22

Verify that $\bigcup\mathscr{C}$ in the proof above does contain e_G and is closed under inverses.

There are other mathematical applications where the maximality criterion is not quite so explicit, for instance:

> every vector space has a basis,

and

> if X is partially ordered by R, then there is a linear order R^* on X extending R, i.e. for all $x, y \in X$, if xRy then xR^*y.

We shall show that every vector space has a basis. This is, of course, a standard result for finite-dimensional vector spaces, but there are plenty of infinite-dimensional spaces for which a special proof might be required. Take, for instance, the set $\mathbb{R}^{\mathbb{R}}$ of all functions from $\mathbb{R}$ to itself, with addition and scalar multiplication defined by $f + g$ and αf being the functions

Observe again the distinction between a result for finite dimension, where AC is not needed, and infinite dimension, where it might be needed.

$$f + g\colon x \longmapsto f(x) + g(x), \text{ for all } x \in \mathbb{R},$$

and

$$\alpha f\colon x \longmapsto \alpha \cdot f(x), \text{ for all } x \in \mathbb{R},$$

for all functions $f, g \in \mathbb{R}^{\mathbb{R}}$ and scalars $\alpha \in \mathbb{R}$. It is easy to check that this makes $\mathbb{R}^{\mathbb{R}}$ a vector space over $\mathbb{R}$, and it is fairly easy to convince oneself that it does not have a finite basis. And it turns out that there is no obvious basis – we really need to explain what is meant by a basis for infinite-dimensional vector spaces! A *basis* of a vector space V over a field F is a linearly independent subset B of V such that every vector of V is a finite linear combination of vectors in B; and B is *linearly independent* means that no finite subset of B is linearly dependent. Another piece of standard terminology from linear algebra is that of the *span* of a set B of vectors, written as $\langle B\rangle$, which is the set of all finite linear combinations of vectors in B. Thus B is a basis of V if it is linearly independent and its span is V.

One way to show that $\mathbb{R}^{\mathbb{R}}$ doesn't have a finite basis is to use cardinalities, which we deal with later. Any finitely generated subspace of $\mathbb{R}^{\mathbb{R}}$ has smaller cardinality than $\mathbb{R}^{\mathbb{R}}$.

Note that we don't allow infinite linear combinations of vectors.

> ***Theorem 5.6***
>
> Let V be a vector space. Then, assuming Zorn's lemma, V has a basis.

Proof

Zorn's lemma is relevant here because a basis of V is also a maximal linearly independent subset of V. If B is a maximal linearly independent subset and there is a vector $v \in V$ which is not in the span of B, then $B \cup \{v\}$ is a larger linearly independent set than B, contradicting the maximality of B. So we define a set P of all linearly independent subsets of V, partially order P by $\subseteq$, and try to show that P satisfies the conditions of Zorn's lemma, so that P has a maximal element.

If V consists of only the zero vector, the result is trivial. Otherwise V contains at least one non-zero vector v, so that P is non-empty, as it contains the linearly independent subset $\{v\}$. Now we must show that any chain $\mathscr{C}$ has an upper bound in P. As with the previous example, the natural candidate for an upper bound is $\bigcup \mathscr{C}$. Is it a linearly independent set of vectors? We must show that every finite subset of it is linearly independent. Take finitely many vectors

The elements of $\mathscr{C}$ are sets of vectors, so that $\bigcup \mathscr{C}$ is a set of vectors.

$$v_1, v_2, \ldots, v_n \in \bigcup \mathscr{C}.$$

Then there are elements

$$B_1, B_2, \ldots, B_n \in \mathscr{C}$$

such that

$$v_i \in B_i$$

for each $i = 1, 2, \ldots, n$. As $\mathscr{C}$ is a chain, either $B_i \subseteq B_j$ or $B_j \subseteq B_i$ for each i, j in $\{1, 2, \ldots, n\}$, and as n is a natural number, and therefore finite, one of the B_is must contain all the other B_js as subsets. Then this B_i contains all of $v_1, v_2, \ldots, v_n$; and as $B_i \in P$, so that B_i is linearly independent, the subset $\{v_1, v_2, \ldots, v_n\}$ is also linearly independent, as required.

Thus $\bigcup \mathscr{C}$ is an upper bound for $\mathscr{C}$ in P, so that by Zorn's lemma P contains a maximal element, meaning that V has a basis. ■

We shall leave some of the mathematical consequences of Theorem 5.6 as exercises at the end of this section.

Let us now turn to some of the more specifically set-theoretic applications of Zorn's lemma. First of all we shall follow up the results of Exercises 5.17 and 5.21 above with an important result about functions between sets A and B.

> ***Theorem 5.7***
>
> Let A and B be sets. Then, assuming Zorn's lemma, there is a one–one function $f\colon A \longrightarrow B$ or a one–one function $g\colon B \longrightarrow A$.

Proof

Define a set P of subsets of $A \times B$ by

$$P = \{f \subseteq A \times B : f \text{ is a one–one function with } \mathrm{Dom}(f) \subseteq A\},$$

and partially order P by $\subseteq$. From the solution to Exercise 5.21, if P has a maximal element f, then either f is a one–one function from A to B or $g = f^{-1}$ is a one–one function from B to A, as required. We shall show that P satisfies the conditions of Zorn's lemma, so that a maximal element exists.

As with other results of this section, there is no need for use of Zorn's lemma if A and B are finite, or if the sets have 'enough' structure to enable a definition of f or g.

P is non-empty, as the trivial empty function is in P. Now we must show that any chain $\mathscr{C}$ has an upper bound in P. We shall investigate whether $\bigcup \mathscr{C}$ is in P, as if it is, then it will be such an upper bound.

First, as every element of $\mathscr{C}$ is a subset of $A \times B$, then so is $\bigcup \mathscr{C}$.

Next, we need to check that $\bigcup \mathscr{C}$ is a function, i.e. if (a, b) and (a, b') are both in $\bigcup \mathscr{C}$, then $b = b'$. If $(a, b), (a, b') \in \bigcup \mathscr{C}$, then there are elements $f, g \in \mathscr{C}$ such that

$$(a, b) \in f \text{ and } (a, b') \in g.$$

As $\mathscr{C}$ is a chain, either $f \subseteq g$ or $g \subseteq f$. Without loss of generality, we can suppose that $f \subseteq g$. Then both (a, b) and (a, b') are in g, so that as g is a function we have $b = b'$, as required.

We shall leave checking that $\bigcup \mathscr{C}$ is one–one as an exercise for you.

Thus, assuming Zorn's lemma, P has a maximal element, so that the required result follows. ■

Exercise 5.23

Complete the proof of Theorem 5.7 by showing that $\bigcup \mathscr{C}$ is one–one.

Solution

Suppose that (a, b) and (a', b) are both in $\bigcup \mathscr{C}$. We must show that $a = a'$. As $(a, b), (a', b) \in \bigcup \mathscr{C}$ there are elements $f, g \in \mathscr{C}$ with $(a, b) \in f$ and $(a', b) \in g$. As $\mathscr{C}$ is a chain, either $f \subseteq g$ or $g \subseteq f$. Without loss of generality, we can suppose that $f \subseteq g$, so that both (a, b) and (a', b) are in g. But g is one–one, so that $a = a'$, as required.

The importance of this result will be seen in the context of cardinalities later in the book. In the terminology of cardinality, we have proved that Zorn's lemma implies the dichotomy principle. Likewise, later we shall show that Zorn's lemma, the dichotomy principle and the axiom of choice are in fact equivalent to each other. As the final result of this section and chapter, we would like you to prove another part of this equivalence, namely that Zorn's lemma implies the axiom of choice.

We shall discuss the dichotomy principle in Chapter 6.

Exercise 5.24

Prove that Zorn's lemma implies the axiom of choice. [Hints: consider the equivalent to AC given by Exercise 5.11:

> Suppose that R is a relation between non-empty sets A and B. Then there is a function F with the same domain as R such that $F \subseteq R$.

Put $P = \{f \subseteq A \times B : f \text{ is a function and } f \subseteq R\}$ and partially order P by $\subseteq$.]

In this chapter, we have tried to show how the axiom of choice and an equivalent, though seemingly different, principle, Zorn's lemma, have various important mathematical consequences for dealing with important sorts of infinite set. In a later chapter we shall see how AC is intimately involved with Cantor's theory of infinite numbers, both cardinals and ordinals, which is both the main historical starting point of set theory and the goal of this, and other similar, books.

Further exercises

Exercise 5.25

Let X be a set weakly partially ordered by R. Prove, assuming Zorn's lemma, that R can be extended to a linear order R^* on X, i.e. such that R^* is a linear order on X and for all $x, y \in X$, if xRy then xR^*y. [Hints: put $P = \{S \subseteq X \times X : S \text{ is a weak partial order on } X \text{ and } R \subseteq S\}$ and partially order P by $\subseteq$. Use Zorn's lemma to show that P has a maximal element R^* and show that R^*, which is automatically a partial order with $R \subseteq R^*$, has the linear property – this involves a bit more effort than in previous examples!]

Equivalently $R \subseteq R^*$.

As ever, Zorn's lemma would not be required if X was finite.

Exercise 5.26

Let R be a commutative ring with a multiplicative identity 1. Show, using Zorn's lemma, that R has a maximal proper ideal.
(Hints: you should be able to invent a suitable partially ordered set P to which to apply Zorn's lemma! When checking that any chain has an upper bound in P, you might need the result for commutative rings with identity that, for all ideals I of R, I is a proper ideal if and only if $1 \notin I$.)

Exercise 5.27

Show that Zorn's lemma, as a general principle for all partially ordered sets P, is equivalent to each of the following general principles.

(a) Every set P partially ordered by R has a chain which is maximal with respect to $\subseteq$, i.e. has an R-chain $\mathscr{C}$ such that there is no R-chain $\mathscr{D}$ containing $\mathscr{C}$ as a proper subset.

(b) For every set X partially ordered by $\subseteq$, there is a maximal $\subseteq$-chain.

This exercise requires a cool head to cope with potentially confusing notation and the use of $\subseteq$ simultaneously on sets and on sets of these sets!

The following exercises investigate some of the mathematical consequences of Theorem 5.6, that, assuming Zorn's lemma, every vector space has a basis. They require no further use of Zorn's lemma or AC.

Exercise 5.28

Let V, W be vector spaces over the same field, with U a subspace of V. Suppose that $f: U \longrightarrow W$ is a linear transformation. Then f can be extended to a linear transformation g on V, i.e. there is a linear transformation $g: V \longrightarrow W$ such that $g(u) = f(u)$ for all $u \in U$.

Exercise 5.29

$\mathbb{R}$ can be regarded as a vector space over the field $\mathbb{Q}$, by defining the sum of two 'vectors' (i.e. real numbers) as their usual sum in $\mathbb{R}$, and defining the result of scaling the vector $r \in \mathbb{R}$ by the scalar $q \in \mathbb{Q}$ to be the usual product qr in $\mathbb{R}$.

So that, assuming Zorn's lemma, $\mathbb{R}$ has a basis.

(a) Show that $f: \mathbb{R} \longrightarrow \mathbb{R}$ is a linear transformation over $\mathbb{Q}$ if and only if

$$f(x + y) = f(x) + f(y), \text{ for all } x, y \in \mathbb{R}.$$

So that f is linear if and only if just this one aspect of linearity holds.

(b) Hence prove that there is a function $f: \mathbb{R} \longrightarrow \mathbb{R}$ such that

$$f(x + y) = f(x) + f(y), \text{ for all } x, y \in \mathbb{R},$$

which is *not* of the form

$$f(x) = kx, \text{ for all } x \in \mathbb{R},$$

where $k \in \mathbb{R}$ is fixed.

6 CARDINALS (WITHOUT THE AXIOM OF CHOICE)

6.1 Introduction

The main reason for Cantor's introduction of set theory was to provide a framework for tying down notions of infinity. Cantor had been investigating the problem of uniqueness of the representation of a function f by a Fourier series, obtaining results of the form

$$\sum_{n=0}^{\infty}(a_n \sin nx + b_n \cos nx) = \sum_{n=0}^{\infty}(a'_n \sin nx + b'_n \cos nx) \text{ for all } x \in S$$
$$\Rightarrow a_n = a'_n \text{ and } b_n = b'_n \text{ for all } n$$

for various sorts of subset S of $\mathbb{R}$ (really of $[0, 2\pi]$). After proving that the result holds for sets S which exclude finitely many points from $[0, 2\pi]$, he managed to extend it to S which exclude certain *infinite* sets of points. To help explain the structure of such sets (those of the excluded points) Cantor developed a theory of infinite 'magnitudes', extending (less controversial!) ideas about finite magnitudes, i.e. natural numbers. Cantor tried to extend two aspects of the natural numbers: first, the way in they can be used, by counting, to give the size of a (finite) set; and second, the way in which their natural ordering by $<$ can be used to describe processes, e.g. 'first do this, next do that, next do ...', labelling the steps by successive natural numbers. These two aspects gave rise to two different sorts of numbers: *cardinal numbers* to describe sizes of sets; and *ordinal numbers* to describe ways of ordering sets. Furthermore, Cantor endowed both sorts of number with an arithmetic, operations of addition, multiplication and exponentiation agreeing with the usual arithmetic for natural numbers.

For a full discussion of Cantor's ideas about sets and their sizes, see Hallett [23].

In this chapter we shall look at some of the theory of cardinal numbers. In subsequent chapters we shall look at the theory of ordinal numbers and at the links between the two sorts of number – in the modern theory the cardinals are special sorts of ordinals.

We shall often use *cardinal* for cardinal number and *ordinal* for ordinal number.

As the interesting cardinals are something to do with infinite sets, let us recall the definitions of *finite* and *infinite*.

> ***Definitions***
>
> A set X is said to be *finite* if there is a bijection $f\colon \mathbf{n} \longrightarrow X$ for some natural number $\mathbf{n}$. X is said to be *infinite* if there is no such bijection, for any $\mathbf{n}$.

$\mathbf{n}$ as we have defined it in *ZF* has intuitively n elements.

Clearly each $\mathbf{n} \in \mathbb{N}$ is finite. Surely $\mathbb{N}$ is infinite – this was shown in Exercise 3.39 of Section 3.4, exploiting the pigeon-hole principle. So we have at least one infinite set, namely $\mathbb{N}$, and as $\mathbf{n} \subseteq \mathbb{N}$ for each natural number $\mathbf{n}$, this infinite set is intuitively 'bigger' than each finite set $\mathbf{n}$.

Cantor created a theory of infinite cardinal numbers, so that you might reasonably ask to be told what an infinite 'number' looks like. But in this chapter we have to dodge that question! One way of interpreting what Cantor thought, made more precise in the work of Frege and Russell, results in the definition of the natural number 1 as

$$\{x : x \text{ is a set and } x \text{ has one element}\}.$$

As we have seen, in Section 4.1, treating this as a set leads to problems; and the way in which (we hope!) we are avoiding these problems, by working within ZF, means that it is not a set within ZF, but a class. Of course we can get round this problem, as far as the number 1 and all natural numbers are concerned, by defining these numbers within ZF as in Chapter 3. A nice feature of our definition of $\mathbf{n} \in \mathbb{N}$ (as opposed to otherwise perfectly reasonable alternatives achievable within ZF) is that, intuitively, $\mathbf{n}$ has n elements, so that e.g. $\mathbf{1}$ is in the class above. But in general we might expect to have to 'choose' a set out of such a class, so the axiom of choice will figure in the discussion.

Cantor's theory turned out with hindsight to have made heavy use of the axiom of choice, hardly surprising if dealings with infinite sets are involved. So it is customary to distinguish between those parts which use AC and those which don't. The theory with AC is richer, but even without it one can derive many of Cantor's most remarkable (and, for some, controversial) results. For instance: there are the same number of rational numbers as natural numbers; and there are many more irrational real numbers than rationals. But surely, one might exclaim, as $\mathbb{N}$ is a proper subset of $\mathbb{Q}$ there must be more rationals than natural numbers. And it is known that between any two distinct irrationals there is a rational, while between any two distinct rationals there's an irrational: so why should there be many more irrationals than rationals? The first thing is to tie down what we mean by one set having 'the same number of elements as', or 'more elements than' another, and that is what we do in the next section.

A key factor behind many of these results is the tying down of what the rational and, especially, the real numbers are by means of the definitions in Chapter 2.

6.2 Comparing sizes

Is $\mathbb{Q}$ really a bigger set than $\mathbb{N}$? The obvious answer is 'Yes', because $\mathbb{N} \subseteq \mathbb{Q}$ and $\mathbb{N} \neq \mathbb{Q}$. In general when $A \subseteq B$ and $A \neq B$ it is natural to say that B is bigger than A. But what would we say if neither of A and B was a subset of the other?

Let $A = \{\mathbf{0}, \mathbf{1}, \mathbf{7}\}$ and $B = \{\mathbf{0}, \mathbf{2}, \mathbf{4}, \mathbf{8}\}$. Is B bigger than A. If so, why? Think about this for a moment before reading on.

In normal life we would surely say that B is bigger than A, despite A not being a subset of B. So as mathematicians we have to make clearer how we use the words 'bigger than'. In some contexts it is clear that we intend 'bigger than' to mean 'contains as a proper subset'. But in the context of sizes of sets this is not what we mean. So what else might we mean? In the case of these particular sets A and B, most of us would probably say that B has 4 elements, A has 3 elements, and 4 is greater than 3.

What's more, it's natural to talk about the 3 and the 4 as the 'sizes' of A and B, respectively. So within ZF we can, for these sets A and B, define their sizes to be the sets **3** and **4** – the 'size' of a set can be described as a set itself. And comparing these numbers makes sense of 'B is bigger than A'.

This all works fine for finite sets, as for any finite set X there is a bijection $f: \mathbf{n} \longrightarrow X$ for some natural number $\mathbf{n}$, by virtue of X being finite. The $\mathbf{n}$ is unique, so we can define the size of X to be this $\mathbf{n}$. Let us write Card(X) for this $\mathbf{n}$. Then in general for finite sets X and Y, we can give a precise meaning to 'Y is bigger than X' by defining it as 'Card(Y) > Card(X)'.

Furthermore, this method works well for explaining when two sets have 'the same size'. For instance if $A = \{\mathbf{0}, \mathbf{1}, \mathbf{7}\}$ and $C = \{\mathbf{2}, \mathbf{4}, \mathbf{5}\}$, then both Card($A$) and Card($C$) equal **3**, so A and C have the same size. In general for finite sets X and Y, we could define 'X has the same size as Y' by 'Card(X) = Card(Y)'.

But what if X is infinite, i.e. there is no natural number $\mathbf{n}$ for which there is a bijection $f: \mathbf{n} \longrightarrow X$? What object can we take to be the size of X? What Cantor did was to identify a suitable class of objects, the $\aleph_\alpha$s, to act as the sizes of infinite sets. These, along with the natural numbers, will make up the class of *cardinal numbers*, which will provide us with a complete collection of 'sizes' of sets. But as many of his results turn out to be equivalent to the axiom of choice, before we deal with them it's instructive to see how much we can do without using AC or an equivalent. It turns out that we cannot do much in the way of identifying a special class of sizes, but we can make progress with *comparing* sizes of sets, without actually committing ourselves to what we mean by the size of any individual set.

$\aleph$, pronounced 'aleph', is the first letter of the Hebrew alphabet.

Exercise 6.1

Let's go back to the sets $A = \{\mathbf{0}, \mathbf{1}, \mathbf{7}\}$ and $C = \{\mathbf{2}, \mathbf{4}, \mathbf{5}\}$. Is there a way of comparing the sizes of these sets that doesn't involve saying what their sizes are?

Solution

The trick is to exploit the idea we have used within set theory to explain when a set X is finite, namely that there is a bijection between X and some natural number $\mathbf{n}$. The bijection pairs off the elements of X with the elements of $\mathbf{n}$, and the existence of the bijection is what underpins the assertion not only that X is finite, but that it has $\mathbf{n}$ elements. The bijection represents our intuition that X and $\mathbf{n}$ have the same size.

For the sets A and C there is likewise a bijection, e.g.

$$\begin{aligned} f\colon A &\longrightarrow B \\ \mathbf{0} &\longmapsto \mathbf{2} \\ \mathbf{1} &\longmapsto \mathbf{4} \\ \mathbf{7} &\longmapsto \mathbf{5} \end{aligned}$$

There are of course several bijections from A to C.

and it seems natural to interpret this as showing that A and C have the same size, without having to say what this size is.

It is this ability to pair off the elements of two sets which we shall take as the definition of sets having the same size.

> **Definition**
>
> Let A and B be sets. We say that A is *equinumerous* with B if there is a bijection $f\colon A \longrightarrow B$; and we write $A \approx B$.

So $\{0,1,7\} \approx \{2,4,5\}$.

'Equinumerous' is just a posher way of saying 'has the same size as', and is standard terminology. Likewise, we can capture the idea of A having size less than or equal to that of B by the following:

> **Definition**
>
> We say that A is *less than or equinumerous* with B if there is a one–one function $f\colon A \longrightarrow B$;
> and we write $A \preceq B$.

Lastly, we can capture the idea that the size of A is strictly smaller than that of B by:

> **Definition**
>
> A is *dominated by* B if $A \preceq B$ and it is not the case that $A \approx B$;
> we then write $A \prec B$.

Let's use these definitions to compare the sizes of a few sets.

Exercise 6.2

For each of the following pairs X, Y of sets, decide which, if any, of the following holds:

$$X \approx Y, \quad X \preceq Y, \quad Y \preceq X, \quad X \prec Y, \quad Y \prec X.$$

(a) $X = \{0,2,4\}$, $Y = \{1,3,5,7\}$

(b) $X = \{0,1,2\}$, $Y = \mathbb{N}$

(c) $X = \mathbb{N}$, $Y = \mathbb{Z}$

Solution

(a) We have $X \preceq Y$ because we can define a one–one function f from X to Y by

$$\begin{aligned} f\colon X &\longrightarrow Y \\ 0 &\longmapsto 1 \\ 2 &\longmapsto 3 \\ 4 &\longmapsto 5 \end{aligned}$$

Clearly we ought to have $X \prec Y$, but as this is our first use of the definition of $\prec$ we ought to prove it. $X \prec Y$ requires showing both that $X \preceq Y$, which we've already done, and that it is not the case that $X \approx Y$ (which we'll write as $X \not\approx Y$). So we have to show that $X \not\approx Y$, i.e. there is no bijection $g\colon X \longrightarrow Y$.

We may also write $A \npreceq B$ for 'not $A \preceq B$' etc.

Assume, hoping to obtain a contradiction, that there *is* such a bijection g. Then its inverse $g^{-1}: Y \to X$, which exists as g is a bijection, is also a bijection. Also there is a bijection h between the natural number $\mathbf{3}$ $(= \{\mathbf{0}, \mathbf{1}, \mathbf{2}\})$ and X defined by

$$\begin{aligned} h\colon \mathbf{3} &\longrightarrow X \\ \mathbf{0} &\longmapsto \mathbf{0} \\ \mathbf{1} &\longmapsto \mathbf{2} \\ \mathbf{2} &\longmapsto \mathbf{4} \end{aligned}$$

and similarly there is a bijection $k\colon \mathbf{4} \longrightarrow Y$. Then the composite function

$$h^{-1} \circ g^{-1} \circ k\colon \mathbf{4} \xrightarrow{k} Y \xrightarrow{g^{-1}} X \xrightarrow{h^{-1}} \mathbf{3}$$

is a bijection from $\mathbf{4}$ to $\mathbf{3}$. But this contradicts the pigeon-hole principle proved for the natural numbers within ZF (Theorem 3.14 of Chapter 3. Thus there is no bijection $g\colon X \longrightarrow Y$. So $X \not\approx Y$, from which we have $X \prec Y$.

Similar arguments by contradiction could be used to show that neither $Y \preceq X$ nor $Y \prec X$ holds.

(b) Clearly $X \preceq Y$ using the inclusion map

$$\begin{aligned} i\colon X &\longrightarrow Y \\ x &\longmapsto x. \end{aligned}$$

Obviously we'd hope to get $X \not\approx Y$, so as before let's assume that $X \approx Y$ and try to obtain a contradiction.

By our assumption there is a bijection $g\colon X \longrightarrow Y$. Then g^{-1} exists and $g^{-1}\colon \mathbb{N} \longrightarrow \mathbf{3}$ is a bijection – note that $X = \{\mathbf{0}, \mathbf{1}, \mathbf{2}\}$ is just the natural number $\mathbf{3}$. Now consider the restriction of g^{-1} to the subset $\mathbf{3}$ of $\mathbb{N}$,

$$g^{-1}|_{\mathbf{3}}\colon \mathbf{3} \longrightarrow \mathbf{3}.$$

As g^{-1} is one–one, so is $g^{-1}|_{\mathbf{3}}$. But, from our work on natural numbers, any one–one function from $\mathbf{3}$ to itself must also be onto, so that $g^{-1}|_{\mathbf{3}}$ is onto. Then $g(\mathbf{3})$, $g(\mathbf{4})$ etc. cannot have been defined as members of $\mathbf{3}$ without contradicting that g is one–one.

The contradiction gives us that $X \not\approx Y$, from which it follows that $X \prec Y$.

Again similar arguments can be used to show that $Y \not\preceq X$ and $Y \not\prec X$. But surely $\approx$, $\preceq$, $\prec$ are sufficiently like $=$, $\leq$, $<$ for us to be able to assert these without lengthy arguments by contradiction. We had better investigate this soon!

(c) Again the inclusion map

$$\begin{aligned} i\colon \mathbb{N} &\longrightarrow \mathbb{Z} \\ n &\longmapsto n \end{aligned}$$

shows that $\mathbb{N} \preceq \mathbb{Z}$. At first sight we might expect that, as $\mathbb{N}$ is a proper subset of $\mathbb{Z}$, $\mathbb{N} \prec \mathbb{Z}$, but, remarkably, this is not so. The function

$$\begin{aligned} f\colon \mathbb{N} &\longrightarrow \mathbb{Z} \\ n &\longmapsto \begin{cases} k, & \text{if } n = 2k, \\ -k, & \text{if } n = 2k+1, \end{cases} \end{aligned}$$

For maps like this which don't really depend on the precise representation of natural numbers within ZF, we shall often not bother to use bold type for natural numbers. We shall usually write bold $\mathbf{n}$, rather than n, when a construction depends on inner properties of natural numbers as sets.

is easily seen to be a bijection, so that in fact $\mathbb{N} \approx \mathbb{Z}$. (So we also have $\mathbb{Z} \preceq \mathbb{N}$, $\mathbb{N} \not\prec \mathbb{Z}$ and $\mathbb{Z} \not\prec \mathbb{N}$.)

The results of these last exercises leave us with a variety of problems. First of all, we ought to investigate the extent to which the definitions of $\approx$, $\preceq$ and $\prec$ capture the properties we expect of $=$, $\leq$ and $<$ for sizes of sets. We shall do this in the next section. Second, we have seen how for infinite sets we do not obtain all of the results which our experience of finite sets might lead us to expect: in particular, for infinite sets A and B with A a proper subset of B, it is not always the case that $A \prec B$. Does this mean that all infinite sets have the same size? Whatever the answer, it is clear that we need to investigate infinite sets with an open mind, which we shall start in the section after the next.

Further exercises

Exercise 6.3

Show that for any sets A and B, if $A \subseteq B$ then $A \preceq B$.

Exercise 6.4

Let $\mathbf{m}$ and $\mathbf{n}$ be natural numbers (in *ZF*). Show that

(a) if $\mathbf{m} \leq \mathbf{n}$ then $\mathbf{m} \preceq \mathbf{n}$;

(b) if $\mathbf{m} < \mathbf{n}$ then $\mathbf{m} \prec \mathbf{n}$.

Exercise 6.5

Let $\mathbf{n} \in \mathbb{N}$. Show that $\mathbf{n} \not\approx \mathbb{N}$.

Exercise 6.6

Suppose that $\mathbb{N} \preceq A$. Show that A is infinite.

Exercise 6.7

Let A and B be finite sets. Show that $A \preceq B$ or $B \preceq B$. [Hint: bear in mind the way we have earlier defined a 'finite' set.]

Exercise 6.8

Decide which, if any, of the following statements hold:

$$2\mathbb{Z} \preceq \mathbb{Z}, \quad 2\mathbb{Z} \approx \mathbb{Z}, \quad 2\mathbb{Z} \prec \mathbb{Z}.$$

Exercise 6.9

A possible alternative definition of 'less than or equinumerous', which we'll write as $\preceq^*$, is given by

$$A \preceq^* B \text{ if there is an onto function } f\colon B \longrightarrow A.$$

Is it the case that $A \preceq^* B$ exactly when $A \preceq B$?

6.3 Basic properties of $\approx$ and $\preceq$

In the previous section you were asked to investigate, for the sets $X = \{0, 2, 4\}$ and $Y = \{1, 3, 5, 7\}$, which of the relationships $X \approx Y$, $X \preceq Y$, $Y \preceq X$, $X \prec Y$ and $Y \prec X$ hold. Our solution showed that $X \preceq Y$ and $X \not\approx Y$, from which it followed by definition that $X \prec Y$; and we said that arguments by contradiction could be used to show that $Y \not\preceq X$ and $Y \not\prec X$. But surely these last two results ought to follow immediately, without any complicated argument, if $\approx$, $\preceq$ and $\prec$ really capture our intuitions about $=$, $\leq$ and $<$ for sizes of sets. In this section we shall investigate the (substantial) extent to which this does happen.

Let us look first of all at the sort of properties we would want $\approx$ to possess to justify that it captures the idea of 'has the same size as'. The usual bare minimum we expect of a mathematical definition of 'the same as' is that it is an equivalence relation.

Exercise 6.10

Show that $\approx$ is an equivalence relation between sets, i.e. that it is

(a) reflexive: for all sets A, $A \approx A$;

(b) symmetric: for all sets A and B, if $A \approx B$ then $B \approx A$;

(c) transitive: for all sets A, B and C, if $A \approx B$ and $B \approx C$ then $A \approx C$.

That $\approx$ is an equivalence relation is a good start. Ideally we would like more, e.g. that performing similar 'size' constructions on equinumerous sets yields equinumerous sets, as in the following exercises.

Exercise 6.11

Suppose that $A \approx B$ and $C \approx D$. Show that $A \times C \approx B \times D$.

Solution

As $A \approx B$ and $C \approx D$ there are bijections $f\colon A \longrightarrow B$ and $g\colon C \longrightarrow D$. Then define a function h by

$$\begin{aligned} h\colon A \times C &\longrightarrow B \times D \\ (a, c) &\longmapsto (f(a), g(c)) \end{aligned}$$

It is easy to show that h is a bijection, so that $A \times C \approx B \times D$.

Exercise 6.12

Suppose that $A \approx B$ and that C is any set. Give an example to show that it is not always the case that $A \cup C \approx B \cup C$. Suggest a condition on C which would guarantee that $A \cup C \approx B \cup C$.

Solution

We can run into trouble if C overlaps with either A or B. For instance, if $A = \{0, 1\}$ and $B = C = \{1, 2\}$, then $A \cup C = \{0, 1, 2\}$ and $B \cup C = \{1, 2\}$, which are not equinumerous.

But if C is disjoint from both A and B, the problem disappears. If f is a bijection from A to B, it is easy to check that the function g defined by

$$g\colon A \cup C \longrightarrow B \cup C$$
$$x \longmapsto \begin{cases} f(x), & \text{if } x \in A, \\ x, & \text{if } x \in C, \end{cases}$$

is a bijection. (Where do you need to know that C is disjoint from both A and B?)

In the next exercise we ask you to show that various pairs of sets are equinumerous. You will have to do this by exhibiting appropriate bijections, which will test your understanding (and memory!) of the definition within ZF of various sets, e.g. $A \times B$ and A^B. You will see later in the section that there is an alternative strategy, which would be less elegant for these particular examples. It might make doing the exercise more interesting to know that you are really proving some of the basic results about the arithmetic of cardinals – more on this soon!

Exercise 6.13

Prove the following statements by exhibiting an appropriate bijection in each case.

(a) $A \times B \approx B \times A$

(b) $A \times (B \times C) \approx (A \times B) \times C$

(c) $A^{\mathbf{2}} \approx A \times A$, where $\mathbf{2} = \{\mathbf{0}, \mathbf{1}\}$.

(d) $A \times (B \cup C) \approx (A \times B) \cup (A \times C)$.

(e) $A^{B \cup C} \approx A^B \times A^C$, where $B \cap C = \varnothing$.

(f) $C^{A \times B} \approx (C^B)^A$

Solution

(a) Define f by

$$f\colon A \times B \longrightarrow B \times A$$
$$(a, b) \longmapsto (b, a)$$

Although this f is clearly a bijection, it is worth bearing in mind in all these exercises how to justify that one's construction works.

Verifying that one has specified a bijection is not always as easy as this example!

First of all, is f actually well-defined as a function? Here this involves noting that given a member x of $A \times B$, the ordered pair construction guarantees both that there are unique $a \in A$, $b \in B$ with $x = (a, b)$ and that the proposed image of x under f is thus a uniquely defined member of $B \times A$.

Next, is f one–one? Here we could argue as follows: if we have $f(a_1, b_1) = f(a_2, b_2)$, then $(b_1, a_1) = (b_2, a_2)$; the ordered pair construction then guarantees that $b_1 = b_2$ and $a_1 = a_2$, so that $(a_1, b_1) = (a_2, b_2)$, as required.

Lastly, we must show that f is onto. Here this is particularly straightforward as (b, a) is the image under f of (a, b).

(b) Not given.

(c) A typical member of A^2 is a function $g\colon \{0,1\} \longrightarrow A$. Define a function θ by

$$\begin{aligned}\theta\colon A^2 &\longrightarrow A \times A\\ g &\longmapsto (g(0), g(1))\end{aligned}$$

You can check that θ is a bijection.

(d) Not given.

(e) Define θ by

$$\begin{aligned}\theta\colon A^{B\cup C} &\longrightarrow A^B \times A^C\\ g &\longmapsto (g|_B, g|_C)\end{aligned}$$

You can check that θ is a bijection. (You need the fact that $B \cap C = \varnothing$ to show that θ is onto: if your bijection is defined the other way round from ours, i.e. from $A^B \times A^C$, you'd need this fact to show your mapping is well-defined.)

(f) Not given. (This is probably the hardest part of this exercise.)

We probably ought only define arithmetic operations on cardinals (in particular addition, multiplication and exponentiation) when we have said what a cardinal number is – this needs the axiom of choice and for the moment we're trying to see how far we can get without using this. But the basic idea can be phrased informally in terms of 'sizes' of sets as follows:

size of A *plus* size of B	size of $(A \times \{0\}) \cup (B \times \{1\})$
size of A *times* size of B	size of $A \times B$
size of A *to the power* size of B	size of A^B

You might like to mull over whether these operations seem reasonable – a good place to start would be with finite sets, e.g. if A has m elements and B has n elements, does the set A^B have the right number of elements? (Note that if we had A and B disjoint, a good measure of the size of A plus the size of B would be the size of $A \cup B$. But in case not, we use the trick of creating disjoint sets equinumerous with A and B.)

With these arithmetic operations we can view the results of the previous exercise as telling us something about the properties of the operations. For instance, $A \times B \approx B \times A$ essentially says that the multiplication is commutative. You can see that Cantor's arithmetic of cardinals is beginning to look as though it not only incorporates the usual arithmetic for natural numbers (regarding $\mathbf{n}$ as a set with n elements), but also extends many of its everyday properties into the realm of more general sets. And our definition of $\approx$ looks like the right way of expressing 'equalities' in this arithmetic.

Let's now look at $\preceq$ and how well it captures what we might expect of $\leq$ for sizes of sets. The most basic requirement of a $\leq$ relation is usually that it satisfies the axioms for a partial order. Does this happen for $\preceq$? We would need $\preceq$ to have the following properties:

Strictly speaking $\preceq$ could never be a relation within ZF as such a relation would have to be a set.

reflexive for all sets A, $A \preceq A$;

anti-symmetric for all sets A and B, if $A \preceq B$ and $B \preceq A$ then $A = B$;

transitive for all sets A, B and C, if $A \preceq B$ and $B \preceq C$ then $A \preceq C$.

Exercise 6.14

In fact $\preceq$ has two of the above properties. Find the property it doesn't have and explain what goes wrong. And prove $\preceq$ does have the other two properties.

Solution

You may have found it easier to spot the two properties which *do* hold, namely reflexivity and transitivity. Proofs of these are very straightforward: the identity function on A is automatically one–one, so $A \preceq A$; and if there are one–one functions from A to B and from B to C, then their composition is a one–one function from A to C, giving the transitivity of $\preceq$.

So by a process of elimination the property which doesn't hold must be anti-symmetry. To justify this we need a counterexample – sets A and B with $A \preceq B$ and $B \preceq A$ for which $A \neq B$. And we have already seen examples of unequal sets A and B for which $A \approx B$, so that both $A \preceq B$ and $B \preceq A$ (why does this follow?), for instance $\{0, 1, 2\}$ and $\{2, 4, 5\}$, or $\mathbb{N}$ and $\mathbb{Z}$. Any such pair of equinumerous but unequal sets provides a counterexample.

How inconvenient of $\preceq$ not to fit into the familiar mould of a partial order! But with hindsight we should not have expected anti-symmetry. If $\preceq$ does live up to our intuition of 'has size less than or equal to' then $A \preceq B$ and $B \preceq A$ should entail that A and B have the same size, not that they are equal. That they do entail this is the content of the following:

> ***Theorem 6.1 Schröder–Bernstein theorem***
>
> If $A \preceq B$ and $B \preceq A$ then $A \approx B$.

The theorem is sometimes called the Cantor–Schröder–Bernstein theorem. Ernst Schröder (1841–1902) and Felix Bernstein (1878–1956) were German mathematicians. Cantor's own proof in 1896 used an equivalent of AC. Schröder's proof at around the same time contained an error which was ultimately corrected. Bernstein's proof, published in 1898, was the first correct proof avoiding use of AC.

The proof of this highly desirable result is sufficiently hard to have challenged several great mathematicians, so it gets a name! We shall leave the proof to the end of this section. The theorem gives us an alternative strategy for showing that $A \approx B$: if we can't find a bijection from one set to the other, we can always try to find a pair of one–one functions, from A to B and from B to A, and use this theorem.

It's worth noting here that Cantor's theory of cardinal numbers defines a set $\mathrm{Card}(X)$ for each set X in such a way that $A \approx B$ if and only if $\mathrm{Card}(A) = \mathrm{Card}(B)$, even when $A \neq B$. And when $\preceq$ is restricted to these cardinals it becomes anti-symmetric in the proper sense, i.e. if $\mathrm{Card}(A) \preceq \mathrm{Card}(B)$ and $\mathrm{Card}(B) \preceq \mathrm{Card}(A)$ then $\mathrm{Card}(A) = \mathrm{Card}(B)$.

This is easy to show, but you need that $\mathrm{Card}(\mathrm{Card}(X)) = \mathrm{Card}(X)$ for each set X. We shall revisit these ideas in Chapter 9.

Thanks to the Schröder–Bernstein theorem, we can regard $\preceq$ as essentially 'less than or equal to' where the corresponding 'equal to' is $\approx$, rather than $=$. In the next exercise you are asked to check that $\preceq$ and $\approx$ indeed have the sort of relationship you'd expect between $\leq$ and $=$.

Exercise 6.15

Using bijections and one–one functions where appropriate, show that:

(a) if $A \preceq B$ and $B \approx C$ then $A \preceq C$;

(b) if $A \approx B$ and $B \preceq C$ then $A \preceq C$.

In the next exercise, you are asked to derive results about $\preceq$ which will eventually translate into inequalities involving the arithmetic operations.

Exercise 6.16

Suppose that $A \preceq B$. Prove the following inequalities by defining appropriate one–one functions.

(a) $A \cup C \preceq B \cup C$, where C is disjoint from both A and B.

(b) $A \times C \preceq B \times C$.

(c) $A^C \preceq B^C$.

(d) $C^A \preceq C^B$.

Solution

(a) As $A \preceq B$ there is a one–one function $f\colon A \longrightarrow B$. Define h by

$$h\colon A \cup C \longrightarrow B \cup C$$
$$x \longmapsto \begin{cases} f(x), & \text{if } x \in A, \\ x, & \text{if } x \in C. \end{cases}$$

h is well-defined because $C \cap A = \varnothing$. The facts that f is one–one and that $C \cap B = \varnothing$ can then be used to show that h is indeed one–one.

(b) Not given.

(c) Not given.

(d) Not given.

These results will translate into statements about cardinal arithmetic, e.g. part (i) above becomes '$\kappa \leq \lambda$ implies $\kappa + \mu \leq \lambda + \mu$'.

Lastly, let's look at some of the properties of $\prec$. By definition $A \prec B$ when $A \preceq B$ and $A \not\approx B$, which is pretty similar to the definition of a strict partial ordering $<$ from a given weak ordering $\leq$. So you might expect to be able to show that $\prec$ is a strict partial ordering on sets. This is your next exercise!

Exercise 6.17

Show that $\prec$ is

(a) irreflexive: for all sets A, $A \not\prec A$;

(b) transitive: for all sets A, B and C, if $A \prec B$ and $B \prec C$ then $A \prec C$.

Solution

(a) By the reflexivity of $\approx$ we have $A \approx A$. So by the definition of $\prec$ we have $A \not\prec A$.

(b) Suppose that $A \prec B$ and $B \prec C$. Then by definition $A \preceq B$ and $B \preceq C$, so by the transitivity of $\preceq$ we have $A \preceq C$.

To conclude that $A \prec C$ we have to show that $A \not\approx C$, so we assume that $A \approx C$ and try to derive a contradiction. But if $A \approx C$, then as $B \preceq C$ we have that $B \preceq A$ (using the result of Exercise 6.15). As we also have $A \preceq B$, the Schröder–Bernstein theorem gives us that $A \approx B$. But this contradicts the original assumption that $A \prec B$. Hence $A \not\approx C$, so that $A \prec C$ as required.

Things are looking so straightforward that it's time to bring ourselves down to earth with a bump by noting some of the properties $\prec$ does *not* have! For instance, you might like to think about the following exercise – no solutions are given as suitable examples will become clearer after reading the next section.

Exercise 6.18

Find counterexamples for each of the following statements.

(a) If $A \prec B$ and C is disjoint from both A and B, then $A \cup C \prec B \cup C$.

(b) If $A \prec B$ and $C \neq \varnothing$, then $A \times C \prec B \times C$.

(c) If $A \prec B$ and $C \neq \varnothing$, then $A^C \prec B^C$.

The point is that cardinal arithmetic will not always preserve *strict* inequalities, even though it will usually preserve weak ones.

And we don't just have problems with $\prec$. We can, and later will, have big problems with $\approx$ and $\preceq$. Many plausible statements involving these – plausible because of our intuition with finite sets or familiar infinite sets – can turn out to be unprovable without some extra principles of set theory, or even wrong! We dodged a major example of such a statement earlier on, when we were looking at the properties we might expect $\preceq$ to have: a nice property possessed by many partial orders in mathematics is that of *linearity*, which for $\preceq$ would be expressed as

$$A \preceq B \text{ or } B \preceq A, \text{ for all sets } A, B.$$

This innocent statement (called the *dichotomy principle*) is in fact equivalent to the axiom of choice. And AC is essentially what is needed to develop Cantor's full theory of cardinal numbers. So we will meet dichotomy again later!

Theorem 5.7 in Section 5.4 shows that the dichotomy principle follows from Zorn's lemma. We shall show that both these principles are equivalent to AC in Chapter 9.

To end the section, let us now give a proof of the Schröder–Bernstein theorem.

Theorem 6.1 Schröder–Bernstein theorem

If $A \preceq B$ and $B \preceq A$ then $A \approx B$.

Proof

As $A \preceq B$ and $B \preceq A$ there are one–one functions $f\colon A \longrightarrow B$ and $g\colon B \longrightarrow A$. Of course, if one of f and g was onto, then we would have the desired bijection to show that $A \approx B$. But in general neither will be onto, so that we have to adapt them in some way to construct a suitable bijection.

First we define subsets A_n of A for all $n \in \mathbb{N}$ by recursion, by setting

$$A_0 = A \setminus \text{Range}(g),$$

and defining A_1 to be the image set of A_0 under the composite function $g \circ f$, A_2 to be the image set of A_1 under $g \circ f$ and so on:

$$A_{n+1} = g(f(A_n)), \text{ for all } n \in \mathbb{N}.$$

We then define subsets B_n of B for all $n \in \mathbb{N}$ by

$$B_n = f(A_n).$$

It's easy to check that $g(B_n) = A_{n+1}$.

Note that if $a \notin A_0$ then $a \in \text{Range}(g)$, so that $a = g(b)$ for some $b \in B$. As g is one–one, this $b \in B$ is unique, so that in this case we can write b as $g^{-1}(a)$.

When a function g is one–one but not onto, it is quite common to use the notation g^{-1} for the 'inverse' function defined on Range(g). Regarding functions as sets of pairs in ZF, g^{-1} is the set $\{\langle a, b\rangle : \langle b, a\rangle \in g\}$.

This means that we can now define a function h from A to B by

$$h: A \longrightarrow B$$
$$h(a) = \begin{cases} f(a), & \text{if } a \in A_n \text{ for some } n \in \mathbb{N}, \\ g^{-1}(a), & \text{otherwise.} \end{cases}$$

This defines a value $h(a)$ for all $a \in A$ as if a is in none of the A_ns, then in particular $a \notin A_0$, so that (as already discussed) $g^{-1}(a)$ exists.

We shall show that h is a bijection, so we must show that it is both one–one and onto. For the one–one property we shall take $a, a' \in A$ such that $h(a) = h(a')$ and show that $a = a'$. If both a and a' are in $\bigcup\{A_n : n \in \mathbb{N}\}$, we have $h(a) = f(a)$ and $h(a') = f(a')$: as f is one–one, this forces $a = a'$. Likewise, if neither a nor a' is in $\bigcup\{A_n : n \in \mathbb{N}\}$, then as g is one–one we also have $a = a'$. The only problem is when one of a, a', say a, is in $\bigcup\{A_n : n \in \mathbb{N}\}$ and the other, a', is not. Suppose that $a \in A_n$. Then

$$h(a) = f(a) \in B_n,$$

while

$$h(a') = g^{-1}(a').$$

Could $g^{-1}(a')$ equal $f(a)$? If so, then

$$\begin{aligned} a' &= g\left(g^{-1}(a')\right) \\ &= g(f(a)) \quad (\text{as we are assuming } g^{-1}(a') = f(a)) \\ &\in A_{n+1}, \end{aligned}$$

contradicting that a' is not in $\bigcup\{A_n : n \in \mathbb{N}\}$. Thus in this case we also have $a = a'$, completing the proof that h is one–one.

Now for the onto property, showing that for each $b \in B$ there is some $a \in A$ with $h(a) = b$. There is no problem if $b \in \bigcup\{B_n : n \in \mathbb{N}\}$ as then $b = f(a)$ for an a in some A_n. But what if $b \notin \bigcup\{B_n : n \in \mathbb{N}\}$? The trick is to look at $g(b)$ and investigate whether $g(b) \in A_n$ for some n. In fact $g(b) \notin A_n$ for any n (which we shall leave as an exercise for you) so that the definition of h gives

$$\begin{aligned} h(g(b)) &= g^{-1}(g(b)) \quad (\text{as } g(b) \notin \textstyle\bigcup\{A_n : n \in \mathbb{N}\}) \\ &= b, \end{aligned}$$

giving that b is indeed in Range(h).

Thus h is onto, so that h is a bijection and giving $A \approx B$, as required. ■

Exercise 6.19

Complete the detail of the proof above, namely that if $b \notin \bigcup\{B_n : n \in \mathbb{N}\}$ then $g(b) \notin A_n$ for all $n \in \mathbb{N}$.

Solution

If $g(b) \in A_0$ this would contradict that $A_0 = A \setminus \text{Range}(g)$. So $g(b) \notin A_0$. Also as $b \notin B_n$ and g is one–one we cannot have $g(b) \in g(B_n)$, i.e. $g(b) \notin A_{n+1}$. Thus $g(b) \notin A_n$ for all $n \in \mathbb{N}$.

Exercise 6.20

In the proof above of the Schröder–Bernstein theorem take $A = B = \mathbb{N}$ and let f and g be the one–one functions defined by $f(n) = 2n$ and $g(n) = 2n + 1$. Describe the rule of the corresponding function h constructed in the proof.

In the next two sections we shall see some impressive applications of the Schröder–Bernstein theorem.

Further exercises

Exercise 6.21

Show that if $A \approx B$ then $A^C \approx B^C$.

Exercise 6.22

Show that if $A \approx B$ then $\mathscr{P}(A) \approx \mathscr{P}(B)$.

Exercise 6.23

Can one define a sensible operation of subtraction on 'sizes' of sets?

Exercise 6.24

Is there a set A such that $A \preceq B$ for all sets B? Likewise is there a set C such that $B \preceq C$ for all sets B?

Exercise 6.25

Prove that the principle of dichotomy holds for *finite* sets, i.e. if A and B are finite sets, then $A \preceq B$ or $B \preceq A$.

Exercise 6.26

Suppose that X, Y are sets each with at least two elements. Show that

$$X \cup Y \preceq X \times Y.$$

Exercise 6.27

Show that if $A \subseteq B \subseteq C \subseteq D$ and $A \approx D$, then $B \approx C$.

6.4 Infinite sets without AC – countable sets

Let us now look more closely at infinite sets and how they are related to each other using $\approx$ etc. We know that $\mathbb{N}$ is infinite, so presumably any 'bigger' set is also infinite – verify this in the following exercise.

$\mathbb{N}$ was shown to be infinite in Section 3.4.

Exercise 6.28

Suppose that $\mathbb{N} \preceq X$. Show that X is infinite. [Hint: assume that X is finite and derive a contradiction.]

Thus the set $\mathbb{Z}$ of all integers is indeed infinite, as $\mathbb{N} \subseteq \mathbb{Z}$, so that $\mathbb{N} \preceq \mathbb{Z}$. But we have already seen the first of Cantor's surprises, namely that $\mathbb{N} \approx \mathbb{Z}$, even though $\mathbb{N}$ is a proper subset of $\mathbb{Z}$. So we have to educate our intuitions about $\approx$ for infinite sets. Let's look first at $\mathbb{N}$ and its subsets.

Exercise 6.29

Give a bijection to show that $\mathbb{N} \approx 2\mathbb{N}$, where $2\mathbb{N} = \{2n : n \in \mathbb{N}\}$, i.e. the set of even integers.

We continue the convention of using bold type for natural numbers, $\mathbf{n}$ rather than n, $\mathbf{0}$ rather than 0 etc., only when we think it might be important to remember that natural numbers are particular sets within ZF.

Solution

Define f by

$$\begin{aligned} f\colon \mathbb{N} &\longrightarrow 2\mathbb{N} \\ n &\longmapsto 2n \end{aligned}$$

It is easy to show that f is a bijection.

It is easy to show similarly that $\mathbb{N} \approx k\mathbb{N}$ $(= \{kn : k \in \mathbb{N}\})$. So $\mathbb{N}$ has the same size as quite a few of its proper subsets. The question arises of what are the possible sizes of subsets of $\mathbb{N}$. Clearly some of the subsets A are finite, i.e. there is some natural number $\mathbf{n}$ such that $\mathbf{n} \approx A$. But is there some subset A of $\mathbb{N}$ which is neither finite, so by definition is infinite, nor equinumerous with $\mathbb{N}$? If such an A exists, then $A \prec \mathbb{N}$, so we have a smaller degree of infinity than $\mathbb{N}$. To put you out of your misery, no such A exists, as the next theorem shows.

We use bold $\mathbf{n}$ here, not n, because in the definition of 'finite', it does matter how natural numbers are represented as sets.

Theorem 6.2

Suppose that $A \subseteq \mathbb{N}$. Then either A is finite or $A \approx \mathbb{N}$.

We've shown that for $\mathbf{n} \in \mathbb{N}$, $\mathbf{n} \not\approx \mathbb{N}$, so that A cannot be simultaneously finite and equinumerous with $\mathbb{N}$.

Proof

For a posh proof we need to set up a bijection between A and either $\mathbb{N}$ or a natural number $\mathbf{n}$. What structure does A have which we can exploit to define such a bijection? Our best bet is surely the ordering of the numbers in A by the usual $\leq$, which arranges the members of A into a list in ascending order – first member, second member, etc. And the most natural way of setting up a bijection is to define one, f say, from $\mathbb{N}$ to A by:

$\mathbf{0} \longmapsto$ first member of A
$\mathbf{1} \longmapsto$ second member of A
$\mathbf{2} \longmapsto$ third member of A
and so on.

The author is suspicious of most arguments that try to define a bijection from A to $\mathbb{N}$: they usually turn out to presuppose an enumeration of A as $a_0, a_1, \ldots$, which really means that our f has already been constructed!

We can smarten up our intuition of the elements of A listed in ascending order by using the language of well-order. So 'first member of A' becomes $\min(A)$, 'second member' becomes $\min(A \setminus \{f(\mathbf{0})\})$ and so on. So it looks as though we could define f by

$$\begin{aligned} f: \mathbb{N} &\longrightarrow A \\ \mathbf{n} &\longmapsto \min\,(A \setminus \text{Range}(f|_{\mathbf{n}})) \end{aligned}$$

This is a rather flash definition, yielding for instance that

$$\begin{aligned} f(\mathbf{0}) &= \min\,(A \setminus \text{Range}(f|_{\mathbf{0}})) \\ &= \min(A \setminus \varnothing) \text{ as } \mathbf{0} = \varnothing \text{ and } \text{Range}(f|_{\varnothing}) = \varnothing \\ &= \min(A), \\ f(\mathbf{1}) &= \min\,(A \setminus \text{Range}(f|_{\mathbf{1}})) \\ &= \min\,(A \setminus \text{Range}(f|_{\{\mathbf{0}\}})) \\ &= \min(A \setminus \{f(\mathbf{0})\}) \end{aligned}$$

and intuitively

$$f(\mathbf{n}) = \min\,(A \setminus \{f(\mathbf{0}), f(\mathbf{1}), \ldots, f(\mathbf{n}-\mathbf{1})\}).$$

Remember that, for a general $\mathbf{n} \in \mathbb{N}$, $\{f(\mathbf{0}), f(\mathbf{1}), \ldots, f(\mathbf{n}-\mathbf{1})\}$ is not a proper expression within *ZF*: however, $\text{Range}(f|_{\mathbf{n}})$ is.

Will this definition do? Could anything go wrong? Stop and have a think!

There is always a potential problem when finding the minimum element of a subset of a well-ordered set: what if the subset is *empty*? So can $A \setminus \text{Range}(f|_{\mathbf{n}})$ ever be empty? Yes, of course! If A is finite, surely there's some $\mathbf{n}$ for which, intuitively, $\{f(\mathbf{0}), f(\mathbf{1}), \ldots, f(\mathbf{n}-\mathbf{1})\}$ lists all of the elements of A. So our definition of f must somehow cope with the case that A is finite. Although one could structure the argument by defining f in two ways, once for the case that A is finite and once for when it isn't, there is something attractive about the following construction, which covers both cases. The trick is to introduce a new set c not in A to be the image of $\mathbf{n}$, should $A \setminus \text{Range}(f|_{\mathbf{n}})$ be empty. Then our definition of f becomes

For example, $c = \langle \mathbf{0}, \mathbf{0} \rangle$ will do, as $\langle \mathbf{0}, \mathbf{0} \rangle \notin \mathbb{N}$.

$$\begin{aligned} f: \mathbb{N} &\longrightarrow A \cup \{c\} \\ \mathbf{n} &\longmapsto \begin{cases} \min\,(A \setminus \text{Range}(f|_{\mathbf{n}})), & \text{if } A \setminus \text{Range}(f|_{\mathbf{n}}) \neq \varnothing, \\ c, & \text{otherwise.} \end{cases} \end{aligned}$$

Unlike our earlier, tentative, definition, this f is well-defined, whatever subset A is of $\mathbb{N}$. We now have two cases: when c is in $\text{Range}(f)$ and when it isn't (intuitively when A is finite and when it is infinite).

Case: $c \in \text{Range}(f)$

Intuitively, if $f(\mathbf{n}) = c$, then $f(\mathbf{m}) = c$ for all $\mathbf{m} \geq \mathbf{n}$; and the first $\mathbf{n}$ such that $f(\mathbf{n})$ hits c is the size of A. To define this $\mathbf{n}$ formally, note that $\{\mathbf{m} : f(\mathbf{m}) = c\}$ is non-empty (as $c \in \text{Range}(f)$) and is a subset of $\mathbb{N}$, so has a least element: let $\mathbf{n}$ be this least element. We claim that $f|_{\mathbf{n}}$ is a bijection from $\mathbf{n}$ to A, so that A is finite.

$f|_{\mathbf{n}}$ is onto A, as follows. By the minimality of $\mathbf{n}$, we have that $f(\mathbf{i}) \neq c$ for all $\mathbf{i} < \mathbf{n}$, so that each such $f(\mathbf{i})$ is in A. Thus $\text{Range}(f|_{\mathbf{n}}) \subseteq A$. And by definition of $\mathbf{n}$, $f(\mathbf{n}) = c$, so that $A \setminus \text{Range}(f|_{\mathbf{n}}) = \varnothing$. But then $A \subseteq \text{Range}(f|_{\mathbf{n}})$, so that $f|_{\mathbf{n}}$ is onto A.

The details of showing that f is one–one are left to you as part of the next exercise. Note for the moment that our solution to the exercise will actually show more than that f is one–one. It will show that f is strictly increasing, i.e. if $\mathbf{i} < \mathbf{j}$ for members $\mathbf{i}, \mathbf{j}$ of $\mathbf{n}$, then $f(\mathbf{i}) < f(\mathbf{j})$.

f strictly increasing implies that f is one–one.

Case: $c \notin \text{Range}(f)$

In this case we have $f(\mathbf{n}) \in A$ for all $\mathbf{n}$, so that $\text{Range}(f) \subseteq A$. The same argument as that showing $f|_{\mathbf{n}}$ to be one–one will show that f is both one–one and strictly increasing. It remains to show that f is onto. It will help to have the following lemma:

Lemma: $f(\mathbf{n}) \geq \mathbf{n}$ for all $\mathbf{n} \in \mathbb{N}$.

Proof of Lemma: We shall use the fact that f is strictly increasing, and proceed by induction.
For $\mathbf{n} = \mathbf{0}$ we have that $f(\mathbf{0})$ is some member of $A \subseteq \mathbb{N}$, so that $f(\mathbf{0}) \geq \mathbf{0}$. Suppose that $f(\mathbf{k}) \geq \mathbf{k}$ for some $\mathbf{k} \in \mathbb{N}$. Then as f is strictly increasing, $f(\mathbf{k}+\mathbf{1}) > f(\mathbf{k})$, so that

$$f(\mathbf{k}+\mathbf{1}) \geq f(\mathbf{k}) + \mathbf{1}$$
$$\geq \mathbf{k}+\mathbf{1} \text{ using the inductive hypothesis.}$$

The result of the lemma follows by induction. □

We will now show that f is onto, arguing by contradiction. Suppose that f is not onto. Then there is some $\mathbf{a} \in A$ with $\mathbf{a} \notin \text{Range}(f)$. Let's investigate the value of $f(\mathbf{a})$. By the lemma above, $f(\mathbf{a}) \geq \mathbf{a}$. Also $f(\mathbf{a}) = \min(A \setminus \text{Range}(f|_{\mathbf{a}}))$, so that $f(\mathbf{a}) \leq b$ for any $b \in A \setminus \text{Range}(f|_{\mathbf{a}})$. However, $\text{Range}(f|_{\mathbf{a}}) \subseteq \text{Range}(f)$ and $\mathbf{a} \notin \text{Range}(f)$, so that $\mathbf{a} \notin \text{Range}(f|_{\mathbf{a}})$. Of course $\mathbf{a} \in A$, by definition of $\mathbf{a}$, so that $\mathbf{a} \in A \setminus \text{Range}(f|_{\mathbf{a}})$, giving that $f(\mathbf{a}) \leq \mathbf{a}$. Putting the two inequalities together, we get $f(\mathbf{a}) = \mathbf{a}$. But then $\mathbf{a} \in \text{Range}(f)$, contradicting our assumption that $\mathbf{a} \notin \text{Range}(f)$.

Thus f is onto, and hence f is a bijection, so that $A \approx \mathbb{N}$. ■

Exercise 6.30

Prove the claim in the proof above that $f|_{\mathbf{n}}$ is strictly order-preserving. (It may help to remember that for natural numbers $\mathbf{i}, \mathbf{j}$ within *ZF*, $\mathbf{i} < \mathbf{j}$ if and only if $\mathbf{i} \in \mathbf{j}$ if and only if $\mathbf{i}$ is a proper subset of $\mathbf{j}$.)

See Exercise 3.11.

Solution

Take $\mathbf{i}, \mathbf{j}$ in $\mathbf{n}$ such that $\mathbf{i} \neq \mathbf{j}$. Without loss of generality $\mathbf{i} < \mathbf{j}$, so that as sets in *ZF*, $\mathbf{i} \subseteq \mathbf{j}$. Then $\text{Range}(f|_{\mathbf{i}}) \subseteq \text{Range}(f|_{\mathbf{j}})$, so that

$$A \setminus \text{Range}(f|_{\mathbf{j}}) \subseteq A \setminus \text{Range}(f|_{\mathbf{i}}).$$

Now $f(\mathbf{j}) \in A \setminus \text{Range}(f|_{\mathbf{j}})$ so $f(\mathbf{j}) \in A \setminus \text{Range}(f|_{\mathbf{i}})$. But by definition $f(\mathbf{i}) = \min(A \setminus \text{Range}(f|_{\mathbf{i}}))$, so that $f(\mathbf{i}) \leq f(\mathbf{j})$. Also, as sets in *ZF*, $\mathbf{i} \in \mathbf{j}$, so that $f(\mathbf{i}) \in \text{Range}(f|_{\mathbf{j}})$, while from the definition $f(\mathbf{j}) \notin \text{Range}(f|_{\mathbf{j}})$. Thus $f(\mathbf{j})$ cannot equal $f(\mathbf{i})$. Hence $f(\mathbf{i}) < f(\mathbf{j})$.

Definitions

A set A is said to be *countable* if it is finite or equinumerous with $\mathbb{N}$. We say that A is *countably infinite* if A is equinumerous with $\mathbb{N}$.

If $\mathbf{n} \in \mathbb{N}$, then $\mathbf{n} \not\approx \mathbb{N}$ (Exercise 3.6). So A cannot be both finite and equinumerous with $\mathbb{N}$.

By our theorem above, any subset of $\mathbb{N}$ is countable; and we have shown that $\mathbb{N} \approx \mathbb{Z}$, so that $\mathbb{Z}$ is countable. How about some other famous sets, like $\mathbb{Q}$, which are clearly infinite? Before we investigate these, let's review the ways in which we might show that a set A is countably infinite. First of all we can construct a bijection from $\mathbb{N}$ to A (or vice versa). We could also exploit the Schröder–Bernstein theorem, by constructing two one–one functions, one from $\mathbb{N}$ to A and one in the reverse direction – it's sometimes easier to do this than to find a bijection. And with both methods we can of course replace $\mathbb{N}$ by another known countably infinite set. In the next exercise we'd like you to justify a further method.

Exercise 6.31

Suppose that the function $f\colon \mathbb{N} \longrightarrow A$ is onto and that $\mathbb{N} \preceq A$. Prove that A is countable. (Can you avoid using any form of AC? The set A will of course be countably infinite, but as this is such a mouthful, it seems customary to be somewhat sloppy with one's terminology!)

A restatement of Exercise 6.31 is that if one can list the members of a set X as $x_0, x_1, \ldots, x_n, \ldots$, where each member of X appears at least once as an x_n for some $n \in \mathbb{N}$, but possibly appears several times, then X is countable.

Let's now use each of these methods to prove one of the first of Cantor's surprising results, that $\mathbb{Q}_0^+$, the set of non-negative rationals, is countable. (It hardly seems necessary to say that the result, at least the first time one meets it, *is* usually surprising – there seem to be vastly more rationals than natural numbers!) As $\mathbb{N} \subseteq \mathbb{Q}_0^+$, we have $\mathbb{N} \preceq \mathbb{Q}_0^+$. Let's define an onto function from $\mathbb{N}$ onto $\mathbb{Q}_0^+$, so that by the result of the previous exercise, $\mathbb{Q}_0^+$ is countable. List all the positive rationals in an array as follows

$$\begin{array}{ccccc}
\frac{1}{1} & \frac{2}{1} & \frac{3}{1} & \frac{4}{1} & \cdots \\[2em]
\frac{1}{2} & \frac{2}{2} & \frac{3}{2} & \frac{4}{2} & \cdots \\[2em]
\frac{1}{3} & \frac{2}{3} & \frac{3}{3} & \frac{4}{3} & \cdots \\[2em]
\frac{1}{4} & \frac{2}{4} & \frac{3}{4} & \frac{4}{4} & \cdots \\[1em]
\vdots & \vdots & \vdots & \vdots & \ddots
\end{array}$$

so that the rational $\frac{m}{n}$ appears as the entry in the nth row and mth column.

Now count through the array, starting at the top left corner and proceeding up successive 'anti-diagonals' as follows:

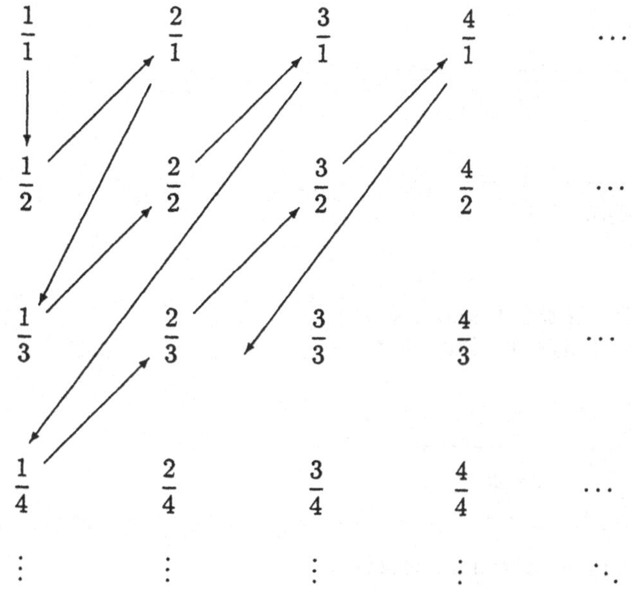

In this counting process $\frac{1}{1}$ is counted as item 1, $\frac{1}{2}$ as item 2, $\frac{2}{1}$ as item 3, and you can check that the entry $\frac{m}{n}$ appears as the $\left(\frac{1}{2}(m+n-1)(m+n-2)+m\right)$th item.

Explain this formula!

This counting process defines a function f from the set of positive integers to the set of positive rationals, and by setting $f(0) = 0$ we obtain a function from $\mathbb{N}$ to $\mathbb{Q}_0^+$. Clearly every non-negative rational is $f(n)$ for some n, and in fact any positive rational is the image of infinitely many n: for instance $\frac{1}{2}$ appears in the array as $\frac{2}{4}$, $\frac{3}{6}$ etc. So f is onto, and the result of Exercise 6.31 allows us to conclude that $\mathbb{N} \approx \mathbb{Q}_0^+$.

An alternative method is to define, in a direct way, a one–one function from $\mathbb{Q}_0^+$ to $\mathbb{N}$ and, given that $\mathbb{N} \preceq \mathbb{Q}_0^+$, use the Schröder–Bernstein theorem to deduce the result. An attractive trick is to exploit the fundamental theorem of arithmetic as follows. Define a function f by

$$f\colon \mathbb{Q}_0^+ \longrightarrow \mathbb{N}$$
$$q \longmapsto \begin{cases} 0, & \text{if } q = 0, \\ 2^m 3^n, & \text{if } q = \frac{m}{n} \text{ and } \gcd(m,n) = 1. \end{cases}$$

The fundamental theorem of arithmetic says that every positive integer greater than 1 has a *unique* decomposition into a product of primes of the form $p_1^{k_1} p_2^{k_2} \ldots p_r^{k_r}$ where $p_1, p_2, \ldots, p_r$ are primes with $p_1 < p_2 < \ldots < p_r$, and $k_1, k_2, \ldots, k_r$ are positive integers.

The condition that $\gcd(m,n) = 1$ is of course needed to give a well-defined image for q. And it is easy to check that f is one–one as required.

How about the most direct way of showing that $\mathbb{N}$ and $\mathbb{Q}_0^+$ are equinumerous, namely defining a bijection from one to the other? Well, it can be done, grubbily!

See Exercise 6.42.

Now that we have shown $\mathbb{Q}_0^+$ to be countable, it's relatively easy to show that $\mathbb{Q}$ is countable. First show that the set $\mathbb{Q}^-$ of negative rationals is countable. Then prove the following theorem:

> **Theorem 6.3**
>
> Let A and B be countable sets. Then $A \cup B$ is countable.

Applying this theorem to $\mathbb{Q}^-$ and $\mathbb{Q}_0^+$ then gives that $\mathbb{Q}$ is countable.

Exercise 6.32

Show that $\mathbb{Q}^-$ is countable. (You have a choice of methods!)

Exercise 6.33

Prove Theorem 6.3. [Hint: as $A \cup B$ equals the union of the disjoint sets A and $B \setminus A$, it is enough to prove the result for A and B disjoint. Only do the case where both A and B are infinite.]

Exercise 6.34

Prove that the union of finitely many countable sets is countable.

Solution

This is just a straightforward induction on the finite number of countable sets.

The next exercise asks you to prove the following theorem in a couple of ways:

> **Theorem 6.4**
>
> Let A and B be countable sets. Then $A \times B$ is countable.

Exercise 6.35

(a) (i) Give a one–one function to show that $\mathbb{N} \preceq \mathbb{N} \times \mathbb{N}$.

(ii) Give a one–one function which shows that $\mathbb{N} \times \mathbb{N} \preceq \mathbb{N}$. [Hint: exploit the trick with prime factorization used for $\mathbb{Q}_0^+$.]

(iii) Prove Theorem 6.4. [How do you prove the result in the case that one or both of A and B are finite?]

(b) Exploit the construction earlier involving an array (of rationals, in that case) to define a bijection between $\mathbb{N}$ and $\mathbb{N} \times \mathbb{N}$.

Solution

(a) (i) Define f by

$$\begin{aligned} f : \mathbb{N} &\longrightarrow \mathbb{N} \times \mathbb{N} \\ n &\longmapsto (n, 0) \end{aligned}$$

This f is clearly one–one.

(ii) Define g by

$$\begin{aligned} g : \mathbb{N} \times \mathbb{N} &\longrightarrow \mathbb{N} \\ (m, n) &\longmapsto 2^m 3^n \end{aligned}$$

Again it is easy to show that g is one–one.

(iii) From the first two parts, we can use the Schröder–Bernstein theorem to conclude that $\mathbb{N} \approx \mathbb{N} \times \mathbb{N}$. Thus if A and B are both countably infinite, i.e. $A \approx B \approx \mathbb{N}$, we can use the result of Exercise 6.11 to deduce that $A \times B \approx \mathbb{N} \times \mathbb{N}$. Thus $A \times B \approx \mathbb{N}$, so is countable.

The case where one of the sets, A say, is countably infinite and the other, B, is finite, can be tackled in a variety of ways. For instance, as B is finite there is some $\mathbf{n} \in \mathbb{N}$ for which $B \approx \mathbf{n}$. Suppose that $\mathbf{n} \geq \mathbf{1}$. Then

What happens if $\mathbf{n} = \mathbf{0}$?

$$\begin{aligned} B &\approx \mathbf{n} \\ &\preceq \mathbb{N} \qquad \text{as } \mathbf{n} \subseteq \mathbb{N}, \end{aligned}$$

so that

$$\begin{aligned} A \times B &\preceq A \times \mathbb{N} \\ &\approx \mathbb{N} \times \mathbb{N} \\ &\approx \mathbb{N}, \end{aligned}$$

while also

$$\begin{aligned} \mathbb{N} &\approx A \\ &\approx A \times \mathbf{1} \qquad \text{(remember that } \mathbf{1} = \{\mathbf{0}\}) \\ &\preceq A \times \mathbf{n} \qquad \text{as } \mathbf{1} \preceq \mathbf{n} \\ &\approx A \times B. \end{aligned}$$

So $\mathbb{N} \preceq A \times B$ and $A \times B \preceq \mathbb{N}$, from which we can conclude that $A \times B \approx \mathbb{N}$, so that $A \times B$ is countable.

Finally, if both A and B are finite, say $A \approx \mathbf{m}$ and $B \approx \mathbf{n}$, for $\mathbf{m}, \mathbf{n} \in \mathbb{N}$, the result follows from the fact that $\mathbf{m} \times \mathbf{n} \approx \mathbf{m} \cdot \mathbf{n}$, where $\mathbf{m} \cdot \mathbf{n}$ is the product of $\mathbf{m}$ and $\mathbf{n}$ within ZF.

(b) This time let us use the familiar picture to define a bijection h from $\mathbb{N} \times \mathbb{N}$ to $\mathbb{N}$, as follows:

$$\begin{aligned} h \colon \mathbb{N} \times \mathbb{N} &\longrightarrow \mathbb{N} \\ (m, n) &\longmapsto \tfrac{1}{2}(m + n + 1)(m + n) + n + 1 \end{aligned}$$

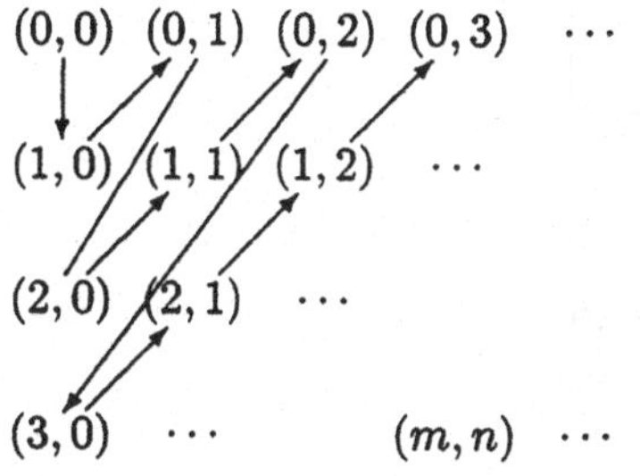

Note that the diagonal containing (m, n) has $m + n + 1$ entries. The preceding diagonals have $1 + 2 + \ldots + (m + n)$ entries. And (m, n) is the $(n + 1)$th entry along its diagonal.

It is easy to check that h is indeed a bijection.

The bijection in the last solution above is often helpful in constructing bijections showing that other sets are countable. Likewise, the fundamental theorem of arithmetic can be very helpful.

So far in this section, all our infinite sets have been countable. Is this the case for all infinite sets? The answer is a resounding 'No', as is given by the following theorem of Cantor – in some sense it is this theorem which makes the subject really take off!

> ***Theorem 6.5 Cantor's theorem***
>
> For any set X, $X \prec \mathscr{P}(X)$.

Think about the consequences of this for a moment! For instance, starting with $\mathbb{N}$, we have a sequence of infinite sets

$$\mathbb{N} \prec \mathscr{P}(\mathbb{N}) \prec \mathscr{P}(\mathscr{P}(\mathbb{N})) \prec \ldots$$

of strictly increasing 'size'. So there are lots of different 'sizes' of infinite sets. Furthermore, there is no set X of 'maximal size', as X is dominated by the set $\mathscr{P}(X)$.

Recall that $\prec$ is irreflexive and transitive, so the sequence cannot loop back on itself.

Proof of Cantor's theorem
We need to show both $X \preceq \mathscr{P}(X)$ and $X \not\approx \mathscr{P}(X)$.

It is easy to show that the function

$$\begin{aligned} f: X &\longrightarrow \mathscr{P}(X) \\ x &\longmapsto \{x\} \end{aligned}$$

is one–one, so that $X \preceq \mathscr{P}(X)$.

Assume that $X \approx \mathscr{P}(X)$. Then there is a bijection $g: X \longrightarrow \mathscr{P}(X)$. Define a subset Y of X by

We are aiming to derive a contradiction.

$$Y = \{x \in X : x \notin g(x)\}.$$

As g is onto, there must be some $y \in X$ for which $Y = g(y)$. Is $y \in Y$? If so, then by definition of Y we have $y \notin g(y)$; but $g(y) = Y$, so we have a contradiction. So we must have $y \notin Y$; but as $Y = g(y)$ this means that $y \in Y$, so again we have a contradiction. So g is not onto, so not a bijection, contradicting our assumption as required. ■

> ***Definition***
>
> An infinite set which is not countable is said to be *uncountable* or *uncountably infinite*.

In the next section we shall investigate some uncountable sets and start to look at an arithmetic of cardinals. But to end this section, let's look at two final conundrums involving countable sets. First of all, in Exercise 6.34 we showed that the union of finitely many countable sets is countable. Here is a powerful extension of this result and its 'proof'.

> ***Theorem 6.6***
>
> Let X be a countable set, each element of which is itself countable. Then $\bigcup X$ is countable; i.e. a countable union of countable sets is countable.

Proof

Consider the most interesting case where X and all its members are countably infinite. As X is countably infinite, there is a bijection $f\colon \mathbb{N} \longrightarrow X$. Each member of X is thus $f(i)$ for some $i \in \mathbb{N}$. $f(i)$ is countably infinite, so there is a bijection $g_i\colon \mathbb{N} \longrightarrow f(i)$. Then the function h defined by

Why are the other cases less interesting?

$$\begin{aligned} h\colon \mathbb{N} \times \mathbb{N} &\longrightarrow \bigcup X \\ (i,j) &\longmapsto g_i(j) \end{aligned}$$

is a bijection, so that $\mathbb{N} \times \mathbb{N} \approx \bigcup X$. By Theorem 6.4, $\bigcup X$ is countable. ■

What, if anything, is fishy about this proof? It is that the proof needs a mild form of the axiom of choice, and in this section we hoped to avoid use of AC.

Exercise 6.36

(a) Where did we use some form of AC in the proof?

(b) Might we need to use AC to prove that a countable union of 2–element sets is countable?

Solution

(a) The problem occurs when we say that for each $i \in \mathbb{N}$ there is a bijection $g_i\colon \mathbb{N} \longrightarrow f(i)$. In general there will be several bijections between $\mathbb{N}$ and the countable set $f(i)$, so that we need to choose one of these for each $i \in \mathbb{N}$. If the sets $f(i)$, $i \in \mathbb{N}$, have sufficient structure, then there may be some way of specifying each g_i. But if not, then a form of AC is needed. (This is the same sort of problem as encountered when proving that the union of countably many null sets is null, in Section 5.3.)

(b) Alas AC might still be needed! For each 2-element set there are two possible bijections between the natural number **2** and the set. Unless each 2-element set has some nice structure, we will again have to make countably many choices to get all these bijections.

Our final conundrum concerns the relationship between an infinite set X and $\mathbb{N}$. We already know that if $\mathbb{N} \preceq X$ then X is infinite. Surely the converse is true: if X is infinite then $\mathbb{N} \preceq X$. Alas (or not as the mood takes you) this result needs a mild form of AC! Knowing that X is infinite just doesn't give us quite enough structure to define a one–one function from $\mathbb{N}$ to X.

By Theorem 6.28.

Exercise 6.37

Let X be an infinite set. Assuming AC, prove that $\mathbb{N} \preceq X$. [Hints: by AC there is a choice function $f\colon \mathscr{P}(X) \setminus \{\varnothing\} \longrightarrow X$. Exploit this f to define by recursion a one–one function from $\mathbb{N}$ to X, along the lines of the proof of Theorem 6.2 – you will need the fact that X is infinite, i.e. not finite, somewhere!]

Further exercises

Exercise 6.38

We shall call a set *Dedekind infinite* if there is a bijection between it and a proper subset of it.

(a) Let X be an infinite set. Show, assuming AC, that X is Dedekind infinite. [Hint: use the result of Exercise 6.37.]

(b) Show, without using AC, that the converse of the above result holds, i.e. if X is Dedekind infinite then X is infinite.

This was Dedekind's definition of an infinite set. It cleverly avoids reference to 'finite'. It is of interest that showing the equivalence of Dedekind's definition and the one that we use in this book requires use of AC. (In fact only a weak version of AC is required.)

For each of the exercises below, it is possible to avoid use of AC.

Exercise 6.39

For the purpose of this exercise, regard a finite sequence of natural numbers as an ordered n-tuple of natural numbers for some $n \in \mathbb{N}$. Thus the set $\bigcup\{\mathbb{N}^n : n \in \mathbb{N}\}$ can be regarded as the set of all finite sequences of natural numbers. Show that this set is countable.

As opposed to such a sequence being a function from $\mathbf{n}$ into $\mathbb{N}$.

Exercise 6.40

(a) Show that the set of all 2-element subsets of $\mathbb{N}$ is countable.

(b) Show that, for a fixed $n \in \mathbb{N}$, the set of all n-element subsets of $\mathbb{N}$ is countable.

(c) Show that the set of all finite subsets of $\mathbb{N}$ is countable.

(d) Is the set of all infinite subsets of $\mathbb{N}$ countable? Explain your answer.

In the first three parts you might like to try to specify bijections, as well as exploit the Schröder–Bernstein theorem.

Exercise 6.41

A complex number is said to be *algebraic* if it is a root of a polynomial equation $x^n + a_{n-1}x^{n-1} + \ldots + a_1x + a_0 = 0$, where each coefficient a_i is rational.

(a) Show that, for a fixed $n \in \mathbb{N}$, the set of all roots of all polynomials of degree n with rational coefficients is countable.

(b) Show that the set of all algebraic numbers is countable.

Exercise 6.42

Define a bijection from $\mathbb{N}$ to $\mathbb{Q}_0^+$, the set of non-negative rationals. [Hint: exploit the list of rational numbers which we used to construct an onto function from $\mathbb{N}$ to $\mathbb{Q}_0^+$.]

Exercise 6.43

We can take the variable symbols for sets in our formal language for set theory, as in Chapter 4, as those in the countable set $\{x_0, x_1, x_2, \ldots, x_n, \ldots\}$. Thus formulas are built up from these variables, brackets and the symbols $\in$, $=$, $\neg$, $\wedge$, $\vee$, $\rightarrow$, $\leftrightarrow$, $\forall$ and $\exists$.

A formula of the language has to be finitely long, so can involve the use of at most finitely many variables.

Show that there are countably many formulas $\phi(y)$ of one free variable y in the formal language. (This shows that the axiom of separation can be used to describe at most countably many subsets of an infinite set x, whereas Cantor's theorem says that such an x has uncountably many subsets. Our inability to describe all of the subsets of x in a finite way is of course an important limitation of both set theory and of us!)

Exercise 6.44

Hilbert's Hotel has a countable infinity of bedrooms, numbered by the natural numbers. The hotel can be filled up in a straightforward way, putting the first guest to arrive in Room 0, the next guest in Room 1, the next in Room 2, and so on. And when the hotel is full, there's a way of squeezing in an extra guest. All the existing guests move from their current room to the one with the next highest number. This frees up Room 0, into which the new guest can move.

Such a hotel was used as a lecture example by Hilbert. The hotel does have a Room 13, unlike some hotels, and a Room 0, unlike most hotels.

(a) The new night manager has been told about this way of adding a guest when the hotel is full. On his first night on duty the hotel is full and 20 new guests arrive, each hoping for a room. He applies the technique for adding a guest 20 times in succession and everyone is satisfied. The next night a somewhat larger number of new guests, namely a countable infinity of them, turn up successively during the evening (at time intervals diminishing by a factor of a half each time), and the night manager accommodates them on the same basis. At breakfast the next morning, the kitchen staff notice that demand is a bit slow. What has happened? What should the night manager have done the night before to accommodate the new guests (and the old)?

It's mildly irritating for the original guests to have to move rooms so often, but the inconvenience is matched by the price.

(b) Business for the hotel recovers and it is full again. The hotel has warning of the arrival, during the evening, of countably infinitely many coach parties, each coach carrying countably infinitely many people requiring separate rooms.

(i) Suggest a scheme for the arrival time of the nth person in the mth coach, assuming that the first person in the first coach checks in at 6.00 p.m. and that all the new guests have checked in by 8.00 p.m.

(ii) How might the night manager (who has apparently retained his job) accommodate all the guests, new and old?

6.5 Uncountable sets and cardinal arithmetic, without AC

In this section we shall investigate the cardinalities of some interesting sets: many will turn out to be uncountable. Among the tools we shall use are the results of Section 6.3, which are really results of cardinal arithmetic. Many of the results are about $\mathbb{R}$ and we shall exploit the various ways of defining the real numbers in Chapter 2. For this section we shall again try to avoid use of AC.

First of all, let us relate the sizes of two important, but very different looking, sorts of set.

Theorem 6.7

For any set X, $\mathscr{P}(X) \approx \mathbf{2}^X$.

$\mathbf{2}^X = \{\mathbf{0}, \mathbf{1}\}^X$, the set of functions from X to $\{\mathbf{0}, \mathbf{1}\}$.

Proof

Define a function f by

$$f\colon \mathscr{P}(X) \longrightarrow \mathbf{2}^X$$
$$A \longmapsto \chi_A$$

where χ_A is the *characteristic function* of A, i.e.

$$\chi_A\colon A \longrightarrow \mathbf{2}$$
$$\chi(x) = \begin{cases} \mathbf{1}, & \text{if } x \in A, \\ \mathbf{0}, & \text{if } x \notin A. \end{cases}$$

It is easy to show that f is a bijection. ■

So Cantor's theorem (Theorem 6.5) can be equivalently stated as

for any set X, $\quad X \prec \mathbf{2}^X$.

As $X \prec \mathscr{P}(X) \approx \mathbf{2}^X$.

Exercise 6.45

Prove directly that $X \not\approx \mathbf{2}^X$ by assuming that there is a bijection between X and $\mathbf{2}^X$ and deriving a contradiction.

Don't involve $\mathscr{P}(X)$ in your argument!

Let us now investigate $\mathbb{R}$, the set of real numbers. As $\mathbb{N} \subseteq \mathbb{R}$, we certainly know (from Exercise 6.28) that $\mathbb{R}$ is infinite. But is $\mathbb{R}$ countably infinite? Cantor showed not, by the following theorem.

Theorem 6.8

$\mathbb{R}$ is uncountable.

Proof

Suppose that $\mathbb{R}$ is countable. Then we can list the reals in the interval $[0, 1)$ as

$$a_1, a_2, a_3, \ldots, a_n, \ldots$$

How do we know that there are infinitely many reals in $[0, 1)$? See Exercise 6.46 below.

with each real in $[0, 1)$ appearing as a_n for exactly one $n \in \mathbb{N}$, $n \geq 1$. We shall represent each such real r by its decimal expansion,

$$r = 0.r_1r_2r_3 \ldots r_n \ldots,$$

avoiding the use of recurring 9s (so that e.g. we represent 0.2 by 0.2000..., rather than 0.1999...). We can then picture the reals in $[0, 1)$ written out in an array:

$$a_1 = 0.a_{1,1}a_{1,2}a_{1,3} \ldots$$
$$a_2 = 0.a_{2,1}a_{2,2}a_{2,3} \ldots$$
$$a_3 = 0.a_{3,1}a_{3,2}a_{3,3} \ldots$$
$$\vdots$$
$$a_n = 0.a_{n,1}a_{n,2}a_{n,3} \ldots$$
$$\vdots$$

As outlined in Chapter 2 all complete ordered fields are isomorphic. An isomorphism is, among other things, a bijection, so that any two such fields are equinumerous. Thus arguments about the cardinality of $\mathbb{R}$ don't depend on which representation of $\mathbb{R}$ is used. Our argument here is prettier using the decimal expansions of everyday maths, rather than Dedekind cuts or Cauchy sequences!

Now define a real number $r = 0.r_1r_2r_3 \dots r_n \dots$ by

$$r_n = \begin{cases} 4, & \text{if } a_{n,n} \geq 6, \\ 7, & \text{if } a_{n,n} < 6. \end{cases}$$

Then r belongs to $[0,1)$. However, r has been constructed to disagree with each a_n at the nth decimal place, so it cannot equal a_n for any n. Thus r does not appear in the list, contradicting that the list contains all reals in $[0,1)$.

This is called a *diagonal* argument. It is essentially the same sort of argument as used in the proof of Cantor's theorem, $X \prec \mathscr{P}(X)$, and is of great use elsewhere, e.g. in the theory of computable functions.

Thus $[0,1) \not\approx \mathbb{N}$. As $\mathbb{N} \preceq [0,1)$, we must then have $\mathbb{N} \prec [0,1)$, so that, as $[0,1) \subseteq \mathbb{R}$, we have that $\mathbb{R}$ is uncountable. ■

Exercise 6.46

Explain the following steps of the proof of Theorem 6.8.

(a) How do you know that there are infinitely many reals in the interval $[0,1)$?

(b) Assuming that there are countably infinitely many reals in the interval $[0,1)$, why can you list them as

$$a_1, a_2, a_3, \dots, a_n, \dots$$

with each number appearing as a_n for exactly one $n \in \mathbb{N}$, $n \geq 1$?

Solution

(a) $[0,1)$ contains the rationals $\frac{1}{n}$ for each $n \in \mathbb{N}$ with $n \geq 1$. This can be used to show that $\mathbb{N} \preceq [0,1)$, so that $[0,1)$ is infinite by Exercise 6.3.

(b) Assuming that $[0,1)$ is countably infinite, there is a bijection $f\colon \mathbb{N} \longrightarrow [0,1)$. Then each real in $[0,1)$ is $f(n)$ for exactly one $n \in \mathbb{N}$. Putting $a_{n+1} = f(n)$, the a_ns have the claimed property. (So why bother with the a_ns? Why not just use the $f(n)$s? I suppose that the variety of notation makes things easier to read.)

This attractive argument tells us that $\mathbb{R}$ is uncountable, but doesn't relate the size of $\mathbb{R}$ to that of any other set we know. The next theorem, however, does just this!

> ***Theorem 6.9***
>
> $\mathbb{R} \approx \mathscr{P}(\mathbb{Q}) \approx 2^{\mathbb{N}}$

Proof

We shall use the Schröder–Bernstein theorem, treating the reals as given by Dedekind cuts. This makes it easy to show that $\mathbb{R} \preceq \mathscr{P}(\mathbb{Q})$, as the function f defined by

Recall that as a Dedekind cut, the real r is a subset of $\mathbb{Q}$.

$$f\colon \mathbb{R} \longrightarrow \mathscr{P}(\mathbb{Q})$$
$$r \longmapsto r$$

is clearly one–one.

As $\mathbb{Q} \approx \mathbb{N}$, we have $\mathscr{P}(\mathbb{Q}) \approx \mathscr{P}(\mathbb{N})$, so by Theorem 6.7 $\mathscr{P}(\mathbb{N}) \approx 2^{\mathbb{N}}$. So to complete the proof we need to show that $2^{\mathbb{N}} \preceq \mathbb{R}$.

By Exercise 6.22.

Given $f \in \mathbf{2}^{\mathbb{N}}$, define a set f^* of rationals by

$$f^* = \{q \in \mathbb{Q} : q < f(0) + \frac{f(1)}{10} + \frac{f(2)}{10^2} + \ldots + \frac{f(n)}{10^n} \text{ for some } n \in \mathbb{N}\}.$$

The idea is that f^* corresponds to the decimal expansion $f(0).f(1)f(2)f(3)\ldots$.

You can check that f^* is a Dedekind cut. Then define a function θ by

$$\begin{aligned} \theta \colon \mathbf{2}^{\mathbb{N}} &\longrightarrow \mathbb{R} \\ f &\longmapsto f^* \end{aligned}$$

You can also check that θ is one–one, so that $\mathbf{2}^{\mathbb{N}} \preceq \mathbb{R}$. ■

Exercise 6.47

(a) Show that f^* in the above proof is a Dedekind cut.

(b) Show that θ in the above proof is one–one.

Now that we have connected the sizes of two very important sets within mathematics, $\mathbb{N}$ and $\mathbb{R}$, let us investigate how the sizes of other such sets compare. We already know, for instance, about $\mathbb{Z}$, $\mathbb{Q}$ and $\mathbb{N} \times \mathbb{N}$: they are all equinumerous with $\mathbb{N}$; while $\mathscr{P}(\mathbb{N})$ is equinumerous with $\mathbb{R}$. How about, say, $\mathbb{R} \setminus \mathbb{Q}$, the set of irrational numbers: or $\mathbb{C}$, the set of complex numbers: or $\mathbb{R}^{\mathbb{R}}$, the set of all real-valued functions of a real variable? For the next exercise, think about some of the possible bounds on the sizes of these sets.

We can now state one of the earliest, and most famous, of Cantor's problems: is there is a subset X of $\mathbb{R}$ with $\mathbb{N} \prec X \prec \mathbb{R}$? Cantor's earliest version of the *continuum hypothesis* is that no such X exists. We shall return to this problem in Chapter 9.

Exercise 6.48

(a) Could $\mathbb{R} \setminus \mathbb{Q}$ be countable? If not, why not?

(b) Why is $\mathbb{C}$ not countable?

(c) Is it possible that $\mathbb{R}^{\mathbb{R}} \approx \mathbb{R}$? If not, why not?

Solution

(a) $\mathbb{R} \setminus \mathbb{Q}$ cannot be countable. If it were countable, then, as $\mathbb{R}$ equals $(\mathbb{R} \setminus \mathbb{Q}) \cup \mathbb{Q}$, $\mathbb{R}$ would be the union of two countable sets, hence countable. This contradicts that $\mathbb{R}$ is uncountable.

(b) $\mathbb{R}$ is uncountable and $\mathbb{R} \subseteq \mathbb{C}$. Thus $\mathbb{C}$ is uncountable.

(c) By Cantor's theorem $\mathbb{R} \prec \mathbf{2}^{\mathbb{R}}$. But $\mathbf{2} \preceq \mathbb{R}$, so that $\mathbf{2}^{\mathbb{R}} \preceq \mathbb{R}^{\mathbb{R}}$. Thus $\mathbb{R} \prec \mathbb{R}^{\mathbb{R}}$. So $\mathbb{R}^{\mathbb{R}} \not\approx \mathbb{R}$.

If $A \preceq B$ then $A^C \preceq B^C$.

Let us try to tie down the sizes of these sets some more. First consider $\mathbb{R} \setminus \mathbb{Q}$. We have shown that $\mathbb{R} \setminus \mathbb{Q} \not\approx \mathbb{N}$, and as $\mathbb{R} \setminus \mathbb{Q} \subseteq \mathbb{R}$ we have $\mathbb{R} \setminus \mathbb{Q} \preceq \mathbb{R}$. Can we locate the size of $\mathbb{R} \setminus \mathbb{Q}$ somewhere between those of $\mathbb{N}$ and $\mathbb{R}$? It may seem obvious that $\mathbb{N} \preceq \mathbb{R} \setminus \mathbb{Q}$, but even this needs justification, e.g. in the form of a one–one function from $\mathbb{N}$ to $\mathbb{R} \setminus \mathbb{Q}$.

Exercise 6.49

Find a one–one function f from $\mathbb{N}$ to $\mathbb{R} \setminus \mathbb{Q}$.

So we can at least conclude that $\mathbb{R} \setminus \mathbb{Q}$ is infinite (by Exercise 6.28). As $\mathbb{R} \setminus \mathbb{Q}$ is not countable the question then arises of whether it might have a size strictly between those of $\mathbb{N}$ and $\mathbb{R}$. The question of whether $\mathbb{R}$ contains any infinite subset which is equinumerous with neither $\mathbb{N}$ nor $\mathbb{R}$ was, perhaps, the most famous one arising from Cantor's work. His *continuum hypothesis* is that there is no such subset and its resolution was proposed by Hilbert as one of the most important tasks for mathematicians in the 20th century. You might guess from this discussion that $\mathbb{R} \setminus \mathbb{Q}$ does not disprove the continuum hypothesis! So as $\mathbb{R} \setminus \mathbb{Q}$ is not countable, it must be the case that it is equinumerous with $\mathbb{R}$, which you will now show.

The continuum is a name for the real line.

The German David Hilbert (1862–1943) was one of the most influential mathematicians at the beginning of the 20th century.

As a consequence of the previous exercise, we know that $\mathbb{R}$ contains two disjoint countably infinite subsets ($\mathbb{Q}$ and $f(\mathbb{N})$). The result that $\mathbb{R} \setminus \mathbb{Q} \approx \mathbb{R}$ will follow from the next theorem.

By the construction of f, $f(\mathbb{N}) \subseteq \mathbb{R} \setminus \mathbb{Q}$.

Theorem 6.10

Let X be a set with countably infinite subsets A and B such that $A \cap B = \varnothing$. Then $X \setminus A \approx X$.

So X is infinite.

Exercise 6.50

Prove this theorem. [Hints: $X \setminus A \preceq X$ is clear. To show $X \preceq X \setminus A$, try using the identity function on $X \setminus (A \cup B)$: where will you map $A \cup B$?]

So the set of irrationals is not just uncountable: it is equinumerous with $\mathbb{R}$. As there is the challenge of trying to find a subset of $\mathbb{R}$ which disproves the continuum hypothesis, we shall give you some exercises investigating the sizes of various infinite subsets of $\mathbb{R}$. The chances are, however, that they will be equinumerous with $\mathbb{N}$ or $\mathbb{R}$! One immediate class of examples is that of all open intervals (x, y), where $x < y$.

Exercise 6.51

(a) Find a bijection between the interval $(-1, 1)$ and $\mathbb{R}$. [Hint: there are differentiable functions which will work.]

(b) Suppose that $x < y$. Show that $(x, y) \approx \mathbb{R}$.

Next let us look at the set $\mathbb{C}$. $\mathbb{R} \subseteq \mathbb{C}$, so $\mathbb{C}$ is uncountable. Is $\mathbb{R} \approx \mathbb{C}$ or is it the case that $\mathbb{R} \prec \mathbb{C}$? It will help to make the connection between $\mathbb{R}$ and $\mathbb{C}$ more precise: $\mathbb{C} = \{a + bi : a, b \in \mathbb{R}\}$. So there is an obvious bijection between $\mathbb{R} \times \mathbb{R}$ and $\mathbb{C}$.

$$f\colon \mathbb{R} \times \mathbb{R} \longrightarrow \mathbb{C}$$
$$(a, b) \longmapsto a + bi$$

So the problem is equivalent to comparing $\mathbb{R}$ and $\mathbb{R} \times \mathbb{R}$, i.e. the real line and the Euclidean plane. Surely there can be no bijection between them! But there is! We shall show this by two methods, both of which have their attractions.

This result was one of those regarded as particularly dramatic by Cantor's peers. As size, the 'obvious' candidate, fails to distinguish the line from the plane, mathematicians had to look for some other difference. This was found in part in the development of topology, in which the two sets are not homeomorphic.

Theorem 6.11

$\mathbb{R} \approx \mathbb{R} \times \mathbb{R}$

Proof

The first method is by constructions of appropriate one–one functions and use of the Schröder–Bernstein theorem.

Define a function $f\colon (0,1)\times(0,1)\longrightarrow(0,1)$ as follows. Given any two reals a and b in $(0,1)$, we can express both by their decimal expansions:

This argument is a modification of Cantor's first attempt to construct a bijection $f\colon (0,1)\times(0,1)\longrightarrow(0,1)$. See Exercise 6.60.

$$a = 0.a_1a_2a_3\ldots a_i\ldots$$
$$b = 0.b_1b_2b_3\ldots b_i\ldots$$

where the a_is and b_is are digits in $\{0,1,\ldots,9\}$, and both expansions avoid the use of recurring 9s, using recurring 0s instead where necessary. Then define $f(a,b)$ to be the real number with decimal expansion

For instance, if $a=\frac{1}{3}=0.3333\ldots$ and $b=\frac{1}{4}=0.2500\ldots$, then $f(a,b)=0.32353030\ldots$.

$$0.a_1b_1a_2b_2a_3b_3\ldots a_ib_i\ldots,$$

interleaving the digits of the expansions of a and b. It is easy to check that f is one–one, so that $(0,1)\times(0,1)\preceq(0,1)$. By Exercise 6.51 we have $(0,1)\approx\mathbb{R}$, so we can deduce that $\mathbb{R}\times\mathbb{R}\preceq\mathbb{R}$.

It is very straightforward to show that $\mathbb{R}\preceq\mathbb{R}\times\mathbb{R}$, for instance by using the function $g\colon\mathbb{R}\longrightarrow\mathbb{R}\times\mathbb{R}$ defined by $g(r)=(r,0)$, which is clearly one–one. As $\mathbb{R}\times\mathbb{R}\preceq\mathbb{R}$, it follows from the Schröder–Bernstein theorem that $\mathbb{R}\approx\mathbb{R}\times\mathbb{R}$.

The second method is to make cunning use of some of the basic facts about $\approx$ from earlier sections, as follows:

$$\begin{aligned}
\mathbb{R}\times\mathbb{R} &\approx \mathbf{2}^{\mathbb{N}}\times\mathbf{2}^{\mathbb{N}} && \text{as } \mathbb{R}\approx\mathbf{2}^{\mathbb{N}}\\
&\approx \mathbf{2}^{\mathbb{N}}\times\mathbf{2}^{\mathbb{N}\times\{0\}} && \text{as } \mathbb{N}\approx\mathbb{N}\times\{0\}\\
&\approx \mathbf{2}^{\mathbb{N}\cup(\mathbb{N}\times\{0\})} && \text{as } A^{B\cup C}\approx A^B\times A^C \text{ when } B\cap C=\varnothing\\
&\approx \mathbf{2}^{\mathbb{N}} && \text{by Theorem 6.3}\\
&\approx \mathbb{R}
\end{aligned}$$

■

The first method used above, giving explicit one–one functions, gives one the satisfaction of having got inside the structure of the sets involved. The second method has the attraction of looking like an algebra, where one exploits a number of fundamental results about $\mathbb{N}$ and $\mathbb{R}$, like $\mathbb{R}\approx\mathbf{2}^{\mathbb{N}}$ and the union of two countable sets is countable, along with some essentially dull facts about bijections. If this method does not seem to be getting anywhere, then see if you can adapt it to use the Schröder–Bernstein theorem, as in the following exercise.

Another crucial result is $\mathbb{N}\times\mathbb{N}\approx\mathbb{N}$.

Exercise 6.52

Show that $\mathbb{R}\approx\mathbb{N}^{\mathbb{N}}$ by two methods, both giving the result by using the Schröder–Bernstein theorem:

(a) construct one–one functions $f\colon\mathbb{R}\longrightarrow\mathbb{N}^{\mathbb{N}}$ and $g\colon\mathbb{N}^{\mathbb{N}}\longrightarrow\mathbb{R}$;

(b) use 'algebra' and suitable facts about $\preceq$ to show that $\mathbb{R}\preceq\mathbb{N}^{\mathbb{N}}$ and $\mathbb{N}^{\mathbb{N}}\preceq\mathbb{R}$.

Use whichever representation of $\mathbb{R}$ you find most convenient.

Solution

(a) Represent real numbers by their decimal expansions, avoiding recurring 9s, with a minus sign at the front if they happen to be negative:

$$a=(-)A_NA_{N-1}\ldots A_0.a_1a_2\ldots a_n\ldots.$$

Then define $f\colon \mathbb{R} \longrightarrow \mathbb{N}^{\mathbb{N}}$ by $f(a) = a^*$, where for $a \geq 0$,

$$a^*(i) = \begin{cases} A_{N-i}, & \text{if } i \leq N, \\ 27, & \text{if } i = N+1 \text{ (so that 27 represents the decimal point),} \\ a_{i-(N+1)}, & \text{if } i > N+1, \end{cases}$$

so that a^* is the function

$$\begin{array}{ccccccccc} 0 & 1 & \ldots & N & N+1 & N+2 & \ldots & N+1+n \\ \downarrow & \downarrow & & \downarrow & \downarrow & \downarrow & & \downarrow \\ A_N & A_{N-1} & \ldots & A_0 & 27 & a_1 & \ldots & a_n \end{array}$$

For $a < 0$ adapt the definition of a^* by setting $a^*(0) = 93$ to represent the minus sign, and then shifting everything in the definition above along by 1.

The only significance of the mysterious 93 and 27 is, of course, that they are different from $0, 1, \ldots, 9$.

It should be clear that f is one–one.

Now define $g\colon \mathbb{N}^{\mathbb{N}} \longrightarrow \mathbb{R}$ as follows. Given $f \in \mathbb{N}^{\mathbb{N}}$ define a corresponding real number f^+ with decimal expansion consisting of 1s separated by $f(i)$ 0s, so that e.g. the identity function

$$\begin{aligned} i\colon \mathbb{N} &\longrightarrow \mathbb{N} \\ n &\longmapsto n \end{aligned}$$

maps to

one 0 as $i(1) = 1$ ↓

$0 \cdot 1010010001000010 0\ldots$

↑ no 0s as $i(0) = 0$ ↑ four 0s as $i(4) = 4$

Formally f^+ is the real number $0.a_1a_2a_3\ldots a_n\ldots$ where

$$a_n = \begin{cases} 1, & \text{if } n = k + \sum_{i=0}^{k-1} f(i) \quad \text{for some } k \geq 1, \\ 0, & \text{otherwise.} \end{cases}$$

You might like to check that this works!

Then define g by

$$\begin{aligned} g\colon \mathbb{N}^{\mathbb{N}} &\longrightarrow \mathbb{R} \\ f &\longmapsto f^+ \end{aligned}$$

You should check that g is one–one.

(b) First note that $\mathbb{R} \approx \mathbf{2}^{\mathbb{N}}$. Then as

$$\mathbf{2} \preceq \mathbb{N},$$

we can use the result of Exercise 6.16(c) to deduce that

$$\mathbf{2}^{\mathbb{N}} \preceq \mathbb{N}^{\mathbb{N}}.$$

If $A \preceq B$ then $A^C \preceq B^C$.

Thus $\mathbb{R} \preceq \mathbb{N}^{\mathbb{N}}$.

The above is rather unsurprising. It's the other way round which is rather spectacular! We have

$$\mathbb{N} \preceq \mathbf{2}^{\mathbb{N}} \quad \text{(by Cantor's theorem)}$$

so that by using again the result of Exercise 6.16(c),

$$\begin{aligned}\mathbb{N}^{\mathbb{N}} &\preceq (\mathbf{2}^{\mathbb{N}})^{\mathbb{N}} \\ &\approx \mathbf{2}^{\mathbb{N}\times\mathbb{N}} \quad \text{(by Exercise 6.13(f))} \\ &\approx \mathbf{2}^{\mathbb{N}} \quad (\text{as } \mathbb{N}\times\mathbb{N}\approx\mathbb{N}),\end{aligned}$$

$C^{A\times B} \approx (C^B)^A$.

so that

$$\mathbb{N}^{\mathbb{N}} \preceq \mathbf{2}^{\mathbb{N}}.$$

You can now use one or other of these methods to resolve the last issue raised by Exercise 6.48. We had shown that $\mathbb{R}^{\mathbb{R}} \not\approx \mathbb{R}$ by noting that $\mathbf{2}^{\mathbb{R}} \preceq \mathbb{R}^{\mathbb{R}}$. Is $\mathbf{2}^{\mathbb{R}} \prec \mathbb{R}^{\mathbb{R}}$? If you now suspect that the answer is 'No', then you are right!

Exercise 6.53

Show that $\mathbf{2}^{\mathbb{R}} \approx \mathbb{N}^{\mathbb{R}} \approx \mathbb{R}^{\mathbb{R}}$.

By now we have made quite an inroad to cardinal arithmetic. We have lots of results comparing the sizes of sets under the relations $\approx$ and $\preceq$. It is very tempting to regard such results about, respectively, the disjoint union of two sets, their Cartesian product, and the set of functions from one to the other as being about addition, multiplication and exponentiation of their sizes. But of course we have rather dodged saying what the size of a set actually is. Without using AC the best we can really do within ZF is to say that the set representing the size of a set X is X itself. But with AC we shall be able to define Card(X) to be a special sort of set which will enable us to define an arithmetic on the cardinals of all sets, with a linear order – indeed a well-order – obeying familiar number-like rules such as

$$\text{if } \kappa < \lambda \text{ then } \kappa^{\mu} < \lambda^{\mu}.$$

And within this arithmetic, we shall become accustomed to some of its differences from the arithmetic of $\mathbb{N}$, once we involve infinite sets. For instance any 'multiplication' of two infinite sets we have done so far has not produced a set of bigger size than both the ones we started with: e.g. $\mathbb{N}\times\mathbb{N}\approx\mathbb{N}$ and $\mathbb{R}\times\mathbb{R}\approx\mathbb{R}$.

Exercise 6.54

(a) Show that $\mathbf{2}^{\mathbf{2}^{\mathbb{R}}} \times \mathbf{2}^{\mathbf{2}^{\mathbb{R}}} \approx \mathbf{2}^{\mathbf{2}^{\mathbb{R}}}$.

(b) Show that $\mathbb{N}\times\mathbb{R}\approx\mathbb{R}$.

Results like these suggest that in the arithmetic of infinite cardinals we should have

$$\kappa\cdot\lambda = \max\{\kappa,\lambda\},$$

where max means the greatest relative to the order $\preceq$; but without AC we cannot show that there is a 'greatest' of two arbitrary sets. (We have only been successful so far because we have been dealing with sets possessing a rich enough structure, like $\mathbb{N}$ and $\mathbb{R}$.)

In the next chapter we shall investigate the other cornerstone of Cantor's theory, the theory of ordinals. With these and AC, we can then really describe Cantor's cardinal arithmetic.

Further exercises

Several of the following exercises ask you to locate the sizes of sets of more or less mathematical significance. Some require a piece of mathematical knowledge from outside the confines of set theory. For many you will need to use the Schröder–Bernstein theorem to prove your result.

Exercise 6.55

Suppose that X is a set for which $X \times X \approx X$.

(a) Must X be infinite?

(b) Suppose that X is infinite. Show that:

(i) $\mathbf{2}^X \approx X^X$;

(ii) $(X \times \{0\}) \cup (X \times \{1\}) \approx X$.

Exercise 6.56

What are the sizes of the following sets? (That is, are they equinumerous with $\mathbb{N}$, $\mathbf{2}^{\mathbb{N}}$, $\mathbf{2}^{\mathbf{2}^{\mathbb{N}}}$ or some other well-known set?)

(a) The set of all infinite subsets of $\mathbb{Q}$.

(b) The set of all finite subsets of $\mathbb{R}$.

(c) The set of all countably infinite subsets of $\mathbb{C}$.

Exercise 6.57

What are the sizes of the following sets? (That is, are they equinumerous with $\mathbb{N}$, $\mathbf{2}^{\mathbb{N}}$, $\mathbf{2}^{\mathbf{2}^{\mathbb{N}}}$ or some other well-known set?)

(a) The set of all continuous functions $f\colon \mathbb{R} \longrightarrow \mathbb{R}$.

(b) The set of all open subsets of $\mathbb{R}$ with the usual topology.

(c) The set of all convergent series of real numbers $\sum_{n=0}^{\infty} a_n$.

Exercise 6.58

A function $f\colon \mathbb{N} \longrightarrow \mathbb{N}$ (or $f\colon \mathbb{R} \longrightarrow \mathbb{R}$) is said to be *increasing* when for all $x, y \in \mathbb{N}$ (or respectively $x, y \in \mathbb{R}$),

Such a function is sometimes called *weakly increasing* because of the requirement that $f(x) \leq f(y)$ rather than $f(x) < f(y)$.

if $x \leq y$ then $f(x) \leq f(y)$.

Similarly f is said to be *decreasing* when for all x, y,

if $x \leq y$ then $f(x) \geq f(y)$.

What are the sizes of the following sets? (That is, are they equinumerous with $\mathbb{N}$, $\mathbf{2}^{\mathbb{N}}$, $\mathbf{2}^{\mathbf{2}^{\mathbb{N}}}$ or some other well-known set?)

(a) The set of all increasing functions $f\colon \mathbb{N} \longrightarrow \mathbb{N}$.

(b) The set of all decreasing functions $f\colon \mathbb{N} \longrightarrow \mathbb{N}$.

(c) The set of all increasing functions $f\colon \mathbb{R} \longrightarrow \mathbb{R}$.

(d) The set of all decreasing functions $f\colon \mathbb{R} \longrightarrow \mathbb{R}$.

Exercise 6.59

(a) Show that any two points in the subset $\mathbb{R}^2 \setminus \mathbb{Q}^2$ of the plane can be joined by a circular arc which lies entirely in this subset.

This is a throwaway remark of Cantor's!

(b) How many such arcs are there? (That is, is the set of all such arcs joining the two points equinumerous with a well-known set like $\mathbb{N}$ or $\mathbb{R}$?)

Exercise 6.60

In our proof of Theorem 6.11 we exploited a function $f\colon (0,1) \times (0,1) \longrightarrow (0,1)$ defined as follows. Given any two reals a and b in $(0,1)$, express both by their decimal expansions:

Cantor's first attempt to show that $\mathbb{R} \times \mathbb{R} \approx \mathbb{R}$ was essentially by constructing this function f and claiming that it was a bijection. Dedekind, to whom he sent the argument, spotted the error.

$$a = 0.a_1a_2a_3 \ldots a_i \ldots$$
$$b = 0.b_1b_2b_3 \ldots b_i \ldots$$

where the a_is and b_is are digits in $\{0,1,\ldots,9\}$, and both expansions avoid the use of recurring 9s, using recurring 0s instead where necessary. Then define $f(a,b)$ to be the real number with decimal expansion obtained by interleaving the digits of the expansions of a and b:

$$0.a_1b_1a_2b_2a_3b_3 \ldots a_ib_i \ldots.$$

Explain why f is not a bijection. [Hint: there is a problem with recurring 9s.]

Exercise 6.61

(a) Show that $\mathbb{R}^m \approx \mathbb{R}^n$ for all m, n in $\mathbb{N}$.

(b) Cantor's result that R and $\mathbb{R}^2$ are the same 'size' closed one promising avenue for explaining why $\mathbb{R}$ and $\mathbb{R}^2$ are 'different'. However, no bijection between the sets is continuous. Thus the sets are not homeomorphic, so that within point-set topology one can establish that they are 'different'. (Cantor's work was a foundation stone for point-set topology.) If you have done some topology (and in particular have met the concepts of continuity and connectedness) you should be able to do the following exercise.

A *homeomorphism* is a continuous bijection with a continuous inverse.

Show that there is no continuous bijection $f\colon \mathbb{R}^2 \longrightarrow \mathbb{R}$, where $\mathbb{R}$ and $\mathbb{R}^2$ have their usual topologies.

Exercise 6.62

The Cantor set C is defined as the set of all real numbers of the form

$$\sum_{n=1}^{\infty} a_n 3^{-n}$$

where each a_n takes one or other of the values 0 or 2.

(a) Show that C is uncountable.

(b) C is a *null* set within the context of the theory of Lebesgue measure, or, equivalently, a set of measure 0. (See Section 5.3 for a brief description of the Lebesgue measure on $\mathbb{R}$.) Let $\mathscr{M}$ be the set of all Lebesgue measurable subsets of $\mathbb{R}$. What is the size of $\mathscr{M}$?

Exercise 6.63

Cantor's first published proof (in 1874) that $\mathbb{R}$ is uncountable used a different method from his later (1891) diagonal construction. This exercise attempts to take you through this earlier argument. Suppose first that $\mathbb{R}$ is countable, so that all the reals can be listed in a sequence $x_0, x_1, x_2, x_3, \ldots$, without repetitions. Sequences $\langle a_n \rangle$ and $\langle b_n \rangle$ of real numbers are defined as follows.

Find the first two numbers in the list of all the reals which lie in the open interval $(0, 1)$. If there aren't two such numbers, it is easy to construct a real number not in the list, giving a contradiction. Otherwise call these two numbers a_0 and b_0, taking $a_0 < b_0$. Now look for the first two numbers in the list in the open interval (a_0, b_0), calling them a_1 and b_1, taking $a_1 < b_1$. (Again if there aren't two such numbers, it is easy to obtain a contradiction.) Carry on in this way, so that a_n and b_n, with $a_n < b_n$, are the first two numbers in the list in the open interval (a_{n-1}, b_{n-1}), where

$$0 < a_0 < a_1 < a_2 < \ldots < a_{n-1} < b_{n-1} < \ldots < b_2 < b_1 < b_0 < 1,$$

assuming that one doesn't run out of numbers in the list at some finite stage of the process (in which case we can construct a real number not in the original list).

(a) Show that if, at the stage of looking for the first two numbers in the list in the open interval (a_{n-1}, b_{n-1}), one fails to find two numbers, then there is a real number in this interval not in the list, hence contradicting that the list contained all the real numbers.

(b) Suppose now that a_n, b_n have been successfully constructed for all $n \in \mathbb{N}$ in the above manner. The sequence $\langle a_n \rangle$ is strictly increasing and bounded above by any of the b_ns, so has a limit α. Likewise, the sequence $\langle b_n \rangle$ is strictly decreasing and bounded below (by any of the a_ns), so has a limit β. It is straightforward real analysis to show that $\alpha \leq \beta$.

(i) Show that $x_n \notin (a_n, b_n)$ for each $n \in \mathbb{N}$.

(ii) Show that $\alpha \neq a_n$ for any $n \in \mathbb{N}$, thus contradicting the assumption that the x_ns listed all real numbers.

Exercise 6.64

A complex number is said to be *transcendental* if it is not algebraic (as defined in Exercise 6.41). Show that the set of transcendental numbers is equinumerous with $\mathbb{R}$. (With any luck, you have essentially replicated Cantor's argument.)

Liouville had shown in 1844 that transcendental numbers exist (see the next exercise), using a bit of calculus. The number e was shown to be transcendental by Hermite in 1873, using quite a lot more calculus. The transcendence of π had to wait for Lindemann in 1882. So you can imagine the stir caused by Cantor's argument (in 1874), which could be interpreted as saying that almost all real numbers are transcendental, without constructing a single one of them!

Exercise 6.65

The purpose of this exercise is to illustrate Liouville's demonstration that the number L defined by $L = \sum_{n=0}^{\infty} \frac{1}{10^{n!}}$ is transcendental.

This exercise has nothing to do with set theory. It uses instead various standard results of real analysis.

Suppose that α is an irrational algebraic number with $f(x)$ a polynomial with integer coefficients of minimal degree d such that $f(\alpha) = 0$. ($f(x)$ can be obtained from the minimum polynomial $m(x)$ of α over $\mathbb{Q}$ by multiplying $m(x)$ by the least common multiple of the denominators of its coefficients.)

(a) Let $\frac{p}{q}$ be any rational number, where $p, q \in \mathbb{Z}$.

(i) Show that $\left|f\left(\frac{p}{q}\right)\right| \geq \frac{1}{q^d}$.

(ii) Suppose that $\left|\frac{p}{q} - \alpha\right| < 1$. Use Taylor's theorem to show that

$$\left|f\left(\frac{p}{q}\right)\right| \leq \left|\frac{p}{q} - \alpha\right| \cdot \left(\sum_{i=1}^{d} \frac{1}{i!}\left|f^{(i)}(\alpha)\right|\right),$$

where $f^{(i)}$ is the ith derivative of f.

(iii) Show that $f'(\alpha) \neq 0$ and deduce that $k = \sum_{i=1}^{d} \frac{1}{i!}\left|f^{(i)}(\alpha)\right| > 0$. [Hint: recall that f is a polynomial of minimal degree such that $f(\alpha) = 0$.]

(iv) Deduce that if $\left|\frac{p}{q} - \alpha\right| < 1$, then

$$\left|\frac{p}{q} - \alpha\right| < \frac{1}{kq^d}$$

has no solutions in p, q.

(b) Let L be defined by $L = \sum_{n=0}^{\infty} \frac{1}{10^{n!}}$.

(i) Explain why L is not rational.

(ii) Show that $\left|L - \sum_{n=0}^{N} \frac{1}{10^{n!}}\right| < \frac{2}{(10^{N!})^{N+1}}$.

(iii) Use the preceding results to deduce that L is transcendental.

7 ORDERED SETS

7.1 Introduction

We have made considerable progress in investigating one sort of number, namely cardinals. In this chapter we shall lay the ground for the second of Cantor's numbers, the ordinals, by looking in greater detail at ordered sets – an ordinal will be a special sort of ordered set. Much of the theory of ordered sets, like so much of what this book covers, was developed by Cantor himself.

But note that we have not yet said what the cardinal numbers are!

The usual orders on $\mathbb{N}$, $\mathbb{Z}$, $\mathbb{Q}$ and $\mathbb{R}$ are a familiar and important part of everyday mathematics. As $\mathbb{N} \subseteq \mathbb{Z} \subseteq \mathbb{Q} \subseteq \mathbb{R}$, the orders are closely related: for instance, $-2 < 1$ is true whether one regards -2 and 1 as members of $\mathbb{Z}$ or of $\mathbb{Q}$ or of $\mathbb{R}$. But to what extent are the orders on these sets the same?

Exercise 7.1

Find properties which distinguish between the orders on $\mathbb{N}$, $\mathbb{Z}$, $\mathbb{Q}$ and $\mathbb{R}$, i.e. find some property expressible in terms of the order (rather than, say, the arithmetic or size) on each set which does not hold for some of the others.

Solution

There are plenty of ways of distinguishing between these orders, and one purpose of this chapter is to investigate properties of order such as those below. For instance:

$\mathbb{N}$ has a minimum element under its order: the other sets do not have one;

for each element of $\mathbb{Z}$ there is a next biggest element in the order: this is also true for $\mathbb{N}$, but not for $\mathbb{Q}$ or $\mathbb{R}$;

between any two distinct elements of $\mathbb{Q}$ there is another element of $\mathbb{Q}$; this is also true for $\mathbb{R}$, but not for $\mathbb{N}$ or $\mathbb{Z}$;

every subset of $\mathbb{R}$ which is bounded above has a least upper bound in $\mathbb{R}$: this is not true for any of $\mathbb{N}$, $\mathbb{Z}$ or $\mathbb{Q}$.

All of the above properties are expressible in terms of the orders on the sets.

One of the key properties of the order on $\mathbb{N}$, not mentioned in the solution above, is that of *well-order*, and that will form the basis of the construction of the ordinals. So in this chapter we shall devote considerable attention to properties of well-ordered sets. We shall also investigate an arithmetic of ordered sets, i.e. ways in which we might add or multiply two ordered sets to get another ordered set. But first we shall remind ourselves of some of the basic axioms and language of the theory of order, which we do in the next section.

Well-order is intimately related to the principle of mathematical induction, which is a distinctive feature of $\mathbb{N}$.

7.2 Linearly ordered sets

First let us remind ourselves of the definitions of the most basic sorts of order.

> ***Definitions***
>
> Let X be a set and R a subset of $X \times X$. R is said to be a *relation* on X. We shall often write xRy for $(x, y) \in R$.
>
> R is a *weak partial order* on X, or X is *weakly partially ordered* by R, if it has the following properties:
>
> **reflexive** for all x in X, xRx;
>
> **anti-symmetric** for all x and y in X, if xRy and yRx then $x = y$;
>
> **transitive** for all x, y and z in X, if xRy and yRz then xRz.
>
> If a weak partial order R has the following additional property:
>
> **linear** for all x and y in X, xRy or yRx,
>
> it is a *weak linear order* on X, and X is *weakly linearly ordered* by R.

An n-ary relation on X is some subset of $\underbrace{X \times \ldots \times X}_{n \text{ times}}$. So R here is technically a 2-ary, or binary, relation.

Standard examples of such orders are:

(a) the usual $\leq$ on any of $\mathbb{N}$, $\mathbb{Z}$, $\mathbb{Q}$ and $\mathbb{R}$ – these are all linear;

(b) the usual $\geq$ on any of $\mathbb{N}$, $\mathbb{Z}$, $\mathbb{Q}$ and $\mathbb{R}$ – these are all linear;

(c) $\subseteq$ on $\mathscr{P}(\mathbb{N})$ – this is partial but not linear.

Technically, $\leq$ on $\mathbb{N}$ is the subset $\{(m, n) : m \leq n\}$ of $\mathbb{N} \times \mathbb{N}$.

Given an ordered set X, any subset of X is also ordered in a natural way. Formally, we have the following.

> ***Definition***
>
> Let R be a partial order on X, and let A be a subset of X. Then the *restriction* of R to A, written as $R|_{A \times A}$, is the set of pairs
>
> $$\{(a, b) : a, b \in A \text{ and } aRb\}.$$

Exercise 7.2

Let R be a partial (respectively linear) order on X, and let A be a subset of X. Show that $R|_{A \times A}$ is a partial (respectively linear) order on A.

So given an ordered set X and $A \subseteq X$, we shall treat A as an ordered set, where it is understood that the order on A is the restriction to A of the order on X. The point of this is that it gives us a quick way of describing a rich class of examples of ordered sets, namely the subsets of one of the standard ordered sets, like $\mathbb{N}$ or $\mathbb{Q}$.

The usual linear order on $\mathbb{N}$ can be represented both by the subset of $\mathbb{N} \times \mathbb{N}$ corresponding to $\leq$ and by the subset corresponding to $\geq$, which are clearly different as subsets.

For example, $(2, 1) \in \geq$, but $(2, 1) \notin \leq$.

Exercise 7.3

Let R be a weak partial order on X. Show that R^{-1}, defined by

$$(x, y) \in R^{-1} \text{ if and only if } (y, x) \in R,$$

is also a partial order on X.

So where we deal with a set with a familiar order, like $\mathbb{N}$, the result of Exercise 7.3 tells us that we always have a choice of two ways of regarding the order: we shall opt for the one we usually regard as '$\leq$'. And if we have a less familiar (weak) order, we shall still discuss it as though it were $\leq$ rather than $\geq$. For instance, if R is a weak linear order on X and $z \in X$ has the property that xRz for all $x \in X$, we describe z as the *maximum* element in the ordered set.

Our convention from now on is to read xRy as 'x is less than or equal to y'.

Associated with every weak order R on X is a corresponding *strict* order S, where xSy means xRy but $x \neq y$. Axioms for strict order are as follows.

For example, the usual $\leq$ on $\mathbb{N}$ is associated with the usual $<$.

Definitions

Let X be a set and S a subset of $X \times X$. S is a *strict partial order* on X, or X is *strictly partially ordered* by S, if it has the following properties:

irreflexive for all x in X, it is not the case that xSx;

transitive for all x, y and z in X, if xSy and ySz then xSz.

If a strict partial order S has the following additional property:

linear for all x and y in X, xSy or $x = y$ or ySx,

it is a *strict linear order* on X, and X is *strictly linearly ordered* by S.

This is the 'strict' version of the linear property.

Exercise 7.4

(a) Suppose that R is a weak partial order on X, and that the relation S is defined by

$$xSy \text{ if and only if } xRy \text{ and } x \neq y.$$

Show that S is a strict partial order on X, and that if R is, in addition, linear, then so is S.

(b) Suppose that S is a strict partial order on X, and that the relation R is defined by

$$xRy \text{ if and only if } xSy \text{ or } x = y.$$

Show that R is a weak partial order on X, and that if S is, in addition, linear, then so is R.

So each weak order on X is associated with a strict order on X, and vice versa. This means that when we talk about a partially or linearly ordered set X, we shall exploit either the weak or strict form of the order, depending on which is the more convenient for the discussion. And following our earlier convention, we shall treat a strict order S on X as though it is 'less than', rather than 'greater than'.

For the rest of this chapter we shall deal only with linearly ordered sets. We shall often write the order relation on X as $<_X$ or $\leq_X$, depending on whether the order is strict or weak. We shall also use the phrase 'ordered set X' to mean the set X with a particular order on it.

We shall need a way of saying when two ordered sets are essentially 'the same' as ordered sets. As an order $<_X$ on a set X is completely determined by the pairs in $<_X$, we shall say that the orders $<_X$ on X and $<_Y$ on Y are the same if there is a bijection between X and Y which matches the pairs in $<_X$ with the pairs in $<_Y$. Formally, this is as follows.

Definition

Let $<_X$ and $<_Y$ be strict partial orders on X and Y, respectively. Then X and Y are *order-isomorphic*, written $X \cong Y$, if there is a bijection $f: X \longrightarrow Y$ such that

$$\text{for all } x_1, x_2 \in X,\ x_1 <_X x_2 \text{ if and only if } f(x_1) <_Y f(x_2).$$

The map f is said to be an *order-isomorphism.*

The definition in terms of the corresponding weak orders $\leq_X$ and $\leq_Y$ looks essentially the same. See Exercise 7.16.

Exercise 7.5

Show that $\mathbb{N}$ is order-isomorphic to $2\mathbb{N}$ (i.e. the set of all even natural numbers), where each set has the usual order.

Solution

Define f by

$$\begin{aligned} f: \mathbb{N} &\longrightarrow 2\mathbb{N} \\ n &\longmapsto 2n. \end{aligned}$$

It is easy to show that f is a bijection and that, for all $m, n \in \mathbb{N}$,

$$m < n \text{ if and only if } f(m) < f(n),$$

so that f is an order-isomorphism.

You could equivalently show that, for this bijection f, $m \leq n$ if and only if $f(m) \leq f(n)$, for all $m, n \in \mathbb{N}$.

To be sure that we have captured at least some of the essence of being 'the same' as ordered sets with this definition, it is customary to check that being order-isomorphic is an equivalence relation. This is a routine exercise.

Exercise 7.6

Prove that $\cong$ is an equivalence relation on partially ordered sets, i.e.

(a) $\cong$ is reflexive: for all X, $X \cong X$;

(b) $\cong$ is symmetric: for all X and Y, if $X \cong Y$ then $Y \cong X$;

(c) $\cong$ is transitive: for all X, Y and Z, if $X \cong Y$ and $Y \cong Z$, then $X \cong Z$.

Assume that X, Y and Z are strictly partially ordered by $<_X$, $<_Y$ and $<_Z$, respectively.

We shall also need the idea of X being order-isomorphic to a subset of Y. It will be more convenient to have a word for the map that shows this.

Definition

The function $f\colon X \longrightarrow Y$ is said to be an *order-embedding* (of X into Y) if f is one–one and, for all $x_1, x_2 \in X$,

$$x_1 <_X x_2 \text{ if and only if } f(x_1) <_Y f(x_2).$$

f is also said to be *order-preserving*.

Exercise 7.7

Give two different order-embeddings of $\mathbb{N}$ into $\mathbb{Z}$.

Solution

There are a large number of possible answers, for instance $f\colon \mathbb{N} \longrightarrow \mathbb{Z}$ defined by any of the following:

$$f(n) = n,$$
$$f(n) = n - 10,$$
$$f(n) = \begin{cases} 0, & \text{if } n = 0, \\ n+2, & \text{if } n > 0. \end{cases}$$

As the name suggests, we can picture the elements of a linearly ordered set X as strung out on a line.

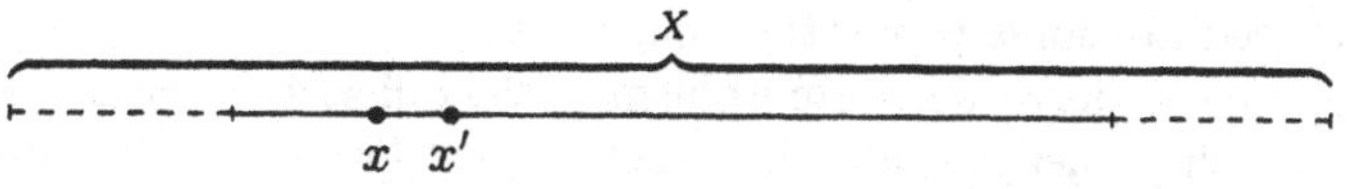

If $<_X$ is the order on X, and the elements x, x' of X are related by $x <_X x'$, we picture x to the *left* of x': so 'smaller' elements lie towards the left and 'larger' ones to the right. Of course, we must not read too much into the picture. The picture is very suggestive of the usual real line, but for instance there might not be any elements of X between x and x' in the picture above; and indeed there might be no order-embedding from X into $\mathbb{R}$ with its usual order. But a picture can help us understand an ordered set. For example, pictures could help us decide whether sets might, or might not, be order-isomorphic.

Can you think of a condition on a linearly ordered X which would guarantee that it could not be order-embedded into $\mathbb{R}$?

Exercise 7.8

Which, if any, pairs of the following three subsets of $\mathbb{Q}$, with the usual order, are order-isomorphic?

$$\mathbb{N}, \qquad A = \{1 - \tfrac{1}{n+1} : n \in \mathbb{N}\}, \qquad B = A \cup \{1\}.$$

Solution

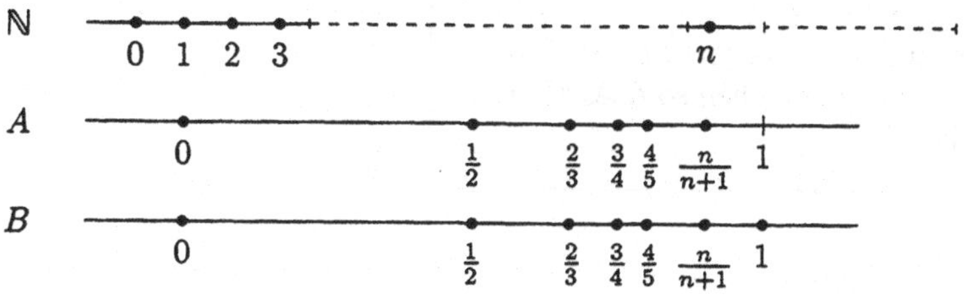

The pictures of the sets make it clear that any candidate for an order-isomorphism from $\mathbb{N}$ to A would have to send the least element of $\mathbb{N}$ to the least element of A (i.e. 0 to 0), the next least element of $\mathbb{N}$ to next least of A (i.e. 1 to $\frac{1}{2}$), and so on. The map f given by this process, i.e.

$$\begin{aligned} f\colon \mathbb{N} &\longrightarrow A \\ n &\longmapsto 1 - \tfrac{1}{n+1}, \end{aligned}$$

is in fact an order-isomorphism. (You can check that f is indeed a bijection matching the orders.) The picture helps by suggesting this process, but might initially mislead one into saying that $\mathbb{N}$ is unbounded as a subset of $\mathbb{R}$, while A is bounded, so that the sets cannot be order-isomorphic – this latter reasoning, referring to the set $\mathbb{R}$, is irrelevant: we must really confine our attention to the order properties of the sets themselves.

The picture also suggests that any candidate for an order-isomorphism from A to B must send 0 to 0, $\frac{1}{2}$ to $\frac{1}{2}, \ldots, \frac{n}{n+1}$ to $\frac{n}{n+1}, \ldots$. But this will exhaust the elements of A, leaving no element which can be mapped to 1, which behaves as a top element of B. (This process in fact defines an order-embedding of A into B.) So it looks as though B is not order-isomorphic to A, and hence not to $\mathbb{N}$ either (as $A \cong \mathbb{N}$).

The argument above suggesting that A is not order-isomorphic to B is, we hope, reasonably convincing. But it would be nice to have a sharper argument establishing that there is no order-isomorphism from A to B – perhaps the picture is misleading in some way. What we need are some properties which an ordered set might possess, that are preserved under order-isomorphism, so that if one set has a particular property and the other doesn't, then they *cannot* be order-isomorphic. Does having a top element, like 1 in B above, constitute such a property? Indeed it does.

We call such a property an *order-theoretic invariant*.

Definitions

Let X be weakly ordered by $\leq_X$ and suppose that $b \in X$. We say that b is the *maximum* element of X if $x \leq_X b$ for all $x \in X$.

Similarly, we say that $a \in X$ is the *minimum* element of X if $a \leq_X x$ for all $x \in X$.

Recall that we are now adopting the convention that all our ordered sets are linearly ordered. So you can show that if X has a maximum element, it is unique.

Thus the set B above has a minimum element, 0, and maximum element, 1, whereas the set A has minimum element, also 0, but no maximum element. We then have the following theorem.

Theorem 7.1

Let $f\colon X \longrightarrow Y$ be an order-isomorphism from X ordered by $\leq_X$ to Y ordered by $\leq_Y$. Then if X has a maximum element then so does Y; and if X has a minimum element then so does Y.

Proof

Suppose that X has a maximum element, b. Then for all $x \in X$ we have $x \leq_X b$. As f is an order-isomorphism, we must then have $f(x) \leq_Y f(b)$, for all $x \in X$. But f is a bijection, so that every element y of Y is of the form $f(x)$ for some $x \in X$. Thus $y \leq_Y f(b)$ for all $y \in Y$, showing that Y has a maximum element, namely $f(b)$.

The argument when X has a minimum element is similar. ■

Use of this theorem confirms that our previous sets A and B are indeed not order-isomorphic, as B has a maximum element while A does not have one.

Exercise 7.9

Show that $\mathbb{N}$ and $\mathbb{Z}$ are not order-isomorphic.

Let us now consider the following subsets of $\mathbb{Q}$ with the usual order:

$$A = \{1 - \tfrac{1}{n+1} : n \in \mathbb{N}\} \quad \text{(as earlier)},$$
$$D = \{1 - \tfrac{1}{n+1} : n \in \mathbb{N}\} \cup \{2 - \tfrac{1}{n+1} : n \in \mathbb{N}\}.$$

Exercise 7.10

Are A and D order-isomorphic? Justify your answer.

Solution

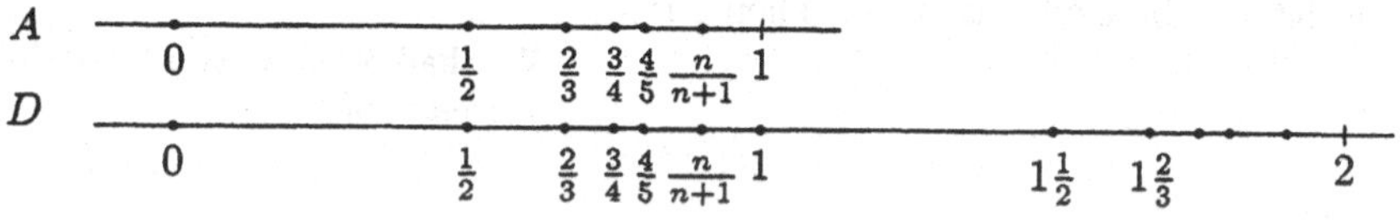

Arguing as in the solution to Exercise 7.8, we can see that any possible order-isomorphism from A to D would simply map the elements of A to themselves so that there would be no element of A left to map to 1, let alone any larger number in D.

This solution is somewhat inelegant, and there is a nicer way of showing that A and D are not isomorphic. The key is that the element 1 of D has a further property which is an order-theoretic invariant. For every element x of D with $x < 1$, there is some element y of D with $x < y$ and $y < 1$. The element 1 is said to be a *limit* point. In general, we have the following.

Definition

Let $<_X$ be a strict order on X. The element c of X is a *limit* point of the ordered set if

> there is some element x of X with $x <_X c$, and
>
> for every element $x \in X$ with $x <_X c$ there is some $y \in X$ with $x <_X y$ and $y <_X c$.

That is, c is not the minimum element of X

> **Theorem 7.2**
>
> Let $f: X \longrightarrow Y$ be an order-isomorphism from X ordered by $<_X$ to Y ordered by $<_Y$. Suppose that c is a limit point of X. Then $f(c)$ is a limit point of Y.

This gives another way of showing that our sets A and D are not isomorphic: D has a limit point, 1, but A has no limit point to correspond to 1 under an order-isomorphism.

Exercise 7.11

Prove Theorem 7.2.

Suppose that c is a point in the ordered set X which is not a minimum element of X. Can we say anything interesting about c if it isn't a limit point? Well, it is *not* the case that for every element $x \in X$ with $x <_X c$ there is *some* $y \in X$ with $x <_X y$ and $y <_X c$. This means that there is some x with $x <_X c$ such that there is no y strictly in between x and c. This leads to the following definition.

> ***Definitions***
>
> Let $<_X$ be a strict order on X and let x, c be elements of X. Then c is the *successor* of x if
>
> $x <_X c$ and
>
> for all y, if $x <_X y$ then $c = y$ or $c <_X y$.
>
> We also say that x is the *predecessor* of c.

We shall often write x^+ for the successor of x.

Thus c is the successor of x if c is the first element of X greater than x.

Exercise 7.12

Let $<_X$ strictly order X and suppose that the element c of X is neither a minimum element of X nor a successor. Show that c is a limit point.

Exercise 7.13

Identify the successors and limit points, if any, of the following ordered sets.

(a) $\mathbb{Q}$ with the usual $<$

(b) $\mathbb{Z}$ with the usual $<$

(c) $\{1 - \frac{1}{n+1} : n \in \mathbb{N}\} \cup \{2 - \frac{1}{n+1} : n \in \mathbb{N}\} \cup \{3\}$ with the usual $<$

Solution

(a) Every rational q is a limit point. For any such q, if $x \in \mathbb{Q}$ with $x < q$, then put $y = \frac{x+q}{2}$. This y is rational (as x and q both are). Also $x < y < q$ (y is in fact half way between x and q). Hence q is a limit point.

Of course, $\mathbb{Q}$ has no minimum element.

As all points in $\mathbb{Q}$ are limit points and a successor cannot be a limit point, $\mathbb{Q}$ has no successors.

(b) Every element of $\mathbb{Z}$ is a successor: n is the successor of $n-1$. So $\mathbb{Z}$ has no limit points.

(c) 0 is the minimum element of the order, so is neither a successor nor a limit point – to be a candidate for either of these, there would have to be some element smaller than 0.

Every element of the form $1 - \frac{1}{n+1}$ with $n > 0$ is a successor (of $1 - \frac{1}{n}$).

The element 1 (which equals $2 - \frac{1}{0+1}$) is a limit point and every element $2 - \frac{1}{n+1}$ with $n > 0$ is a successor.

Lastly, 3 is a limit point of the order. (Note that a limit point in the order-theoretic sense does not have to be a limit point of the set in the topological sense.)

Exercise 7.14

Let c be a limit point of the ordered set X, strictly ordered by $<_X$. Suppose that for each x with $x <_X c$, x has a successor x^+. Show that the set $\{x \in X : x <_X c\}$ is infinite. (Thus X is infinite.)

This set is called the *initial segment* of X determined by c.

Solution

Note first that if $x <_X c$, so that x has a successor x^+, then by definition of successor either $x^+ = c$ or $x^+ <_X c$. But c is a limit point. Thus $x^+ <_X c$.

c is a limit point, so there is a smaller element x_0, i.e. $x_0 <_X c$. Define a function f recursively by

$$\begin{aligned} f\colon \mathbb{N} &\longrightarrow \{x \in X : x <_X c\} \\ f(\mathbf{0}) &= x_0 \\ f(\mathbf{n}+\mathbf{1}) &= (f(\mathbf{n}))^+ \text{ for } \mathbf{n} \geq \mathbf{0}. \end{aligned}$$

We can show by induction that the function is well-defined. For the base step, $f(\mathbf{0}) = x_0$ and we are given that $x_0 <_X c$. For the inductive step, from $f(\mathbf{n}) <_X c$ we know that that $f(\mathbf{n})$ has a successor $(f(\mathbf{n}))^+$ and as c is a limit point, we must have $(f(\mathbf{n}))^+ <_X c$.

Likewise, we can show that f is an order-embedding: for instance, fix $\mathbf{m}$ and show by induction on $\mathbf{n} \geq \mathbf{1}$ that $f(\mathbf{m}) <_X f(\mathbf{m}+\mathbf{n})$. Thus f is one–one, so that $\{x \in X : x <_X c\}$ is infinite.

In general, a linearly ordered set need not have either a maximum or a minimum element. But examples of such sets (other than $\varnothing$) have to be infinite. In the next exercise we ask you to show that any finite linearly ordered set really looks like a natural number in ZF, ordered by $\in$. So words like maximum and minimum can be used without risk for such sets.

Recall from Chapter 3 that $<$ on $\mathbb{N}$ and on any natural number $\mathbf{n}$ is defined as $\in$.

Exercise 7.15

(a) Let $f\colon \mathbf{n} \longrightarrow X$ be a bijection from the natural number $\mathbf{n}$, in its guise as a set in ZF, to the set X linearly ordered by $<_X$. Show that there is an order-isomorphism $g\colon \mathbf{n} \longrightarrow X$.

(b) Why does a finite linearly ordered set X (non-empty!) have minimum and maximum elements?

Solution

(a) As ever, the basic technique is induction, i.e. showing that the set of $\mathbf{n}$ with this property is inductive.

For $\mathbf{n} = \mathbf{0}$, if $f: \mathbf{0} \longrightarrow X$ is a bijection, then X has to be the empty set, so that the result is vacuously true.

If you are uneasy with $\mathbf{n} = \mathbf{0}$, try the case $\mathbf{n} = \mathbf{1}$. Then $X = \{f(\mathbf{0})\}$, so that f is already an order-isomorphism.

Suppose that the result holds for $\mathbf{n} \geq \mathbf{0}$, i.e. for all linearly ordered sets X, if there is a bijection from this $\mathbf{n}$ to X, then $\mathbf{n}$ and X are order-isomorphic. Now take an ordered set A for which there is a bijection $f: \mathbf{n}^+ \longrightarrow A$. To exploit the inductive hypothesis for $\mathbf{n}$, restrict f to the subset $\mathbf{n}$ of $\mathbf{n}^+$. As f is a bijection, the range of $f|_\mathbf{n}$ is clearly $A \setminus \{f(\mathbf{n})\}$. Also $f|_\mathbf{n}$ is one–one, so that

Recall that $\mathbf{n}^+ = \mathbf{n} \cup \{\mathbf{n}\}$.

$$f|_\mathbf{n}: \mathbf{n} \longrightarrow A \setminus \{f(\mathbf{n})\}$$

is a bijection. $A \setminus \{f(\mathbf{n})\}$ is a subset of A, hence linearly ordered by $<_X$. Then by the inductive hypothesis there is an order-isomorphism

$$g: \mathbf{n} \longrightarrow A \setminus \{f(\mathbf{n})\}.$$

We shall exploit this g to construct an order-isomorphism

$$h: \mathbf{n}^+ \longrightarrow A.$$

The trick is to reinsert $f(\mathbf{n})$ into A and shift the images under g accordingly.

Define a subset S of $\mathbf{n}$ by

$$S = \{\mathbf{i} \in \mathbb{N} : g(\mathbf{i}) > f(\mathbf{n})\}.$$

If $S = \varnothing$ then $f(\mathbf{n})$ is in fact the maximum element of A and we can define h by

$$h(\mathbf{i}) = \begin{cases} f(\mathbf{n}), & \text{if } \mathbf{i} = \mathbf{n}, \\ g(\mathbf{i}), & \text{if } \mathbf{i} \in \mathbf{n}. \end{cases}$$

It is easy to check that h is indeed an order-isomorphism.

If $S \neq \varnothing$, then as $\mathbf{n}$ is well-ordered, S has a least element $\mathbf{k}$.

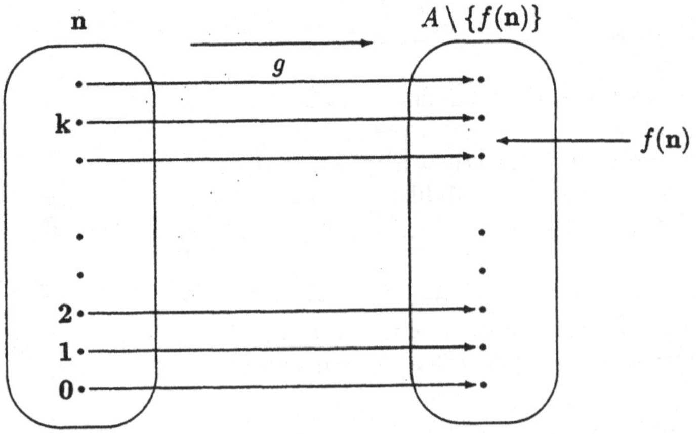

In this case we define h to agree with g for all $\mathbf{i}$ (if any) below $\mathbf{k}$, fit $f(\mathbf{n})$ in as $h(\mathbf{k})$ and make $h(\mathbf{k}+\mathbf{1})$ equal to $g(\mathbf{k})$, etc. Formally, define

$$h(\mathbf{i}) = \begin{cases} g(\mathbf{i}-\mathbf{1}), & \text{if } \mathbf{i} > \mathbf{k}, \\ f(\mathbf{n}), & \text{if } \mathbf{i} = \mathbf{k}, \\ g(\mathbf{i}), & \text{if } \mathbf{i} < \mathbf{k}. \end{cases}$$

Again it is straightforward to show that h is an order-isomorphism.

The result follows by induction.

(b) By virtue of X being finite, there is a bijection $f\colon \mathbf{n} \longrightarrow X$ for some $\mathbf{n} \in \mathbb{N}$. Thus there is an order-isomorphism $g\colon \mathbf{n} \longrightarrow X$. As X is non-empty, we must have $\mathbf{n} \neq \mathbf{0}$, so that $\mathbf{n}$ has a minimum element, $\mathbf{0}$, and a maximum element, $\mathbf{n}-\mathbf{1}$. Then X has minimum element $g(\mathbf{0})$ and maximum element $g(\mathbf{n}-\mathbf{1})$.

Thus a finite linearly ordered set is really just a dressed-up version of some natural number $\mathbf{n}$, ordered by $\in$. We shall primarily be looking at well-ordered sets like $\mathbf{n}$, where the identification and use of successors and limit points will be of great value. But it would be a shame to end this section without investigating countably infinite linearly ordered sets. We shall need to look at an interesting property of the standard order on $\mathbb{Q}$. In the solution to Exercise 7.13(a) we found that each element of $\mathbb{Q}$ is a limit point of the order. But this isn't the most interesting way of looking at the order on $\mathbb{Q}$. The property which is of much more significance is that between any two distinct rationals there is another rational. The corresponding definition is as follows.

Definition

The strict order $<_X$ on X is *dense* if for all $x, y \in X$,

if $x <_X y$, then there is some $z \in X$ with $x <_X z$ and $z <_X y$.

It is easy to show that density is an order-theoretic invariant. The density of the usual $<$ on $\mathbb{Q}$ is a key part of the following theorem, due to Cantor.

Theorem 7.3

Let $<_X$ be a strict linear order on a countable set X. Then there exists an order-embedding of X into $\mathbb{Q}$ (with its usual order).

Proof

We shall deal with the case when X is countably infinite. (The result for finite X will then drop out in an obvious way.) One key to the proof will be that $<$ is a dense order on $\mathbb{Q}$. The other is that the countability of X allows us to list its elements as

$$x_0, x_1, x_2, \ldots, x_n, \ldots$$

without repetitions, so that each element of X appears as x_n for exactly

Technically, if X is countably infinite, there is a bijection $g\colon \mathbb{N} \longrightarrow X$. Then x_n is just $g(n)$.

one $n \in \mathbb{N}$. This will permit us to define an order-embedding $f\colon X \longrightarrow \mathbb{Q}$ recursively on X, as follows.

To start the recursive definition, set $f(x_0)$ to be any rational that you like. For the sake of definiteness in the construction, fix some listing of $\mathbb{Q}$:

This is no problem, as $\mathbb{Q}$ is countable.

$$q_0, q_1, q_2, \ldots, q_n, \ldots$$

Then define $f(x_0)$ to be q_0.

For the recursive step, suppose that we have defined f on $\{x_0, x_1, x_2, \ldots, x_{n-1}\}$ to be order-preserving. How can we now define $f(x_n)$ so that f is order-preserving on $\{x_0, x_1, x_2, \ldots, x_{n-1}, x_n\}$? First note that $x_0, x_1, x_2, \ldots, x_{n-1}$ split the rest of X into $n+1$ disjoint intervals, $I_0, I_1, I_2, \ldots, I_n$ – a picture may help.

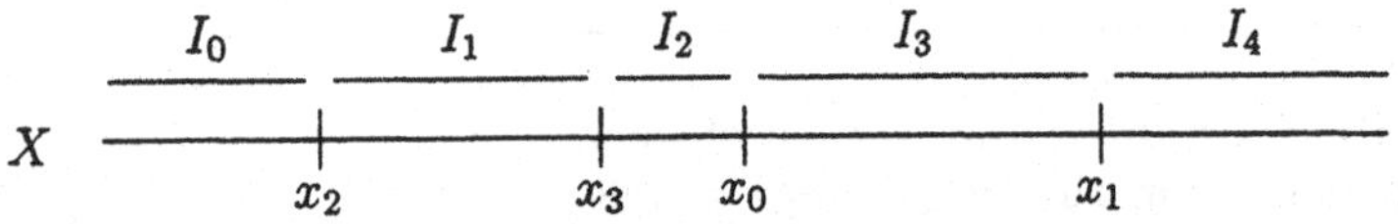

Some of the I_i might in fact be empty, but this won't matter. These I_n are disjoint and their union is $X \setminus \{x_0, x_1, x_2, \ldots, x_{n-1}\}$, so that x_n lies in a unique interval I_k.

The picture illustrates that the list $x_0, x_1, x_2, \ldots$ is usually not in ascending $<_X$ order.

To be somewhat more rigorous, put $X_n = \{x_0, x_1, x_2, \ldots, x_{n-1}\}$. Then the map

$$\begin{aligned} h\colon \mathbf{n} &\longrightarrow X_n \\ \mathbf{i} &\longmapsto x_i \end{aligned}$$

is a bijection from $\mathbf{n}$ to the linearly ordered subset X_n of X. Thus by Exercise 7.15 there is an order-isomorphism

$$\mathbf{k}\colon \mathbf{n} \longrightarrow X_n.$$

Define the subsets I_i, where $i = 0, 1, \ldots, n$, of $X \setminus X_n$ by

$$\begin{aligned} I_0 &= \{x \in X : x <_X k(0)\}, \\ I_i &= \{x \in X : k(i-1) <_X x \text{ and } x <_X k(i)\} \quad \text{for } 1 \leq i \leq n-1, \\ I_n &= \{x \in X : k(n-1) <_X x\}.) \end{aligned}$$

As $f|_{X_n}$ is order-preserving, the rationals $f(x_0), f(x_1), f(x_2), \ldots, f(x_{n-1})$ split the rest of $\mathbb{Q}$ into $n+1$ corresponding disjoint intervals $J_0, J_1, J_2, \ldots, J_n$. Then define $f(x_n)$ to be the first q_j in the listing of $\mathbb{Q}$ that is in the interval J_k, where I_k is the interval containing x_n – as $<$ is dense on $\mathbb{Q}$, and as $\mathbb{Q}$ has no maximum or minimum element, there *are* rationals in J_k, so that $f(x_n)$ is really defined!

Hence we can define f by recursion to give an order-embedding of X into $\mathbb{Q}$. ■

The upshot of this result and that of Exercise 7.15 is that the rather humdrum-looking set $\mathbb{Q}$ contains an example, up to order-isomorphism, of any possible countable linear order, just by finding the right subset. As you will see in the next section, finding such a subset is not always too easy, and we shall look at other ways of describing linear orders.

$\mathbb{Q}$ is said to be *universal* for linear orders.

Further exercises

Exercise 7.16

Let $<_X$ and $<_Y$ be strict partial orders on X and Y respectively, with corresponding weak orders $\leq_X$ and $\leq_Y$, and let $f: X \longrightarrow Y$ be a function. Show that f is an order-isomorphism if and only if

> f is a bijection and for all $x_1, x_2 \in X$, $x_1 \leq_X x_2$ if and only if $f(x_1) \leq_Y f(x_2)$.

This then gives a definition of *order-isomorphism* in terms of weak orders.

Exercise 7.17

Give examples of subsets of $\mathbb{Q}$ which, when ordered by the usual $<$, have the following properties.

(a) A dense order with maximum and minimum elements.

(b) An order with one limit point in which every other element is a successor.

(c) An order with infinitely many limit points, each of which has a successor.

Exercise 7.18

Let X, Y and Z be linearly ordered sets. Decide whether each of the following statements is true or false, and give a proof or counterexample as appropriate.

(a) If X is order-embeddable in Y and Y is order-embeddable in Z, then X is order-embeddable in Z.

(b) If X is order-isomorphic to Y and Y is order-embeddable in Z, then X is order-embeddable in Z.

(c) If X is order-embeddable in Y and Y is order-embeddable in X, then X and Y are order-isomorphic.

Exercise 7.19

Let X be a linearly ordered set and let A be a non-empty subset of X. An element x of X is said to be the *least upper bound* of A if both of the following are satisfied:

The terminology *supremum* is often used instead of least upper bound.

(1) x is an *upper bound* of A, i.e. $y \leq x$ for all $y \in A$;

(2) x is the *least* such upper bound of A, i.e. if z is an upper bound of A, then $x \leq z$.

A is said to be a *bounded* subset of X if it has an upper bound in X.

The notation lub A is sometimes used for the least upper bound of A.

(a) Show that if the non-empty subset A of X has a least upper bound, then lub A is a unique element of X.

(b) Let A be a non-empty subset of X. Show that $x = \text{lub}\, A$ if and only if both the following are satisfied:

   (1) x is an upper bound of A;

   That is, the same as (1) above.

   (2') for all z, if $z < x$ then z is not an upper bound of A, i.e. there is some $y \in A$ with $z < y$.

(c) Show that being a least upper bound is an order-theoretic invariant, as follows. Let $f: X \longrightarrow Y$ be an isomorphism between linearly ordered sets X and Y. Let A be a non-empty subset of X and x an element of X. Show that

$$x = \text{lub}\, A \text{ if and only if } f(x) = \text{lub}\, f(A).$$

Exercise 7.20

(a) Let a, b, c, d be rationals with $a < b$ and $c < d$. Show that there is an order-isomorphism $f\colon \mathbb{Q} \longrightarrow \mathbb{Q}$ such that $f(a) = c$ and $f(b) = d$.

(b) Let a_i, c_i, for $i = 1, 2, \ldots, n$, be rationals with $a_1 < a_2 < a_3 < \ldots < a_n$ and $c_1 < c_2 < c_3 < \ldots < c_n$. Show that there is an order-isomorphism $f\colon \mathbb{Q} \longrightarrow \mathbb{Q}$ such that $f(a_i) = c_i$ for all $i = 1, 2, \ldots, n$.

Exercise 7.21

Is there an order-isomorphism $f\colon \mathbb{Q} \longrightarrow \mathbb{Q}$ such that $f(1 - \frac{1}{n+1}) = 1 - \frac{1}{n+1}$ for all $n \in \mathbb{N}$ and $f(1) = 2$?

Exercise 7.22

(a) Show that there are infinitely many order-isomorphisms $f\colon \mathbb{Q} \longrightarrow \mathbb{Q}$.

(b) With which of the following is the set of all such order-isomorphisms f equinumerous: $\mathbb{N}$, $2^{\mathbb{N}}$, $2^{2^{\mathbb{N}}}$?

Exercise 7.23

Suppose that X and Y are both countably infinite sets, both linearly ordered by dense orders (respectively $<_X$ and $<_Y$), and both with neither maximum nor minimum elements. Show that X and Y are order-isomorphic. [Hint: adapt the proof of Theorem 7.3. It may help you to know that the classic proof, by Cantor, uses what is called a 'back and forth' argument'!]

Exercise 7.24

Prove each of the following.

(a) If the non-empty set X is linearly ordered and has neither a maximum nor minimum element, then X is infinite.

(b) If the linearly ordered set X contains a limit point c, then X is infinite.

(c) If a linear order on X is dense and X has at least two elements, then X is infinite.

Exercise 7.25

Let X be a countably infinite set linearly ordered by $<_X$.

(a) As X is countably infinite, there is a bijection $f\colon \mathbb{N} \longrightarrow X$. Define a relation R_f on $\mathbb{N}$ by

$$mR_f n \text{ if and only if } f(m) <_X f(n), \text{ for all } m, n \in \mathbb{N}.$$

Show that R_f is a linear order on $\mathbb{N}$ and that $\mathbb{N}$ with this order is order-isomorphic to X.

(b) The relation R_f in part (a) is a subset of $\mathbb{N} \times \mathbb{N}$. Consider the set of all subsets of $\mathbb{N} \times \mathbb{N}$ which represent orders on $\mathbb{N}$ order-isomorphic to X linearly ordered by $<_X$. With which of the following is this set equinumerous: $\mathbb{N}$, $2^{\mathbb{N}}$, $2^{2^{\mathbb{N}}}$?

7.3 Order arithmetic

Our goal in the next chapter is the theory of ordinal numbers, along with their arithmetic. Ordinals are special sorts of ordered set, so we shall investigate an arithmetic of ordered sets, to which ordinal arithmetic corresponds.

First, let us look at a way of 'adding' together two ordered sets X and Y. One natural idea is to take all the elements of X in ascending $<_X$ order and then add on, as greater elements, all the elements of Y in ascending $<_Y$ order.

$$\ldots x < x' < \ldots < y < y' \ldots$$

So we could tentatively define the 'sum' of X and Y to be $X \cup Y$ with this order. There's just one technical hitch, which is what to do if X and Y happen to have elements in common. The solution is to create copies of X and Y which are disjoint and then define the sum as follows.

> ***Definition***
>
> Let X and Y be linearly ordered sets. Then define the *order sum* of X and Y, written as
>
> $$X + Y,$$
>
> to be the linearly ordered set
>
> $$(X \times \{\mathbf{0}\}) \cup (Y \times \{\mathbf{1}\})$$
>
> with order $<$ defined by
>
> $$(x, \mathbf{0}) < (x', \mathbf{0}) \text{ if } x, x' \in X \text{ and } x <_X x',$$
> $$(x, \mathbf{0}) < (y, \mathbf{1}) \text{ if } x \in X \text{ and } y \in Y,$$
> $$(y, \mathbf{1}) < (y', \mathbf{1}) \text{ if } y, y' \in Y \text{ and } y <_Y y'.$$

Clearly $X \times \{\mathbf{0}\}$ and $Y \times \{\mathbf{1}\}$ are disjoint.

Exercise 7.26

(a) Check that $X + Y$ with $<$ as defined above is a linearly ordered set.

(b) Show that there are order-embeddings of both X and Y into $X + Y$.

Let us investigate what happens when familiar ordered sets are added in this way. Do we always obtain 'new' ordered sets? By 'new' we mean sets not order-isomorphic to the sets we added together. For finite sets we intuitively do obtain new orders – indeed it is the mental picture for finite sets which motivates the definition. As typical finite linearly ordered sets we can take natural numbers within ZF, by Exercise 7.15. In fact, we would surely expect that, for $\mathbf{m}, \mathbf{n} \in \mathbb{N}$, the sum $\mathbf{m} + \mathbf{n}$ in the order sense of $\mathbf{m}$ and $\mathbf{n}$ should be order-isomorphic to the natural number $\mathbf{m} + \mathbf{n}$. As a temporary piece of notation, let us write the usual $+$ operation on natural numbers in ZF as $\oplus$, so that this expectation becomes the following theorem.

Theorem 7.4

Let $\mathbf{m}, \mathbf{n}$ be natural numbers within ZF, with their usual orders (i.e. $<$ is $\in$). Then their order-theoretic sum $\mathbf{m} + \mathbf{n}$ is order-isomorphic to their arithmetic sum $\mathbf{m} \oplus \mathbf{n}$.

Proof

As is often the case with statements about general natural numbers within ZF, we shall proceed by induction, on $\mathbf{n}$ for fixed $\mathbf{m}$.

For $\mathbf{n} = \mathbf{0}$, we have on the one hand that $\mathbf{m} + \mathbf{0}$ is the ordered set $(\mathbf{m} \times \{\mathbf{0}\}) \cup (\varnothing \times \{\mathbf{1}\})$, i.e. $\mathbf{m} \times \{\mathbf{0}\}$. On the other hand $\mathbf{m} \oplus \mathbf{0}$ equals $\mathbf{m}$, by the definition of $\oplus$. These sets are clearly order-isomorphic, so that $\mathbf{m} + \mathbf{0} \cong \mathbf{m} \oplus \mathbf{0}$.

Remember that $\mathbf{0}$ in ZF is the empty set $\varnothing$.

For the inductive step, suppose that for some $\mathbf{n} \geq \mathbf{0}$ we have $\mathbf{m} + \mathbf{n} \cong \mathbf{m} \oplus \mathbf{n}$: let $f\colon \mathbf{m} + \mathbf{n} \longrightarrow \mathbf{m} \oplus \mathbf{n}$ be a corresponding order-isomorphism. Then on the one hand

$$\begin{aligned} \mathbf{m} + \mathbf{n}^+ &= (\mathbf{m} \times \{\mathbf{0}\}) \cup (\mathbf{n}^+ \times \{\mathbf{1}\}) \quad \text{(appropriately ordered)} \\ &= (\mathbf{m} \times \{\mathbf{0}\}) \cup (\mathbf{n} \times \{\mathbf{1}\}) \cup (\{\mathbf{n}\} \times \{\mathbf{1}\}) \quad (\text{as } \mathbf{n}^+ = \mathbf{n} \cup \{\mathbf{n}\}), \end{aligned}$$

while on the other

$$\begin{aligned} \mathbf{m} \oplus \mathbf{n}^+ &= (\mathbf{m} \oplus \mathbf{n})^+ \\ &= (\mathbf{m} \oplus \mathbf{n}) \cup \{\mathbf{m} \oplus \mathbf{n}\}. \end{aligned}$$

Then it is clear that the function g defined in terms of the order-isomorphism f by

$$\begin{aligned} g\colon \mathbf{m} + \mathbf{n}^+ &\longrightarrow \mathbf{m} \oplus \mathbf{n}^+ \\ (\mathbf{i}, \mathbf{0}) &\longmapsto f((\mathbf{i}, \mathbf{0})) \quad \text{for } \mathbf{i} \in \mathbf{m} \\ (\mathbf{j}, \mathbf{1}) &\longmapsto f((\mathbf{j}, \mathbf{1})) \quad \text{for } \mathbf{j} \in \mathbf{n} \\ (\mathbf{n}, \mathbf{1}) &\longmapsto \mathbf{m} \oplus \mathbf{n} \end{aligned}$$

is an order-isomorphism from $\mathbf{m} + \mathbf{n}^+$ to $\mathbf{m} \oplus \mathbf{n}^+$. The result follows by induction. ∎

As $\oplus$ on $\mathbb{N}$ is commutative and associative, Theorem 7.4 allows us to conclude that order $+$ on finite linearly ordered sets is also commutative and associative, with respect to order-isomorphism, i.e.

$$\mathbf{m} + \mathbf{n} \cong \mathbf{n} + \mathbf{m} \quad \text{and} \quad \mathbf{l} + (\mathbf{m} + \mathbf{n}) \cong (\mathbf{l} + \mathbf{m}) + \mathbf{n},$$

where the $+$ here is the order-theoretic sum. For instance

$$\begin{aligned} \mathbf{m} + \mathbf{n} &\cong \mathbf{m} \oplus \mathbf{n} \quad \text{by Theorem 7.4} \\ &= \mathbf{n} \oplus \mathbf{m} \\ &\cong \mathbf{n} + \mathbf{m} \quad \text{by Theorem 7.4 again.} \end{aligned}$$

What happens when we add infinite sets? As ever, we would expect some surprises! For the exercises below, and the rest of the book, it will help to have a new piece of notation to describe a familiar and important ordered set.

> ***Definition***
>
> Define ω to be the set $\mathbb{N}$ of natural numbers with its usual order $<$ (given by $\in$ in ZF).

Why do we want this notation? Partly it is because we might want to define other orders on $\mathbb{N}$, so that it is convenient to reserve a symbol for the set with its usual order. Indeed one natural way of describing a countable ordered set is by reordering $\mathbb{N}$ to give an order-isomorphic set, as you will see later in this section. And partly it is because it is a standard notation!

This is often more informative than giving an order-isomorphic subset of $\mathbb{Q}$.

Exercise 7.27

Which, if any, of the ordered sets $\omega + \mathbf{2}$, $\mathbf{2} + \omega$ and ω are order-isomorphic?

Recall that in ZF the natural number $\mathbf{2}$ is the set $\{\mathbf{0}, \mathbf{1}\}$ with $<$ defined to be $\in$.

Solution

A sketch might help.

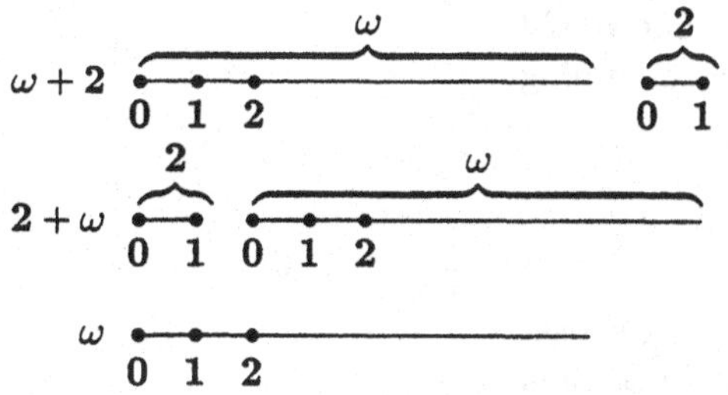

In our sketch above of $\omega + \mathbf{2}$ we should, strictly speaking, have labelled the members of the copy of ω as $(\mathbf{0}, \mathbf{0}), (\mathbf{1}, \mathbf{0}), (\mathbf{2}, \mathbf{0})$, etc., and the members of the copy of $\mathbf{2}$ as $(\mathbf{0}, \mathbf{1})$ and $(\mathbf{1}, \mathbf{1})$. But the sloppy notation conveys a good sense of what is going on, and we shall continue to use it!

Similarly, in the sketch of $\mathbf{2} + \omega$ the members of the copy of $\mathbf{2}$ should be $(\mathbf{0}, \mathbf{0})$ and $(\mathbf{1}, \mathbf{0})$, and those of ω should be $(\mathbf{0}, \mathbf{1}), (\mathbf{1}, \mathbf{1}), (\mathbf{2}, \mathbf{1}), \ldots$.

The sketch should convey that $\omega + \mathbf{2}$ has a limit point, namely the $\mathbf{0}$ in the copy of $\mathbf{2}$. (Technically the point $(\mathbf{0}, \mathbf{1})$ of $\omega + \mathbf{2}$ is a limit point.) Thus $\omega + \mathbf{2}$ cannot be order-isomorphic to ω.

It doesn't look from the sketch as though $\mathbf{2} + \omega$ has a limit point. In fact it not only doesn't have one, but is also order-isomorphic to ω, and the sketch makes it plainer how to define an order-isomorphism. Define f by

$$f\colon \omega \longrightarrow \mathbf{2} + \omega$$
$$\mathbf{i} \longmapsto \begin{cases} (\mathbf{i}, \mathbf{0}), & \text{if } \mathbf{i} = \mathbf{0} \text{ or } \mathbf{1}, \\ (\mathbf{i} - \mathbf{2}, \mathbf{1}), & \text{if } \mathbf{i} \geq \mathbf{2}. \end{cases}$$

This f is an order-isomorphism.

So adding two ordered sets sometimes gives a new order. But we are warned that addition of order is *not commutative* with respect to order-isomorphism, as $\omega + \mathbf{2} \not\cong \mathbf{2} + \omega$. Order addition does, however, have some familiar algebraic properties, for instance associativity, as we ask you to show in the next exercise.

Exercise 7.28

Show that addition of order is associative with respect to order-isomorphism, i.e. if X, Y and Z are linearly ordered sets, then $X + (Y + Z) \cong (X + Y) + Z$.

We have seen how a sketch of $X + Y$ can help us understand the order. But for countable sets there are (at least) two other helpful ways of representing the order. One way is as a subset of $\mathbb{Q}$, using the result of Theorem 7.3. And the other way is to take a well-known countable set, namely $\mathbb{N}$, and reorder it, exploiting its rich structure, so that it is order-isomorphic to the given order. For the latter it will help to know something about the order structure of subsets of $\mathbb{N}$ with the normal $<$.

Exercise 7.29

Let A be a subset of $\mathbb{N}$ ordered by the usual $<$. Show that A is order-isomorphic to either some $\mathbf{n} \in \mathbb{N}$ or to ω.

Solution

If A is finite the result follows from Exercise 7.15. If A is infinite, one would expect to be able to set up an order-isomorphism $f \colon \mathbb{N} \longrightarrow A$ by something like

$$f(\mathbf{0}) = \min(A),$$
$$f(\mathbf{n}) = \min(A \setminus \mathrm{Range}(f|_{\mathbf{n}})), \text{ for } \mathbf{n} > \mathbf{0}.$$

And indeed this works fine! In fact the details of the argument for any subset A, finite or infinite, are precisely as in the proof of Theorem 6.2 in Section 6.4.

You'll apply this in the next batch of exercises, but first have a think about conditions on X and Y which guarantee that $X + Y$ is countable.

Exercise 7.30

Suppose that X and Y are countable linearly ordered sets. Show that $X + Y$ is also a countable set.

Solution

$X \approx X \times \{\mathbf{0}\}$ and $Y \approx Y \times \{\mathbf{1}\}$. Also X and Y are both countable, and the union of two countable sets is countable, so that $X + Y$, which equals $(X \times \{\mathbf{0}\}) \cup (Y \times \{\mathbf{1}\})$, is countable.

Exercise 7.31

For each of the following ordered sets X,

(a) $X = \omega + \mathbf{2}$; (b) $X = \omega + \omega$,

(i) give an order-embedding of X into $\mathbb{Q}$ with the usual order;

(ii) define a new order $<'$ on the set $\mathbb{N}$ so that with this order $\mathbb{N}$ is order-isomorphic to X.

Solution

(a) There are many correct answers, for instance as follows.

(i) Define f by

$$\begin{aligned} f\colon \omega + \mathbf{2} &\longrightarrow \mathbb{Q} \\ (\mathbf{n}, \mathbf{0}) &\longmapsto 1 - \tfrac{1}{n+1} \\ (\mathbf{0}, \mathbf{1}) &\longmapsto 1 \\ (\mathbf{1}, \mathbf{1}) &\longmapsto 2. \end{aligned}$$

In this solution, as elsewhere, we use bold type for natural numbers, e.g. $\mathbf{n}$ and $\mathbf{0}$, rather than the ordinary n and 0, when it might be important to remember that we are dealing with our particular representation of natural numbers within ZF.

Then f is a suitable order-embedding with image set as below.

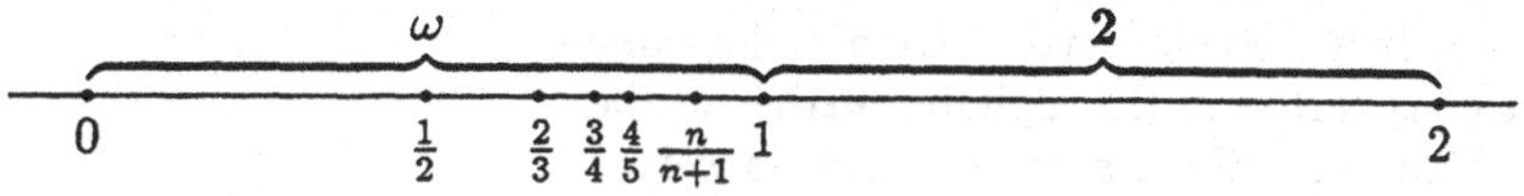

y

(ii) Define $<'$ so that

$$2 <' 3 <' 4 <' \ldots <' n <' \ldots <' 0 <' 1.$$

Clearly, or by Exercise 7.29, $\{2, 3, 4, \ldots\} \cong \omega$!

(b) (i) Define f by

$$\begin{aligned} f\colon \omega + \omega &\longrightarrow \mathbb{Q} \\ (\mathbf{n}, \mathbf{0}) &\longmapsto 1 - \tfrac{1}{n+1} \\ (\mathbf{n}, \mathbf{1}) &\longmapsto 2 - \tfrac{1}{n+1}. \end{aligned}$$

Then f is a suitable order-embedding with image set as below.

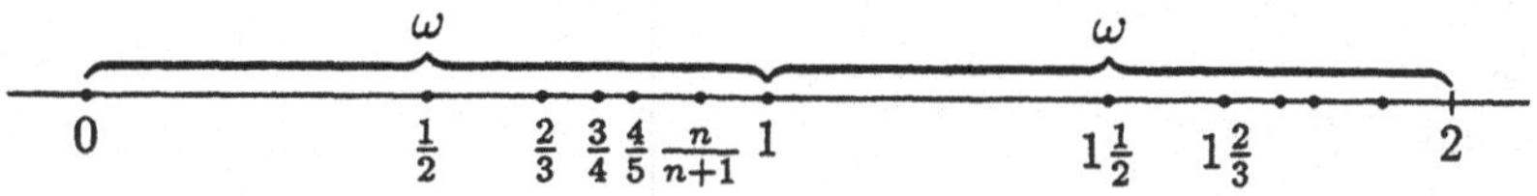

(ii) Define $<'$ so that

$$0 <' 2 <' 4 <' \ldots <' 2n <' \ldots <' 1 <' 3 <' 5 <' \ldots <' 2n+1 <' \ldots$$

Let us now define a multiplication of ordered sets. If we were to look at the natural numbers $\mathbf{n}$ and $\mathbf{m}$ as ordered sets in ZF, we might aim to define their product as ordered sets so that it was at least isomorphic, if not actually equal, to the natural number $\mathbf{m} \cdot \mathbf{n}$. As $\mathbf{m} \cdot \mathbf{n} \approx \mathbf{m} \times \mathbf{n}$, this would suggest defining the order product to be some sort of order on the Cartesian product $\mathbf{m} \times \mathbf{n}$. And this is what we shall do in general, for any linearly ordered sets X and Y. One clever sort of order on $X \times Y$ copies the way one orders words in an English dictionary.

Exercise 7.32

Describe an algorithm for deciding the order in which words are found in an English dictionary. The algorithm would have to sort a list like

feast   beast   bear   bird

into the correct order.

Solution

We are taught the alphabet of 26 letters in a particular order:

a, b, c, d, e, . . . , y, z.

To decide which of a pair of words, like 'feast' and 'beast', comes first in a dictionary, we look at the first letters of the words, i.e. their leftmost characters, and if they are not the same, as with this example, the word starting with the earlier letter in the alphabet comes earlier in the dictionary: so here 'beast' comes before 'feast'. But if the words start with the *same* letter, as with 'bear' and 'bird', we look at the second letters of each word, again working from left to right, and if the second letters are not the same we put the word whose second letter is earlier in the alphabet earlier in the dictionary: so 'bear' comes before 'bird'. And if the second letters, as well as the first, are the same, we look at the third letters; and if they are the same, we look at the fourth letters, and so on. So 'bear' comes before 'beast'.

If we confined ourselves to words with exactly 2 letters, we would have a special case of the following definition.

> ***Definition***
>
> Let X and Y be linearly ordered sets. Then the *lexicographic order* on $X \times Y$ is defined by
>
> $$(x, y) < (x', y') \text{ if (i) } x <_X x', \text{ or (ii) } x = x' \text{ and } y <_Y y'.$$

One meaning of 'lexicon' is 'dictionary'.

Before you get too excited by this entirely reasonable definition, let us define the product of two ordered sets by a different, but clearly very similar order, called the *anti*-lexicographic order. This differs from the dictionary order by reading words from *right to left* rather than from left to right.

Like in a Hebrew dictionary.

> ***Definition***
>
> Let X and Y be linearly ordered sets. Then the *product* of these orders is the set $X \times Y$ with the *anti-lexicographic* order defined by
>
> $$(x, y) < (x', y') \text{ if (i) } y <_Y y', \text{ or (ii) } y = y' \text{ and } x <_X x'.$$

There is a feeble reason for preferring the anti-lexicographic to the lexicographic order, which will become apparent only later, but they are clearly strongly related, as you will show in one of the subsequent exercises. The following exercises are designed to give some feel for the product order.

Exercise 7.33

Show that the anti-lexicographic order on $X \times Y$ is indeed a linear order.

Exercise 7.34

(a) Show that $X \times Y$ with the anti-lexicographic order is order-isomorphic to $Y \times X$ with the lexicographic order.

(b) Is it true in general that $X \times Y$ with the anti-lexicographic order is order-isomorphic to $X \times Y$ with the lexicographic order? Prove it, if true; give a counterexample, if false.

From now on, when we talk about the ordered set $X \times Y$ we shall assume that it has the anti-lexicographic order. It will also sometimes help to have a mental picture of $X \times Y$.

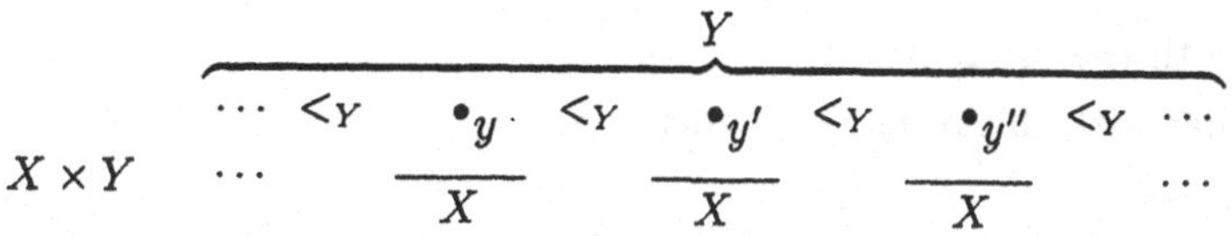

It is as though a copy of X is attached to each $y \in Y$, with the order increasing from left to right (as usual) through these copies of X.

Exercise 7.35

(a) Show that $\mathbf{2} \times \mathbf{3} \cong \mathbf{6}$, where we are treating natural numbers as sets in ZF.

(b) Show that $\mathbf{m} \times \mathbf{n} \cong \mathbf{m} \cdot \mathbf{n}$ for all $\mathbf{m}, \mathbf{n} \in \mathbb{N}$.

Exercise 7.36

Which, if any, of the ordered sets

$\omega \times \mathbf{2} \quad \omega + \omega \quad \mathbf{2} \times \omega$

are order-isomorphic?

Solution

A sketch of each set may help us.

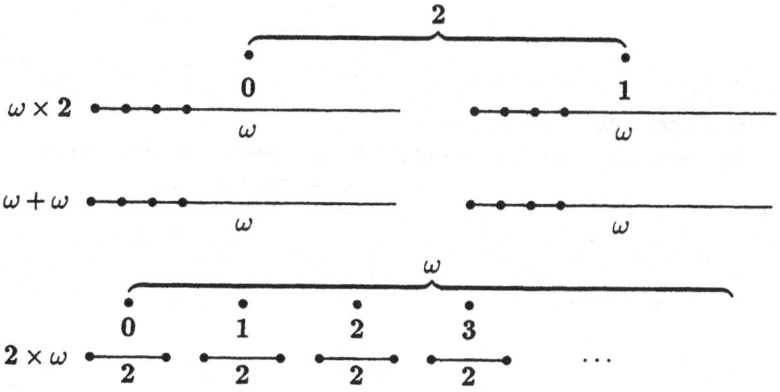

It is clear that $\omega \times \mathbf{2}$ and $\omega + \omega$ must be order-isomorphic. In fact, a look at how they are defined shows that they are actually the same sets with the same order!

In general $X \times \mathbf{2} = X + X$, but e.g. $X \times \mathbf{3} \neq X + X + X$. However, $X \times \mathbf{3} \cong X + X + X$.

However, $\mathbf{2} \times \omega$ looks different – for instance, it looks as though every element, except the minimum, has a predecessor, so that there are no limit points. And indeed it is the case that $\mathbf{2} \times \omega$ is order-isomorphic to ω. So multiplication of orders is, like addition, not commutative with respect to order-isomorphism.

Construct a suitable order-isomorphism between $\mathbf{2} \times \omega$ and ω!

Exercise 7.37

Is $X \times (Y \times Z)$ order-isomorphic to $(X \times Y) \times Z$, for ordered sets X, Y and Z?

Exercise 7.38

For each of the following ordered sets X,

(i) give an order-embedding of X into $\mathbb{Q}$ with the usual order;

(ii) define a new order $<'$ on the set $\mathbb{N}$ so that with this order $\mathbb{N}$ is order-isomorphic to X.

(a) $X = \omega \times \mathbf{3}$

(b) $X = (\omega \times \mathbf{3}) + \mathbf{4}$

(c) $X = \omega \times \omega$

(d) $X = (\omega \times \omega) + \omega$

Solution

(a) A picture of $\omega \times \mathbf{3}$ might help.

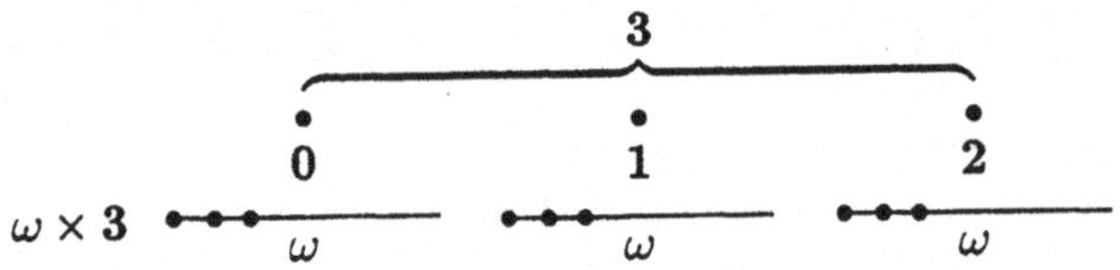

(i) We can mimic the order-isomorphism from ω to $\{1 - \frac{1}{n+1} : n \in \mathbb{N}\}$ which we have used often above. Map the copy of ω attached to $\mathbf{0}$ to $\{1 - \frac{1}{n+1} : n \in \mathbb{N}\}$, the copy of ω attached to $\mathbf{1}$ to $\{2 - \frac{1}{n+1} : n \in \mathbb{N}\}$, and so on. Formally define a function f by

This copy of ω is the subset $\{(\mathbf{0},\mathbf{0}), (\mathbf{1},\mathbf{0}), (\mathbf{2},\mathbf{0}), \ldots, (\mathbf{n},\mathbf{0}), \ldots\}$ of $\omega \times \mathbf{3}$.

$$f : \omega \times \mathbf{3} \longrightarrow \mathbb{Q}$$
$$(\mathbf{n}, \mathbf{m}) \longmapsto m - \tfrac{1}{n+1}.$$

It is easy to check that f is an order-embedding.

(ii) We just need to split $\mathbb{N}$ into three infinite subsets, each ordered by the usual $<$, and order these subsets sensibly. For instance, take the subsets

Each subset is order-isomorphic to ω, by Exercise 7.29.

$$\{3n : n \in \mathbb{N}\},\ \{3n+1 : n \in \mathbb{N}\},\ \{3n+2 : n \in \mathbb{N}\},$$

in this order, and define $<'$ by

$$0 <' 3 <' 6 <' \ldots <' 1 <' 4 <' 7 <' \ldots <' 2 <' 5 <' 8 <' \ldots.$$

(b)

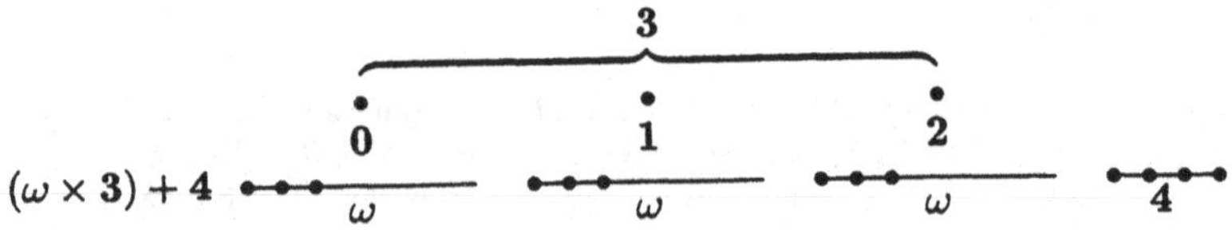

(i) We can easily adjust our solution to part (a)(i) above by selecting four rationals greater than or equal to 3 to which to map the copy of **4** in $(\omega \times \mathbf{3}) + \mathbf{4}$, for instance 3, 4, 5 and 6. The description of the corresponding order-embedding, g say, is a bit messy, mainly because the description of $(\omega \times \mathbf{3}) + \mathbf{4}$ is so messy!

$$\begin{aligned} g\colon (\omega \times \mathbf{3}) + \mathbf{4} &\longrightarrow \mathbb{Q} \\ ((\mathbf{n}, \mathbf{m}), \mathbf{0}) &\longmapsto m - \tfrac{1}{n+1} \\ (\mathbf{i}, \mathbf{1}) &\longmapsto 3 + i. \end{aligned}$$

(ii) Likewise, we can pluck out four natural numbers, say **0**, **1**, **2** and **3**, from the subsets in our answer to part (a)(ii), and place them at the top of our new order $<'$:

$$6 <' 9 <' 12 <' 15 <' \ldots <' 4 <' 7 <' 10 <' 13 <' \ldots$$
$$\ldots <' 5 <' 8 <' 11 <' 14 <' \ldots <' 0 <' 1 <' 2 <' 3.$$

(c)

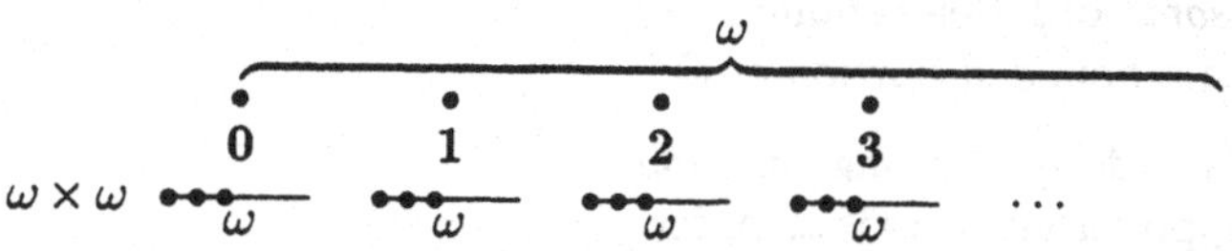

(i) Again we can exploit the idea of the order-isomorphism between ω and $\{1 - \frac{1}{n+1} : n \in \mathbb{N}\}$. Put

$$A_m = \{m - \tfrac{1}{n+1} : n \in \mathbb{N}\},$$

for each $m \in \mathbb{N}$. Clearly $A_m \cong \omega$. Note that if $m < m'$ then any number in A_m is less than any number in $A_{m'}$. (Check this!) Thus making the copy of ω corresponding to $\mathbf{m}$ in $\omega \times \omega$ map to A_m in the obvious way will give an order-embedding h. Formally this order-embedding (there are many correct alternatives) is defined by

This copy of ω is the subset $\{(\mathbf{0}, \mathbf{m}), (\mathbf{1}, \mathbf{m}), \ldots, (\mathbf{n}, \mathbf{m}), \ldots\}$ of $\omega \times \omega$.

$$\begin{aligned} h\colon \omega \times \omega &\longrightarrow \mathbb{Q} \\ (\mathbf{n}, \mathbf{m}) &\longmapsto m - \tfrac{1}{n+1}. \end{aligned}$$

(ii) This time we need to split $\mathbb{N}$ into infinitely many infinite subsets, $\{B_n : n \in \mathbb{N}\}$. Order each B_n by the usual $<$, so that $B_n \cong \omega$, and arrange the B_n order-isomorphically to ω.

There are many ways of defining these B_n. For instance, one can exploit the fact that there are infinitely many prime numbers. Let p_n, for $n \geq 1$, be the nth prime, ordering the primes in the usual way. Then for each $n \geq 1$ put

So $p_1 = 2$, $p_2 = 3$, $p_3 = 5$, etc.

$$B_n = \{p_n^i : i > 0\}.$$

The B_n are clearly infinite and disjoint, but their union is not the whole of $\mathbb{N}$ – it doesn't include 0 or 1 or any of the infinitely many composite numbers. So just put all these numbers into a set B_0, ordering them by the usual $<$ to obtain $B_0 \cong \omega$.

Alternatively, one can exploit any bijection $h\colon \mathbb{N} \times \mathbb{N} \longrightarrow \mathbb{N}$ by putting

$$B_n = \{h(n, i) : i \in \mathbb{N}\}, \quad \text{for each } n.$$

For instance, the bijection h used in our proof of Theorem 6.4 in Section 6.4.

However we have defined the B_n, we can then re-order $\mathbb{N}$ by $<'$ to obtain a set order-isomorphic to $\omega \times \omega$ by, for example,

$$i <' j \text{ if } i, j \in B_n \text{ for some } n \text{ and } i < j,$$
$$\text{or } i \in B_m \text{ and } j \in B_n \text{ and } m < n.$$

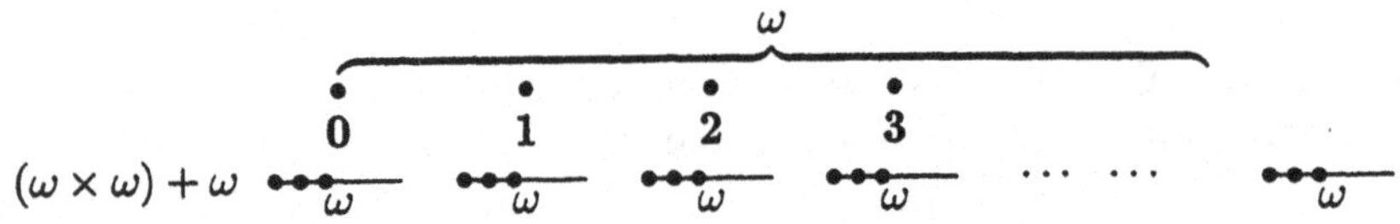

(d)

(i) Unfortunately, we cannot easily adapt our solution to part (c)(i) by putting some subset of $\mathbb{Q}$ order-isomorphic to ω to the right of those used in that solution – the image set of h is unbounded above! But of course Theorem 7.3 guarantees that there must be *some* order-embedding. We must somehow squash up a copy of $\omega \times \omega$ into a bounded portion of $\mathbb{Q}$.

It would be handy if there were some sort of order-preserving function squashing $\mathbb{Q}$, or part of it. For the moment, suppose that there is an order-preserving function $\theta\colon [0, \infty) \longrightarrow [0, 1)$ which maps rationals to rationals. Then applying θ to the subset $\mathbb{N}$ of $[0, \infty)$, which of course is ω, gives a subset of $[0, 1)$ order-isomorphic to ω. For each $m \in \mathbb{N}$ construct an order-isomorphic copy of this lying in the interval $[m, m+1)$, say C_m, where

The function $x \longmapsto \frac{2}{\pi}\arctan(x)$ would do if only it mapped rationals to rationals – which it doesn't!

$$C_m = \{m + \theta(n) : n \in \mathbb{N}\}.$$

So that if $m < m'$ then any number in C_m is less than any number in $C_{m'}$. Then the set C is order-isomorphic to $\omega \times \omega$, where

So far this is like the solution to part (c)(i) above.

$$C = \bigcup\{C_m : m \in \mathbb{N}\}.$$

The trick is to now use θ again to shrink C into a subset of $[0, 1)$ order-isomorphic to $\omega \times \omega$ and add a suitable subset of $[1, \infty)$ order-isomorphic to ω to give a subset of $\mathbb{Q}$ order-isomorphic to $(\omega \times \omega) + \omega$. Formally this amounts to defining an order-embedding k by something like

$$\begin{aligned} k\colon (\omega \times \omega) + \omega &\longrightarrow \mathbb{Q} \\ ((\mathbf{n}, \mathbf{m}), \mathbf{0}) &\longmapsto \theta(m + \theta(n)) \\ (\mathbf{i}, \mathbf{1}) &\longmapsto 1 + i. \end{aligned}$$

Obviously rather a lot of the above depends on finding the function θ. We ask you to do this in the next exercise.

(ii) Take the infinite subsets B_n as in part (c)(ii), but this time arrange them so that everything in B_0 is greater than the members of the other B_n: i.e. define $<'$ by

We are really arranging the subscript ns in the order $\omega + \mathbf{1}$ and exploiting the fact that $(\omega \times \omega) + \omega \cong \omega \times (\omega + \mathbf{1})$.

$$i <' j \text{ if } i, j \in B_n \text{ for some } n,$$
$$\text{or } i \in B_m \text{ and } j \in B_n \text{ and } 0 < m < n,$$
$$\text{or } i \in B_m \text{ for some } m > 0 \text{ and } j \in B_0.$$

Exercise 7.39

Find an order-preserving function $\theta\colon [0,\infty) \longrightarrow [0,1)$ which maps rationals to rationals.

Solution

One such function, essentially already exploited in much of the above, is

$$\begin{aligned}\theta\colon [0,\infty) &\longrightarrow [0,1)\\ x &\longmapsto 1-\tfrac{1}{x+1}.\end{aligned}$$

You can check that θ is in fact an order-isomorphism from $[0,\infty)$ to $[0,1)$. Clearly if x is a rational, then so is $\frac{1}{x+1}$, hence also $\theta(x)$.

To round off this excursion into the arithmetic of order, we should ask if there is any further nice connection between order sums and products. Our experience of ordinary arithmetic makes us look at the distributivity of $+$ over $\times$.

Exercise 7.40

Let X, Y and Z be ordered sets.

(a) Show that $X \times (Y+Z) \cong (X \times Y) + (X \times Z)$.

(b) Is it always the case that $(X+Y) \times Z \cong (X \times Z) + (Y \times Z)$?

It is now time to leave the arithmetic of sets with any old linear order, and to concentrate on more specialized ordered sets, namely well-ordered sets, in particular ordinals. Happily, many of the examples investigated above are well-ordered sets. Indeed, all natural numbers $\mathbf{n}$ and ω itself are actually ordinals. We shall define an arithmetic on ordinals, using a sum and product which are different from the order-theoretic operations of this chapter; but the results of applying the ordinal operations will give sets order-isomorphic to those obtained by using the order-theoretic operations.

Sum and product on ordinals will extend the definitions for the members of $\mathbb{N}$ in ZF.

Further exercise

Exercise 7.41

Let X be a linearly ordered set. Let X^n stand for the ordered set $\underbrace{X \times X \times \ldots \times X}_{n}$, for $n \in \mathbb{N}$, $n > 0$.

As order product is associative relative to order-isomorphism, we don't have to worry about brackets.

For each of the following ordered sets X,

(i) give an order-embedding of X into $\mathbb{Q}$ with the usual order;

(ii) define a new order $<'$ on the set $\mathbb{N}$ so that with this order $\mathbb{N}$ is order-isomorphic to X.

(a) ω^3

(b) $\mathbb{Z}^3$

(c) $\omega^3 + (\omega^2 \times 2) + \omega$

7.4 Well-ordered sets

We shall now look at a special sort of linear order, called well-order. Ordinals, which are what we aiming at in the next chapter, are well-ordered sets, so that the results of this section will help us later on, as well as being of mathematical interest in their own right.

Cantor devised the ordinals as a way of extending beyond the finite the idea of numbering the stages of a process – 'first do this, next do that, next do that, ...'. If we represent these stages by the elements of an ordered set X, then we require that for a given stage x there *is* a 'next' stage x^+ (unless x happens to be the 'final' stage, i.e. x is the maximum element of X). So if x is not the maximum element of X, in which case $\{x' \in X : x < x'\} \neq \varnothing$, we want this latter set to contain a least element x^+. As a bit of overkill, for a well-ordered set we shall require that *every* non-empty subset of X has a least element.

$x < x'$ means that stage x comes earlier than stage x'.

x^+ is the successor of x.

> ***Definition***
>
> A set X is said to be *well-ordered* by $<$ if X is linearly ordered by $<$ and in addition has the *well-ordering property*:
>
> if A is any non-empty subset of X, then A contains a least element, i.e. there is some element $x \in A$ such that $x \leq a$ for all $a \in A$.

$<$ is said to be a *well-order.*

The classic example of a well-ordered set is ω, i.e. the set $\mathbb{N}$ with the usual $<$ ($\in$ in *ZF*). See whether you can spot other examples in the next exercise.

Exercise 7.42

Decide which of the following ordered sets are well-ordered.

(a) $\mathbf{5}$ in *ZF*, i.e. $\{\mathbf{0}, \mathbf{1}, \mathbf{2}, \mathbf{3}, \mathbf{4}\}$ ordered by $\in$.

(b) $\mathbb{Z}$ with the usual order.

(c) The closed interval $[0, 1]$ of $\mathbb{R}$ with the usual order.

(d) The subset $\{m + \frac{1}{n+1} : m, n \in \mathbb{N}\} \cup \{0\}$ of $\mathbb{Q}$ with the usual order.

(e) The subset $\{m - \frac{1}{n+1} : m, n \in \mathbb{N}\}$ of $\mathbb{Q}$ with the usual order.

(f) ω^2 with the product order.

Recall that ω^2 is $\omega \times \omega$.

Solution

(a) $\mathbf{5}$ is well-ordered. $\mathbf{5}$ is a subset of the well-ordered set ω, so as well as being linearly ordered by $\in$, any non-empty subset of $\mathbf{5}$ is a subset of ω, so has a least element. In general any subset of a well-ordered set is well-ordered.

Similarly every natural number in *ZF* is well-ordered.

(b) $\mathbb{Z}$ is not well-ordered, as e.g. the subset $\mathbb{Z}$ itself has no least element. The key feature here of $\mathbb{Z}$ is that the negative integers form an infinite descending $<$-chain. An ordered set with such a descending chain cannot be well-ordered.

We define 'infinite descending $<$-chain' in Exercise 7.45 below.

(c) $[0, 1]$ is not well-ordered. For instance, the subset given by the open interval $(0, 1)$ contains no least element. Similarly any subset of $\mathbb{R}$ containing a non-trivial interval (a, b) is not well-ordered by the usual $<$, in particular $\mathbb{R}$ itself.

(d) This set is not well-ordered. For instance, it includes the numbers $1 + \frac{1}{n+1}$, $n \in \mathbb{N}$, which form a descending $<$-chain. As with $\mathbb{R}$, one would expect $\mathbb{Q}$ to contain lots of subsets which are not well-ordered.

(e) The set $\{m - \frac{1}{n+1} : m, n \in \mathbb{N}\}$ is well-ordered. As we already know that it is linearly ordered, all we need to show is that it has the well-ordering property. So take any non-empty subset A and show that A contains a least element. To guess what this least element should be, it helps to realize that the m in each $m - \frac{1}{n+1}$ is more important than the n, in the sense that

$$m - \tfrac{1}{n+1} < m' - \tfrac{1}{n'+1} \text{ if and only if } m < m' \text{ or } m = m' \text{ and } n < n'.$$

Thus the least element of A has to have the least m available among the elements of A and the smallest n for this m. More precisely, put

$$B = \{m \in \mathbb{N} : m - \tfrac{1}{n+1} \in A \text{ for some } n \in \mathbb{N}\}.$$

As $A \neq \varnothing$, B is a non-empty subset of $\mathbb{N}$ and so has a least element m_0. As $m_0 \in B$ there is at least one n such that $m_0 - \frac{1}{n+1} \in A$, so that the set

$$C = \{n \in \mathbb{N} : m_0 - \tfrac{1}{n+1} \in A\}$$

is a non-empty subset of $\mathbb{N}$ and thus has a least element n_0. As $n_0 \in C$, we have that $m_0 - \frac{1}{n_0+1}$ is an element of A. You can check that $m_0 - \frac{1}{n_0+1}$ is indeed the least element of A.

(f) ω^2 is well-ordered. You could show this directly by an argument similar to the one we used in the preceding solution, showing how to find the least element of any non-empty subset A of ω^2. In fact this sort of argument can be used to show that the product of any two well-ordered sets is well ordered. Or you could show that being well-ordered is an order-theoretic invariant, and exploit the function

That is, being well-ordered is preserved under order-isomorphism.

$$f \colon \{m - \tfrac{1}{n+1} : m, n \in \mathbb{N}\} \longrightarrow \omega^2$$
$$m - \tfrac{1}{n+1} \longmapsto (n, m)$$

which is an order-isomorphism. As $\{m - \frac{1}{n+1} : m, n \in \mathbb{N}\}$ is well-ordered, so is ω^2.

Now prove for yourself some of the general results arising from the examples above.

Exercise 7.43

Show that any finite linearly ordered set is well-ordered. [Hint: look at the exercises in Section 7.2.]

Exercise 7.44

Suppose that X is well-ordered by $<$ and that Y is subset of X. Show that Y is also well-ordered by $<$.

Technically the order on Y is $< |_{Y \times Y}$.

Exercise 7.45

Suppose that the set X, ordered by $<$, contains an infinite descending $<$-chain, i.e. contains elements x_n, $n \in \mathbb{N}$, such that

$$\ldots < x_n < \ldots < x_2 < x_1 < x_0.$$

Show that X is not well-ordered by $<$.

Exercise 7.46

Suppose that X and Y are ordered sets and that $f\colon X \longrightarrow Y$ is an order-isomorphism. Show that if X is well-ordered then so is Y.

Exercise 7.47

Suppose that X and Y are well-ordered sets. Show that the ordered sets $X + Y$ and $X \times Y$ are both well-ordered.

So e.g. $\omega + \mathbf{2}$ and $\omega^2 + \omega$ are well-ordered.

Why are well-ordered sets worth studying? First of all, they provided Cantor with a way of extending counting beyond the natural numbers. Second, in the light of historically later concerns over the use of the axiom of choice, a well-order on a set X provides a rich enough structure to define a choice function on X, with no need for AC.

Exercise 7.48

Suppose that X is well-ordered by $<$. Show how to define a choice function on X of the form

$$f\colon \mathscr{P}(X) \setminus \{\varnothing\} \longrightarrow X$$

such that

$$f(A) \in A \text{ for all non-empty subsets } A \text{ of } X.$$

Solution

Just define $f(A)$ to be the least element of A!

As we have seen, $\mathbb{R}$ is not well-ordered by the usual $<$. What can we say about subsets of $\mathbb{R}$? If we could find a subset A equinumerous with $\mathbb{R}$ which was well-ordered by the usual $<$, we could exploit the bijection between A and $\mathbb{R}$ to define a well-ordering of $\mathbb{R}$, settling one of Cantor's main problems. Clearly no subset of $\mathbb{R}$ containing a non-trivial open interval (a, b) is well-ordered – a shame, as such a subset would be equinumerous with $\mathbb{R}$! In fact any well-ordered subset of $\mathbb{R}$ must be a long way short of equinumerous with $\mathbb{R}$, as the next exercise shows.

Exercise 7.49

Let A be a subset of $\mathbb{R}$ which is well-ordered by the usual $<$ on $\mathbb{R}$. Show that there is an order-embedding $f\colon A \longrightarrow \mathbb{Q}$, where $\mathbb{Q}$ has the usual order and hence deduce that A is countable. [Hints: exploit the fact that between any two distinct reals there is a rational, and for each $a \in A$ define $f(a)$ to be a suitable rational greater than a.]

Solution

For each $a \in A$, as long as a doesn't happen to be the maximum element of A, a will have a successor a^+ in A, i.e. the least element of A bigger than a. Between a and a^+ there is at least one rational, so define $f(a)$ to be such a rational. As the open interval (a, a^+) of $\mathbb{R}$ contains no elements of A, there is no danger that the rational $f(a)$ could be associated with another element b of A, so that f will be one–one (and in fact order-preserving as well).

A need not have a maximum element.

In fact there are infinitely many rationals in (a, a^+).

More formally, list the rationals in some way as $q_0, q_1, q_2, \ldots, q_n, \ldots$. Provided that the set $\{b \in A : a < b\}$ is non-empty, define a^+ to be the least element of this set. Now define $f\colon A \longrightarrow \mathbb{Q}$ by:

We are making the choice of the rational $f(a)$ more definite and coping with the possibility that a might be $\max(A)$.

$$f(a) = \begin{cases} \text{the first } q_i \text{ in the list in } (a, a^+), & \text{if } \{b \in A : a < b\} \neq \varnothing, \\ \text{the first } q_i \text{ in the list in } (a, \infty), & \text{otherwise, i.e. } a \text{ is } \max A. \end{cases}$$

To check that f is an order-embedding, suppose that $a, a' \in A$ with $a < a'$. Then

$$\{b \in A : a < b\} \neq \varnothing,$$

so that a^+ exists, and by definition $a^+ \leq a'$. Then as $f(a) \in (a, a^+)$ and $f(a') > a'$, we have

$$a < f(a) < a^+ \leq a' < f(a'),$$

giving $f(a) < f(a')$ as required.

In particular f is one–one, so that $A \preceq \mathbb{Q}$, from which we can deduce that A is countable.

In the rest of this section we shall derive a result about the comparison of well-ordered sets with each other, i.e. whether, given two well-ordered sets X and Y, one is order-embeddable in the other. For linearly ordered, but not necessarily well-ordered, sets X and Y, this isn't always true: take for instance $\mathbb{Z}$ and $\{1 - \frac{1}{n+1} : n \in \mathbb{N}\} \cup \{1\}$, both with the usual $<$.

Exercise 7.50

(a) Explain why there can be no order-embedding $f\colon \mathbb{Z} \longrightarrow \{1 - \frac{1}{n+1} : n \in \mathbb{N}\} \cup \{1\}$.

(b) Explain why there can be no order-embedding $f\colon \{1 - \frac{1}{n+1} : n \in \mathbb{N}\} \cup \{1\} \longrightarrow \mathbb{Z}$.

Solution

(a) The set $\{1 - \frac{1}{n+1} : n \in \mathbb{N}\} \cup \{1\}$ is well-ordered so doesn't contain an infinite descending $<$-chain. Thus there is no way of order-embedding the negative integers into the set.

(b) Whichever integers we chose for $f(0)$ and $f(1)$, there would only be finitely many possible images between them for the infinitely many $1 - \frac{1}{n+1}$, $n > 0$. So we could not define f to be both one–one and order-preserving.

$0 = 1 - \frac{1}{0+1}$.

It isn't even the case that for linearly ordered X and Y with X order-embeddable in Y and Y order-embeddable in X, we are then guaranteed that X and Y are order-isomorphic.

You might have worked this out for yourself when doing Exercise 7.18 in Section 7.2.

Well-ordered sets are, however, comparable with each other via order-embedding: given any two well-ordered sets, X and Y, then either they are order-isomorphic or one can be order-embedded into the other.

The basic idea is to pair off the least elements of X and Y, then their next least elements, and so on. If the process has worked for all the elements of X less than some limit point x of X, it turns out that the least available element of Y to be paired up with x also has to be a limit point. And if in this process we run out of elements of both X and Y at the same stage, we have an order-isomorphism; while if we run out of elements of one of the sets before we've used up all the elements of the other, then we have an order-embedding of one set into the other.

As X and Y are well-ordered, 'least' and 'next least' make sense.

Consider, for instance, the subsets

$$X = \{1 - \tfrac{1}{2^n} : n \in \mathbb{N}\} \cup \{2 - \tfrac{1}{2^n} : n \in \mathbb{N}\} \cup \{2\}$$

and

$$Y = \{1 - \tfrac{1}{n+1} : n \in \mathbb{N}\} \cup \{1\}$$

of $\mathbb{Q}$ with the usual order.

Check that X and Y are well-ordered!

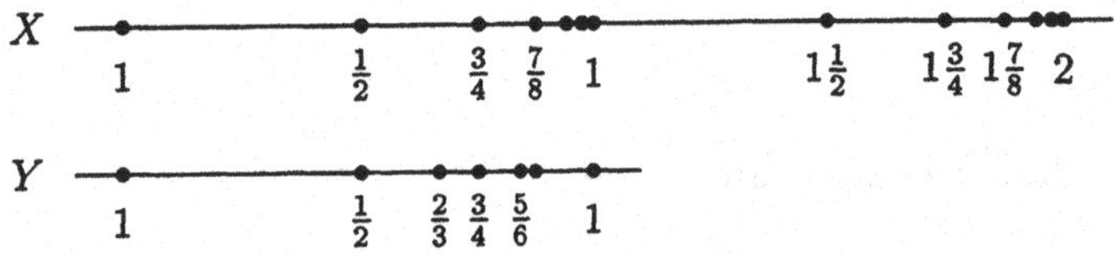

Both sets are well-ordered. The process would give the function

$$\begin{aligned} f\colon Y &\longrightarrow X \\ 1 - \tfrac{1}{n+1} &\longmapsto 1 - \tfrac{1}{2^n} \quad n \in \mathbb{N} \\ 1 &\longmapsto 1 \; (= 2 - \tfrac{1}{2^0}) \end{aligned}$$

which is clearly an order-embedding.

Y runs out of elements before X does.

Exercise 7.51

For the X and Y above, find an order-embedding $g\colon Y \longrightarrow X$ mapping $\frac{5}{6}$ $(= 1 - \frac{1}{6} \in Y)$ to $1\frac{3}{4}$ $(= 2 - \frac{1}{2^2} \in X)$.

Solution

There are several correct answers, for instance

$$\begin{aligned} g\colon Y &\longrightarrow X \\ 1 - \tfrac{1}{n+1} &\longmapsto 1 - \tfrac{1}{2^n} \qquad n = 0,1,2,3,4 \\ 1 - \tfrac{1}{n+1} &\longmapsto 2 - \tfrac{1}{2^{n-3}} \quad n = 5,6,\ldots \\ 1 &\longmapsto 2. \end{aligned}$$

Although there are plenty of order-embeddings from Y to X like g above, the one (f) given by our earlier process is special in that it maps Y onto the corresponding initial portion of X. In some sense X is just Y with some extra elements added at the (right-hand) end. The initial portion of X, involving all the elements of X up to a particular element, is worth a general definition:

Definition

Let X be a well-ordered set with $a \in X$. Then the *initial segment* of X determined by a, written as $\mathrm{Seg}_X(a)$, is the set

$$\mathrm{Seg}_X(a) = \{x \in X : x < a\}.$$

Our order-embedding f above maps Y onto the initial segment $\mathrm{Seg}_X(2\frac{1}{2})$ of X. And, for instance, with $X = \omega$ and $a = \mathbf{3}$,

$$\begin{aligned}\mathrm{Seg}_\omega(\mathbf{3}) &= \{\mathbf{n} \in \omega : \mathbf{n} < \mathbf{3}\}\\ &= \{\mathbf{0}, \mathbf{1}, \mathbf{2}\}\\ &= \mathbf{3}.\end{aligned}$$

We can now state the main theorem about comparing well-ordered sets.

Theorem 7.5

Let X and Y be well-ordered sets. Then exactly one of the following holds:

1. X is order-isomorphic to an initial segment of Y;
2. X is order-isomorphic to Y;
3. Y is order-isomorphic to an initial segment of X.

Exercise 7.52

Let X be $\mathbb{N}$ and Y be $\{m - \frac{1}{n+1} : m = 1, 2 \text{ and } n \in \mathbb{N}\}$, both with their usual orders. One of X and Y is order-isomorphic to an initial segment $\mathrm{Seg}(a)$ of the other. Decide which set is which and identify the element a.

Solution

X is order-isomorphic to $\mathrm{Seg}_Y(1)$.

Exercise 7.53

According to Theorem 7.5, given an infinite well-ordered set X and ω, either they are order-isomorphic or one is order-isomorphic to an initial segment of the other. Suppose that they are not order-isomorphic. Decide which one is order-isomorphic to an initial segment of the other and give the corresponding order-isomorphism.

Solution

As ω is such a basic infinite set, it seems likeliest that it is ω that is order-isomorphic to an initial segment of X, rather than the other way round; and this is indeed the case. Let c be a set not in X and define a function g by

$$\begin{aligned} g : \omega &\longrightarrow X \cup \{c\}\\ \mathbf{n} &\longmapsto \begin{cases} \min(X \setminus \mathrm{Range}(g|_{\mathbf{n}})), & \text{if } \min(X \setminus \mathrm{Range}(g|_{\mathbf{n}})) \neq \varnothing,\\ c, & \text{otherwise.}\end{cases}\end{aligned}$$

You should by now recognize this sort of construction! Firstly we can show that c is not in Range(g), arguing by contradiction. If c is in Range(g), put $\mathbf{n} = \min\{\mathbf{m} \in \mathbb{N} : g(\mathbf{m}) = c\}$. You can check that $g|_{\mathbf{n}}$ is an order-isomorphism from $\mathbf{n}$ to X, contradicting that X is infinite.

We can conclude that Range(g) is a subset of X. As ω and X are not order-isomorphic, Range(g) must be a proper subset of X. That means that $X \setminus$ Range(g) is a non-empty subset of X, so it has a least element, y say. We'll leave you to show that g gives an order-isomorphism from ω to the initial segment $\text{Seg}_X(y)$.

You can check that g is an order-embedding.

It's worth comparing this result to that of Exercise 6.37 of Section 6.4. Knowing that an infinite set X is well-ordered enables one to show that $\mathbb{N} \preceq X$ without assuming AC. We shall explore the strong relationship between well-ordering and AC in Chapter 9.

Theorem 7.5 is a very strong result. It says that any two well-orders are essentially the same or one of them is just the other with some extra elements tacked on at the end. It also suggests that one can organize well-ordered sets in terms of their 'length' as orders – we shall do this properly in the next section by looking at ordinals, which are just special well-ordered sets.

Of course 'same' means order-isomorphic.

In the rest of this section we shall prove Theorem 7.5. The proof, not surprisingly, will make essential use of the well-ordering property, which lends itself to very powerful proof techniques. We shall leave some of the details for you to verify – they don't require special tricks! First of all, we shall need some results about initial segments.

The well-ordering property for $\mathbb{N}$ is equivalent to the principle of mathematical induction, which is pretty powerful!

Exercise 7.54

Let X be a well-ordered set and let $x, x' \in X$. Show the following.

(a) $\text{Seg}_X(x) \subseteq \text{Seg}_X(x')$ if and only if $x \leq x'$.

(b) x is the least element of $X \setminus \text{Seg}_X(x)$.

(c) If $x < x'$ then $\text{Seg}_{\text{Seg}_X(x')}(x) = \text{Seg}_X(x)$.

Exercise 7.55

Let $f\colon X \longrightarrow Y$ be an order-isomorphism between well-ordered sets X and Y. Show that for any $x \in X$, $\text{Seg}_X(x) \cong \text{Seg}_Y(f(x))$.

This result really only needs that X and Y are linearly ordered. However, our definition of initial segment applies only to well-ordered sets.

Solution

It is easy to show that the restriction of f to $\text{Seg}_X(x)$, i.e. $f|_{\text{Seg}_X(x)}$, is onto the subset $\text{Seg}_Y(f(x))$ of Y. The restriction inherits being one–one and order-preserving from f.

The terminology of initial segments makes it more straightforward to state a pleasing property of well-ordered sets, namely that one can prove things about them by induction.

> ***Theorem 7.6 Principle of transfinite induction***
>
> Let A be a subset of the well-ordered set X such that
>
> (i) the least element of X is in A;
>
> (ii) for all other $x \in X$, if $\mathrm{Seg}_X(x) \subseteq A$ then $x \in A$.
>
> Then $A = X$.

'Transfinite' because the induction might go beyond the finite, e.g. if $X = \omega + \omega$.

The base step.

The inductive step.

If you think of A as the subset of X consisting of all elements with a particular property P, then the inductive step can be read as saying that

> if every element less than x has property P, then so does x itself.

And $A = X$ just says that every element of X has property P.

Let us now prove Theorem 7.6.

Proof

Let A be a subset satisfying (i) and (ii). Suppose that A does not equal X. Then the subset $X \setminus A$ of X is non-empty, so as X is well-ordered it has a least element, x say. By (i) x cannot be the least element of X, so that $\mathrm{Seg}_X(x)$ is non-empty. By the minimality of this x, for any $y < x$ we must have $y \notin X \setminus A$, i.e. $y \in A$. This means that $\mathrm{Seg}_X(x) \subseteq A$, so that by (ii) $x \in A$, contradicting that x is in $X \setminus A$.
Thus $A = X$. ■

We shall try to obtain a contradiction.

The method used in this proof could be described as that of the 'minimal counterexample' – assume that the sought-for result does not hold for all the elements of X, so that as X is well-ordered there is a least element of X for which it doesn't hold: then exploit this least element in some way or other to try to obtain a counterexample.

We shall prove the next result needed in our proof of Theorem 7.5 by induction. But the proof can also be done by the method of the minimal counterexample, which we shall give you as an exercise.

> ***Theorem 7.7***
>
> If $f\colon X \longrightarrow X$ is an order-embedding of a well-ordered set X into itself, then $x \leq f(x)$ for all $x \in X$.

Proof

Define the subset A of X by

$$A = \{x \in X : x \leq f(x)\}.$$

We shall show by induction (Theorem 7.6) that $A = X$.

Assume that X is non-empty, so that X has a least element x_0. Then as $f(x_0)$ has to be some element of X, we have $x_0 \leq f(x_0)$ by the minimality of x_0. Thus the least element of X is in A.

For $X = \varnothing$, the result is trivial.

Now for the inductive step. Suppose that x is an element of X for which $\mathrm{Seg}_X(x) \subseteq A$. We must show that $x \in A$, i.e. $x \leq f(x)$.

For each $y < x$ we have both

$$y \leq f(y) \quad (\text{as } y \in \mathrm{Seg}_X(x) \subseteq A)$$

and

$$f(y) < f(x) \quad (\text{as } f \text{ is an order-embedding}),$$

so that

$$y < f(x).$$

Thus $\mathrm{Seg}_X(x) \subseteq \mathrm{Seg}_X(f(x))$, so that by the result of Exercise 7.54(a) $x \leq f(x)$, as required for the inductive step.

Hence by Theorem 7.6 $A = X$. ■

Exercise 7.56

Prove the last result using the method of the minimal counterexample, i.e. assume that there are elements of X for which it is not the case that $x \leq f(x)$, so that there is a least such element x_0, and try to derive a contradiction.

This technical result is used in the proof of the next, somewhat more surprising, theorem.

Theorem 7.8

Let $f\colon X \longrightarrow X$ be an order-isomorphism of the well-ordered set X with itself. Then f is the identity function.

Proof

Take any $x \in X$. Then as f is an order-embedding, the previous result gives

$$x \leq f(x).$$

But f^{-1} is also an order-embedding, so the same result (applied to the element $f(x)$ of X) also gives

$$f(x) \leq f^{-1}(f(x)),$$

i.e.

$$f(x) \leq x.$$

Hence $f(x) = x$ for all x, so that f is the identity function. ■

This result is quite impressive as it does *not* apply to linearly ordered sets which are not well-ordered.

Exercise 7.57

Give an example of a linearly ordered set X and an order-isomorphism of X with itself which is *not* the identity function.

The last result can be used to prove something slightly more general: the content of the next exercise.

Exercise 7.58

Suppose that $f\colon X \longrightarrow Y$ and $g\colon X \longrightarrow Y$ are both order-isomorphisms between the well-ordered sets X and Y. Then $f = g$.

So that if $X \cong Y$, there's exactly one order-isomorphism from X to Y. This is not true in general if X and Y are merely linearly ordered rather than well-ordered.

Solution

The composite function $g^{-1} \circ f$ will be an order-isomorphism of X with itself, so by the result of the last exercise must be the identity. Hence $f = g$.

The last result we need before we prove the big theorem comes out of the next set of exercises.

Exercise 7.59

Let X be a well-ordered set. Show that X cannot be order-isomorphic to $\mathrm{Seg}_X(x)$ for any $x \in X$.

Exercise 7.60

Let Y be a well-ordered set with $y, y' \in Y$. Show that if $\mathrm{Seg}_Y(y) \cong \mathrm{Seg}_Y(y')$, then $y = y'$.

We are now in a position to prove Theorem 7.5.

Theorem 7.5

Let X and Y be well-ordered sets. Then exactly one of the following holds:

1. X is order-isomorphic to an initial segment of Y;
2. X is order-isomorphic to Y;
3. Y is order-isomorphic to an initial segment of X.

We shall essentially formalize our informal process of matching the least elements of the sets, then the next least elements, and so on. Define a subset E of $X \times Y$ by

$$E = \{(x, y) : \mathrm{Seg}_X(x) \cong \mathrm{Seg}_Y(y)\}.$$

It is easy to show that E is an order-isomorphism from a subset of X to a subset of Y. What is only a bit harder is showing that either

$$\mathrm{Dom}(E) = X \text{ and } \mathrm{Range}(E) = Y,$$

in which case X and Y are order-isomorphic, or

$$\mathrm{Dom}(E) = X \text{ and } \mathrm{Range}(E) \text{ is an initial segment of } Y,$$

or

$$\mathrm{Range}(E) = Y \text{ and } \mathrm{Dom}(E) \text{ is an initial segment of } X,$$

in which case E^{-1} is an order-isomorphism of Y with an initial segment of X.

First of all, let us show that for each $x \in \text{Dom}(E)$ there is a *unique* $y \in Y$ such that $(x, y) \in E$. Given such an x there is at least one y with $(x, y) \in E$, as we are assuming that $x \in \text{Dom}(E)$. If there was another $y' \in Y$ such that $(x, y') \in E$, then we would have both

$$\text{Seg}_X(x) \cong \text{Seg}_Y(y)$$

and

$$\text{Seg}_X(x) \cong \text{Seg}_Y(y'),$$

so that

$$\text{Seg}_Y(y) \cong \text{Seg}_Y(y').$$

It follows from the result of Exercise 7.60 that $y = y'$, giving the uniqueness of the y such that $(x, y) \in E$. Thus E is a function.

A similar argument exploiting the result of Exercise 7.60 shows that for any $y \in \text{Range}(E)$, if $(x, y), (x', y) \in E$ then $x = x'$. Thus E is one–one.

To show that E is an order-embedding, suppose that $x, x' \in \text{Dom}(E)$ with $x <_X x'$. As $x <_X x'$ we have that $x \in \text{Seg}_X(x')$. As $x' \in \text{Dom}(E)$, there is a $y' \in Y$ such that

y' is $E(x)$.

$$\text{Seg}_X(x') \cong \text{Seg}_Y(y'),$$

so that under this isomorphism x must map to some $y \in \text{Seg}_Y(y')$. Then by the result of Exercise 7.55

$$\text{Seg}_{\text{Seg}_X(x')}(x) \cong \text{Seg}_{\text{Seg}_Y(y')}(y),$$

which by the result of Exercise 7.54 simplifies to

$$\text{Seg}_X(x) \cong \text{Seg}_Y(y).$$

This means that $(x, y) \in E$, i.e. $y = E(x)$ (now that we know E is a function).

What we have shown above in a somewhat roundabout way is that if $x <_X x'$ then

We have also effectively shown that if $x' \in \text{Dom}(E)$ then $\text{Seg}_X(x') \subseteq \text{Dom}(E)$.

$$\begin{aligned} E(x) &= y \\ &<_Y y' \quad \text{as } y \in \text{Seg}_Y(y') \\ &= E(x'). \end{aligned}$$

Thus E is an order-embedding and gives an order-isomorphism of the subset $\text{Dom}(E)$ of X with the subset $\text{Range}(E)$ of Y.

Next we shall show that at least one of the subsets $\text{Dom}(E)$ and $\text{Range}(E)$ is all of, respectively, X and Y. Suppose that this is not the case. Then both $X \setminus \text{Dom}(E)$ and $Y \setminus \text{Range}(E)$ are non-empty, so have least elements x' and y', respectively. It is a straightforward exercise to show that

We'll use the method of the minimal counterexample.

1. $\text{Dom}(E) = \text{Seg}_X(x')$,
2. $\text{Range}(E) = \text{Seg}_Y(y')$,

and

3. $\text{Seg}_X(x') \cong \text{Seg}_Y(y')$.

Exercise 7.61

Prove 1, 2 and 3 above.

From 3 above $(x', y') \in E$, contradicting the definitions of x' and y'. Thus either $\mathrm{Dom}(E) = X$ or $\mathrm{Range}(E) = Y$ or both.

If both $\mathrm{Dom}(E) = X$ and $\mathrm{Range}(E) = Y$, then without further ado E is an order-isomorphism from X to Y.

Exercise 7.62

Suppose that $\mathrm{Dom}(E) = X$ and $\mathrm{Range}(E)$ is a proper subset of Y. Show that $\mathrm{Range}(E)$ is an initial segment of Y. [Hint: for which $y' \in Y$ will $\mathrm{Range}(E)$ equal $\mathrm{Seg}_Y(y')$?]

So if $\mathrm{Dom}(E) = X$ and $\mathrm{Range}(E)$ is a proper subset of Y, then X is order-isomorphic to an initial segment of Y.

As E is an order-isomorphism from $\mathrm{Dom}(E)$ to $\mathrm{Range}(E)$.

Similarly if $\mathrm{Range}(E) = Y$ and $\mathrm{Dom}(E)$ is a proper subset of X, then Y is order-isomorphic to an initial segment of X.

Consider the order-isomorphism E^{-1} from $\mathrm{Range}(E)$ to $\mathrm{Dom}(E)$.

We have almost completed the proof of Theorem 7.5. We have shown that *at least* one of the following holds:

1. X is order-isomorphic to an initial segment of Y;
2. X is order-isomorphic to Y;
3. Y is order-isomorphic to an initial segment of X.

We must now tidy up the detail of Theorem 7.5 that *exactly* one of the above holds.

Exercise 7.63

Complete the proof of Theorem 7.5 by showing that at most one of 1, 2 and 3 above can hold.

Solution

Make use of Exercise 7.59.

Now that we have finally proved Theorem 7.5 let's look at what might follow from its result.

Clearly order-isomorphism acts as an equivalence relation on well-ordered sets. Unfortunately the equivalence classes are indeed classes – they are too big to be sets! But if we can find a 'nice' member of each class, then these 'nice' members will give examples of all possible well-orders. By Theorem 7.5 we can then order these by the relation 'is order-isomorphic to an initial segment of': it would not be surprising if this turned out to be a well-ordering! In the next chapter we shall find such 'nice' well-ordered sets.

The equivalence is of course not a relation within ZF.

Indeed this is the effect of Exercise 7.66 below.

Further exercises

Exercise 7.64

Explain why in general the collection of well-ordered sets order-isomorphic to a given well-ordered set is a proper class, not a set. Is there an example of a well-ordered set for which this collection is indeed a set?

Exercise 7.65

Let X be a well-ordered set. Show that X is order-isomorphic to the set $\{\mathrm{Seg}_X(x) : x \in X\}$ of all initial segments of X, where these are ordered by $\subseteq$.

Exercise 7.66

Let $\mathscr{W}$ be a set of well-ordered sets.

(a) Show that $\mathscr{W}$ contains a member A with the property that for each $B \in \mathscr{W}$ either $A \cong B$ or A is order-isomorphic to an initial segment of B.

(b) Does $\mathscr{W}$ contain a member C with the property that for each $B \in \mathscr{W}$ either $B \cong C$ or B is order-isomorphic to an initial segment of C?

Exercise 7.67

Let X be a linearly ordered set. In Exercise 7.45 you were effectively asked to show that if X is well-ordered then there is no infinite descending $<$-chain of elements of X, i.e. there is no sequence $\{x_n\}_{n \in \mathbb{N}}$ of elements of X such that

$$\ldots < x_{n+1} < x_n < \ldots < x_2 < x_1 < x_0.$$

Does the converse to this result hold, i.e. if X contains no infinite descending chain then X is well-ordered?

Exercise 7.68

For each of the following linearly ordered sets, decide whether the order is also a well-order. For each set which is well-ordered, how does it compare, in the sense of Theorem 7.5, with the sets ω^n, for $n \in \mathbb{N}$?

(a) The subset $\{\frac{1}{2^m} - \frac{1}{3^n} : m, n \in \mathbb{N}\}$ of $\mathbb{Q}$, with the usual order.

(b) The subset $\{2^m - \frac{1}{3^n} : m, n \in \mathbb{N}\}$ of $\mathbb{Q}$, with the usual order.

(c) The set $\mathbb{N}^+$ of positive integers ordered by $<'$ defined by

$m <' n$ if either m has fewer divisors than n
or m and n have the same number of divisors and $m < n$,

where $<$ is the usual order on $\mathbb{N}^+$.

(d) The set S of all sequences of natural numbers which are zero almost everywhere, i.e. regarding such a sequence as a function from $\mathbb{N}$ to $\mathbb{N}$,

$$S = \{f \in \mathbb{N}^{\mathbb{N}} : \text{ for some } n \in \mathbb{N}, f(i) = 0 \text{ for all } i > n\},$$

ordered by $<'$, where

Check that this *is* a linear order!

$f <' g$ if there is some $m \in \mathbb{N}$ such that $f(m) < g(m)$
and $f(i) = g(i)$ for all $i > m$.

(e) The set S as in part (d) ordered by $<^*$, where

$$f <^* g \text{ if there is some } m \in \mathbb{N} \text{ such that } f(m) < g(m) \text{ and } f(i) = g(i) \text{ for all } i < m.$$

(f) The set T of all sequences of $\mathbf{0}$s and $\mathbf{1}$s which are zero almost everywhere, i.e.

$\{\mathbf{0}, \mathbf{1}\} = \mathbf{2}$ is ordered by the usual $<$.

$$T = \{f \in \mathbf{2}^{\mathbb{N}} : \text{ for some } n \in \mathbb{N}, f(i) = \mathbf{0} \text{ for all } i > n\},$$

ordered by the same $<'$ as earlier, i.e.

$$f <' g \text{ if there is some } m \in \mathbb{N} \text{ such that } f(m) < g(m) \text{ and } f(i) = g(i) \text{ for all } i > m.$$

(g) The set T as in part (f) ordered by $<^*$ as earlier, i.e.

$$f <^* g \text{ if there is some } m \in \mathbb{N} \text{ such that } f(m) < g(m) \text{ and } f(i) = g(i) \text{ for all } i < m.$$

Exercise 7.69

Let X be well-ordered by $<$. Define a relation R on the set of pairs $X \times X$ as follows:

$$\begin{aligned}(x, y)R(x', y') \text{ if (i) } & \max\{x, y\} < \max\{x', y'\}, \\ \text{or (ii) } & \max\{x, y\} = \max\{x', y'\} \text{ and } y < y', \\ \text{or (iii) } & \max\{x, y\} = \max\{x', y'\} \text{ and } y = y' \text{ and } x < x'.\end{aligned}$$

(a) Show that R is a well-order on $X \times X$.

R will usually not be the same as the more usual anti-lexicographic order on $X \times X$.

(b) (i) Suppose that X is set ω. Show that $\omega \times \omega$ ordered by R is order-isomorphic to ω.

(ii) Suppose that X is $\omega + \mathbf{1}$. Find $\mathbf{m}, \mathbf{n} \in \mathbb{N}$ such that the well-ordered set $(\omega + \mathbf{1}) \times (\omega + \mathbf{1})$ ordered by R is order-isomorphic to $(\omega \times \mathbf{m}) + \mathbf{n}$ with its usual order.

(c) Put $x_0 = \min X$ and suppose that $x \in X$. Show that

$$\mathrm{Seg}_X(x) \times \mathrm{Seg}_X(x) \cong \mathrm{Seg}_{X \times X}((x, x_0)),$$

where both sets are ordered by R.

8 ORDINAL NUMBERS

8.1 Introduction

Cantor's work on Fourier series (see Section 6.1) led him to a construction on subsets A of $\mathbb{R}$ described as finding the *derived set* of A. The derived set of A, written as A', is the set of all *limit* points of A, where in this context x is a limit point of A if every neighbourhood of x contains infinitely many points of A. Cantor considered the sequence of sets obtained by starting with A, finding its derived set A', then the derived set A'' of A', and so on:

$$A, A', A'', A''', \ldots.$$

The definitions of A' and limit are not of direct importance here. But they, and Cantor's work on subsets of $\mathbb{R}$, were seminal for a variety of mathematical developments. It is not important for you to understand all of this introduction!

Recall that Cantor was interested in the uniqueness of representation of a function f on $[0, 2\pi]$ by a Fourier series and, in some sense, how far one could push the subset S of $[0, 2\pi]$ on which two Fourier series for f had to agree to ensure that the series were in fact identical. Putting A as the complement of S in $[0, 2\pi]$, he proved uniqueness not only in the case when A is empty, but also when one of the derived sets $A', A'', A''', \ldots$ is empty. The case when $A \neq \varnothing$ and $A' = \varnothing$ corresponds to A being finite. And when $A' \neq \varnothing$ and $A'' = \varnothing$, this means that A is infinite, but somehow with a very sparse structure – note that if A included a non-trivial closed interval $[a, b]$, then each of $A', A'', A''', \ldots$ would also include it, so that none of the latter could be empty.

Cantor wished to continue his construction in the case when none of A', A'', A''', etc., was empty. He ultimately hit on the idea of constructing a set $A^{(\infty)}$ as the intersection of $A', A'', A''', \ldots$. He then made repeated use of the derived set construction on this set to give $A^{(\infty+1)}, A^{(\infty+2)}, \ldots$, and in the case when none of these was empty, he again took their intersection, to give $A^{(\infty+\infty)}$, and so on.

This was where Cantor explicitly described the intersection of infinitely many sets, another example of consciously incorporating the infinite into mathematics.

Ultimately Cantor studied the superscripts

$$\infty, \infty+1, \infty+2, \ldots, \infty+\infty, \ldots$$

in their own right, the *ordinal numbers*, as an extension of the natural numbers. He introduced the ω notation, rather than ∞, to describe the first 'infinite' one. He focused on their order properties – in some sense what is important about them is that each one has a successor and that there is some process of taking limits. This he finally encapsulated in terms of well-order. He then defined an ordinal as an order-type, meaning what was in common as an order to a whole class of order-isomorphic well-ordered sets. Thus ω was the order-type of all ordered sets order-isomorphic to $\mathbb{N}$ with its usual order. And this order-type was 'greater than' any finite order-type, e.g. 1, which is the order-type of any one-element linearly ordered set.

We have seen another example in this book of where we wish to continue an iterative process beyond the natural numbers, namely the iterative hierarchy of sets $\mathscr{V}_n$ discussed in Section 4.5.

This, however, gives rise to a considerable problem when trying to represent such mathematical objects in terms of sets – remember that the main purpose of *ZF* or any other form of set theory is precisely to validate Cantor's use of infinity! Can a set represent an order-type? The natural set to take is the collection of all ordered sets of a given order-type. But, as we hope you showed in Exercise 7.64 at the end of the previous chapter, this usually gives you a proper class, not a set. For instance, the order-type 1 would be represented by the class of all singletons, which is not a set.

Recall from Chapter 3 that if the collection of all singletons was a set, then the universe $\mathscr{U}$ of sets would also be a set, ultimately leading to Russell's paradox.

In this chapter, we shall look at a construction developed by von Neumann (arising out of work by the Russian and Swiss mathematician Dimitry Mirimanov) which gives a particular set out of each order-type of well-ordered sets in a systematic way, without any need for axioms beyond those in ZF. We shall then reconstruct Cantor's arithmetic on his ordinals using the von Neumann ordinals, which from here on we shall just refer to as the ordinals. Cantor's arithmetic on ordinals gives an insight into what infinite ordinals look like. At the end of the chapter, we shall start to see how Cantor's two sorts of infinite number, ordinals and cardinals, relate to each other.

Actually, some of the axioms in ZF are there precisely so that von Neumann's construction works!

Exercise 8.1

Which set might we take as the representative of all n-element well-ordered sets?

Solution

Within the context of this book, the obvious choice is the natural number $\mathbf{n}$ within ZF. Our intuition is that $\mathbf{n}$ has n elements.

8.2 Ordinal numbers

In the previous chapter we saw that well-ordered sets can be compared with each other in a particularly strong way. By Theorem 7.5 in the Section 7.4, given any two well-ordered sets, either they are order-isomorphic or (exactly) one is not merely order-embeddable in the other, but is order-isomorphic to an initial segment of the other. It would be nice to be able to pick from each class of order-isomorphic well-ordered sets a single such set to represent them all. All these representatives could then be linearly ordered by the relation 'is order-isomorphic to an initial segment of'. It would be even nicer if the 'order-isomorphism to an initial segment' took some very simple form – simplest of all would surely be the identity function, which would entail that given any two such representative sets, one would be a subset of the other.

'Pick' suggests that we might have to use AC. Will we?

This relation would even be a well-order, by Exercise 7.66 in Section 7.4.

So how might we chose the representatives? Let's look at some of the well-ordered sets about which we know something. The classic interesting well-ordered set is surely the set of natural numbers $\mathbb{N}$ with its usual order – what we've also called the set ω. Another interesting example is any natural number $\mathbf{n}$ within ZF – what makes this particularly interesting for us now is that by Exercises 7.15 and 7.43 of Chapter 7 every finite well-ordered set is order-isomorphic to some $\mathbf{n} \in \mathbb{N}$. Thus, given that we know lots about natural numbers within ZF, we might as well choose these to be the representatives of all finite well-ordered sets. And as these are all elements of the well-ordered set ω, we might as well take ω as the representative of its order-isomorphism class of well-ordered sets. How do ω and its members shape up as representatives? Certainly there's no duplication: distinct natural numbers aren't order-isomorphic, and no natural number is order-isomorphic to ω – there are no bijections connecting them. And there are no well-ordered sets that 'fit in between' ω and its members: every finite well-ordered set is represented by a

You've been set up! Historically this representation of $\mathbb{N}$ within ZF stemmed from von Neumann's representation of the ordinals, which is what we're working up to.

member of ω; and by Exercise 7.53 any well-ordered set that isn't finite, i.e. is infinite, is, or contains an initial segment, order-isomorphic to ω, so cannot fit 'below' ω.

Exercise 8.2

What does an 'order-isomorphism to an initial segment of' look like when comparing distinct elements of $\{\omega\} \cup \omega$?

Solution

It will simply be the identity function. Each $\mathbf{n} \in \omega$ is an initial segment of ω, namely $\mathrm{Seg}_\omega(\mathbf{n})$. And similarly if $\mathbf{m} \neq \mathbf{n}$, so that without loss of generality $\mathbf{m} < \mathbf{n}$, we have $\mathbf{m} = \mathrm{Seg}_{\mathbf{n}}(\mathbf{m})$.

Of course we've not exhausted all order-isomorphism classes of well-ordered sets by taking ω and its members as representatives. For instance, none of these is order-isomorphic to $\omega + \omega$. So what aspects of ω and its members shall we exploit to extend our class of representative well-ordered sets? First of all, note that for all of them the $<$ relation is actually $\in$, i.e. set membership. As $\in$ is one of the fundamental starting points of *ZF*, there is something satisfying in using it as the order relation. And second, note that for each of these sets, any member of the set is also a *subset* of it – the $\in$ relation gives them a very strong structure as well-ordered sets. These aspects become the key features of our definition of the special, representative, well-ordered sets, namely the *ordinals*.

Definition

A set α is said to be an *ordinal* if

(i) α is well-ordered by $\in$, and

(ii) if $\beta \in \alpha$ then $\beta \subseteq \alpha$.

It's common to use Greek letters to stand for ordinals.

Equivalently, if $\gamma \in \beta$ and $\beta \in \alpha$ then $\gamma \in \alpha$. So α is $\in$-transitive.

All $\mathbf{n} \in \omega$ and ω itself are ordinals – our definition was made to ensure this. Are there any other ordinals?

There better had be some more!

Exercise 8.3

(a) Let $\mathbf{n} \in \omega$. Show that $\omega \notin \mathbf{n}$ and that $\omega \notin \omega$. [Hint: use the definition of ordinal and properties of well-ordered sets, rather than the axiom of foundation or any of its consequences.]

(b) Show that $\omega \cup \{\omega\}$ $(= \omega^+)$ is an ordinal.

(c) Is ω^+ equal to ω or some $n \in \omega$?

As before, we shall use the notation α^+ to stand for $\alpha \cup \{\alpha\}$.

Solution

(a) If $\omega \in \mathbf{n}$ then, as $\mathbf{n}$ is an ordinal, all the elements of ω are elements of $\mathbf{n}$. In particular we have $\mathbf{n} \in \mathbf{n}$, which is not true. Hence $\omega \notin \mathbf{n}$.

If $\omega \in \omega$, we have that $\{\omega\}$ is a subset of ω. But the only element of this subset fails to obey the requirement of being its *least* element, as the subset contains an element less than it – $\omega \in \omega$ is the same as $\omega < \omega$. This contradicts that ω is well-ordered. Hence $\omega \notin \omega$.

As an alternative to the above, we could have exploited the consequence of the axiom of foundation, *ZF9*, that there are no infinite descending $\in$-chains of sets. If $\omega \in \mathbf{n}$ then, as $\mathbf{n} \in \omega$, we would have the $\in$-chain

$$\ldots \in \mathbf{n} \in \omega \in \mathbf{n} \in \ldots \in \omega \in \mathbf{n} \in \omega,$$

giving a contradiction. Similarly if $\omega \in \omega$, the chain

$$\ldots \in \omega \in \omega \in \ldots \in \omega \in \omega$$

gives a contradiction. Our preferred argument essentially uses only well-ordering, which is intrinsic to the whole point of ordinals. The whole theory of ordinals can be developed entirely without the axiom of foundation, so it is customary to avoid its use, as we shall do in this chapter.

Note that all the exercises in this chapter expect you to avoid use of ZF9 in your answers. As *ZF9* is one of the more technical, and less intuitive, of the axioms, it is nice to know it is not needed for this part of the theory.

(b) First we must show that $\omega \cup \{\omega\}$ is well-ordered by $\in$. We need to check that $\in$ is a linear order on this set and then check the well-ordering property.

How about irreflexivity? We know from our work on natural numbers in *ZF* that $\mathbf{n} \notin \mathbf{n}$ for all $\mathbf{n} \in \omega$. And from the previous part we have $\omega \notin \omega$. Thus $\in$ is irreflexive on $\omega \cup \{\omega\}$.

For transitivity, the only cases we need to consider are

As we cannot have $\omega \in \mathbf{n}$ or $\omega \in \omega$.

$$\mathbf{l} \in \mathbf{m} \text{ and } \mathbf{m} \in \mathbf{n} \text{ for } \mathbf{n} \text{ in } \omega$$

and

$$\mathbf{l} \in \mathbf{m} \text{ and } \mathbf{m} \in \omega.$$

In the first case $\mathbf{l} \in \mathbf{n}$ as $\in$ is transitive on $\mathbf{n}$; and in the second case $\mathbf{l} \in \omega$ as all elements of a natural number are natural numbers.

We already know that $\in$ is linear on the elements of ω. And we can compare any element of ω with the other member of $\omega \cup \{\omega\}$, namely ω itself – we have $\mathbf{n} \in \omega$! Hence the order is linear.

To show the well-ordering property suppose that A is a non-empty subset of $\omega \cup \{\omega\}$. If A is simply the subset $\{\omega\}$, then as $\omega \notin \omega$ we have that ω is the least (and only!) element of A. Otherwise A contains at least one element of ω: as ω is well-ordered by $\in$, the least such element in A, $\mathbf{n}$ say, is also the least element of A – we have $\mathbf{n} \in \mathbf{m}$ for all other $\mathbf{m}$s in A, and if A also includes ω then automatically $\mathbf{n} \in \omega$.

We must also show that if $\beta \in \omega \cup \{\omega\}$ then $\beta \subseteq \omega \cup \{\omega\}$. This is straightforward: as β is either a natural number or ω we have

$$\beta \subseteq \omega \subseteq \omega \cup \{\omega\}.$$

(c) If ω^+ equalled ω then we would have

$$\omega \in \omega \cup \{\omega\} = \omega,$$

so that $\omega \in \omega$, which is not true. Likewise if ω^+ equalled $\mathbf{n}$ for some $\mathbf{n} \in \omega$, we would have

$$\mathbf{n} \in \omega \subseteq \omega^+ = \mathbf{n},$$

so that $\mathbf{n} \in \mathbf{n}$, which is not true.

Another way of approaching this is by comparing the order structure of ω^+ with the ordered sets considered in previous sections. It is easy to show that ω^+ is order-isomorphic to the order sum $\omega + \mathbf{1}$, which is known not to be order-isomorphic to ω or any finite set. So that ω^+ cannot actually equal any of these latter sets.

We shall look at how order sums and products relate to ordinals in the next section.

So ω^+ is an ordinal we hadn't listed before. We shall show that there are further ordinals, enough to represent all well-ordered sets. But first let's establish some of the basic properties of ordinals.

Exercise 8.4

Let α be an ordinal and let $\beta \in \alpha$.

(a) Show that β is an ordinal.

(b) Show that $\beta = \mathrm{Seg}_\alpha(\beta)$.

So that any member of an ordinal is also an ordinal; and an initial segment is an ordinal.

Solution

(a) First of all, we must check that β is well-ordered by $\in$. As $\beta \in \alpha$ and α is an ordinal, we have that $\beta \subseteq \alpha$. Any subset of a well-ordered set is clearly well-ordered by the same order, so that β is well-ordered by $\in$.

We must also show that if $\gamma \in \beta$ then $\gamma \subseteq \beta$, i.e. that for all δ,

if $\delta \in \gamma$ then $\delta \in \beta$.

So suppose that $\gamma \in \beta$ and take any $\delta \in \gamma$. If we knew that δ, γ and β were all members of α, then as $\in$ is a linear order on α, the transitivity of the order would allow us to conclude that $\delta \in \beta$, which is what we want. We are given that β is in α. And as α is an ordinal this means $\beta \subseteq \alpha$. Now $\gamma \in \beta$ and $\beta \subseteq \alpha$, so that $\gamma \in \alpha$. Similarly, this gives that $\gamma \subseteq \alpha$, so that as $\delta \in \gamma$ we also have $\delta \in \alpha$. Thus δ, γ and β *are* all in α, and our argument is complete.

(b) Left for you!

We can use the results of this exercise to rephrase the principle of transfinite induction (Theorem 7.6) when the well-ordered X to which it is applied is an ordinal, α say. We leave it for you to check that it is merely a rephrasing!

Theorem 8.1 Principle of transfinite induction for ordinals

Let A be a subset of the ordinal α such that

(i) $\mathbf{0} \in A$;

(ii) for all $\beta \in \alpha$, if $\beta \subseteq A$ then $\beta \in A$.

Then $A = \alpha$.

Theorem 8.2

Suppose that α and β are order-isomorphic ordinals. Then $\alpha = \beta$.

Proof

We shall argue by contradiction. Suppose that $f\colon \alpha \longrightarrow \beta$ is an order-isomorphism and that f is not the identity map. Then $\{\gamma \in \alpha : f(\gamma) \neq \gamma\}$ is a non-empty subset of α, so has a least element γ_0. We shall consider the restriction, $f|_{\text{Seg}_\alpha(\gamma_0)}$, of f to the initial segment of α determined by γ_0.

From the solution to Exercise 7.55 in Section 7.4 we know that $f|_{\text{Seg}_\alpha(\gamma_0)}$ is an order-isomorphism between $\text{Seg}_\alpha(\gamma_0)$ and $\text{Seg}_\beta(f(\gamma_0))$. We can use the result of Exercise 8.4(b) to rephrase this by saying

$$f|_{\gamma_0} : \gamma_0 \longrightarrow f(\gamma_0) \text{ is an order-isomorphism.}$$

By the minimality of γ_0 we have

$$f|_{\gamma_0}(\delta) = f(\delta) = \delta \text{ for all } \delta \in \gamma_0,$$

so that the sets γ_0 and $f(\gamma_0)$ have precisely the same members. But this means that $\gamma_0 = f(\gamma_0)$, contradicting the definition of γ_0.

By the axiom of extensionality.

Hence f is the identity map, so that $\alpha = \beta$. ■

This theorem tells us that if, as we hope, each class of order-isomorphic well-ordered sets contains an ordinal, then this ordinal is unique. We have yet to show that there is such an ordinal! The theorem will also help us prove that the class of ordinals, not just any single ordinal α, is well-ordered by $\in$, which we do now.

Why not the *set* of ordinals?

> ***Theorem 8.3***
>
> The class of ordinals is well-ordered by $\in$, in the following sense:
>
> (i) for all ordinals α, $\alpha \notin \alpha$;
>
> (ii) for all α, β and γ, if $\alpha \in \beta$ and $\beta \in \gamma$ then $\alpha \in \gamma$;
>
> (iii) for all ordinals α and β, $\alpha \in \beta$ or $\alpha = \beta$ or $\beta \in \alpha$;
>
> (iv) any non-empty set X of ordinals contains a least element.

Proof

The proofs of (i) and (ii) are left as an exercise for you: they can both be done using the basic definition of ordinals. (Although (i) is an easy consequence of the axiom of foundation, avoid its use here!)

For (iii), suppose that α and β are ordinals and that $\alpha \neq \beta$. Then by Theorem 8.2 α and β are not isomorphic. So by the big result on comparing well-ordered sets, Theorem 7.5, one of α and β is order-isomorphic to an initial segment of the other. Without loss of generality β is order-isomorphic to an initial segment of α, so that

$$\beta \cong \text{Seg}_\alpha(\gamma), \text{ for some } \gamma \in \alpha.$$

By Exercise 8.4(b) $\text{Seg}_\alpha(\gamma)$ is just γ itself, so that

$$\beta \cong \gamma \text{ for some } \gamma \in \alpha.$$

But then, by Theorem 8.2, β equals γ for this element γ of α, i.e. $\beta \in \alpha$.

(iv) follows from Exercises 7.66 and 8.4(b). We can also prove this directly as follows. If X is a non-empty set of ordinals, then it contains at least one ordinal, α say. Consider the set $Y = \{\beta \in X : \beta \in \alpha\}$. If Y is empty then α is the least element of X. And if $Y \neq \varnothing$ then, as Y is a non-empty subset of the well-ordered set α, Y has a least element γ, which must also be the least element of X. ■

The same proof shows that any non-empty *class* of ordinals contains a least element.

Exercise 8.5

Prove (i) and (ii) of Theorem 8.3 above. (For (i) in particular, avoid use of the axiom of foundation or any of its consequences, especially that $x \notin x$ for any set x, not just an ordinal.)

Exercise 8.6

Suppose that α and β are ordinals. Show that

$\alpha \leq \beta$ if and only if $\alpha \subseteq \beta$.

Solution

Suppose first that $\alpha \leq \beta$ and show that $\alpha \subseteq \beta$. If $\alpha = \beta$ there is nothing more needed, so suppose that $\alpha < \beta$, i.e. $\alpha \in \beta$. Then as β is an ordinal, we have $\alpha \subseteq \beta$ by the definition of ordinal.

Conversely, suppose that it is not the case that $\alpha \leq \beta$, so that $\alpha \notin \beta$ and $\alpha \neq \beta$, and show that it is not the case that $\alpha \subseteq \beta$. Then as $\in$ linearly orders the ordinals, we have $\beta \in \alpha$, so that, as in the first part of this solution, $\beta \subseteq \alpha$. As $\alpha \neq \beta$ we must in fact have that β is a proper subset of α, so that α cannot be a subset of β.

Using Theorem 8.3.

We have been careful to avoid saying that there is a *set* of all ordinals – there isn't such a set and we are now in a position to prove this.

Exercise 8.7

Suppose that there is a set $\mathscr{O}$ whose members are precisely the ordinals.

(a) Show that $\mathscr{O}$ is an ordinal.

(b) Hence derive a contradiction.

Solution

(a) Theorem 8.3 shows that $\mathscr{O}$ is well-ordered by $\in$. And if $\alpha \in \mathscr{O}$, then as every member of α is an ordinal (by Exercise 8.4(a)) we have $\alpha \subseteq \mathscr{O}$. Thus $\mathscr{O}$ is an ordinal.

(b) As we are supposing that $\mathscr{O}$ is the set of all ordinals, we must have $\mathscr{O} \in \mathscr{O}$. But this contradicts Theorem 8.3(i).

This contradiction is called *Burali-Forti's paradox*, published in 1897 by the Italian Cesare Burali-Forti (see [11]). It was the first key paradox to arise from Cantor's work and sparked the proper evaluation of the foundations of set theory. Its significance in the context of ZF is that the ordinals form a proper class. To help us towards our goal of showing that each isomorphism class of well-ordered sets does contain an ordinal, the following theorem gives a couple of very useful ways of constructing an ordinal from other, known, ordinals.

It is often convenient to use the notation $\mathscr{O}$ for the class of all ordinals. (Some books use Cantor's notation, Ω, the last letter of the Greek alphabet.)

Theorem 8.4

(i) Let α be an ordinal. Then α^+ is

(a) an ordinal;

(b) the successor of α.

(ii) Let X be a set of ordinals. Then $\bigcup X$ is

(a) an ordinal;

(b) the least upper bound of X.

(Thus we can write $\bigcup X$ as lub X.)

α^+ is $\alpha \cup \{\alpha\}$.

Least in the ordinals.

Proof

(i) is left as an exercise for you.

To prove (ii)(a), suppose that X is a set of ordinals. To show that $\bigcup X$ is an ordinal, we have, as ever, to show that it is well-ordered by $\in$ and that each of its members is a subset.

Each member of $\bigcup X$ is in some ordinal α in the set X, so is itself an ordinal. Thus $\bigcup X$ is a set of ordinals and is then, by Theorem 8.3, well-ordered by $\in$.

Now suppose that $\beta \in \bigcup X$. We shall show that $\beta \subseteq \bigcup X$. As $\beta \in \bigcup X$, we have $\beta \in \alpha$ for some $\alpha \in X$. As $\alpha \in X$ then $\alpha \subseteq \bigcup X$ and as α is an ordinal we then have $\beta \subseteq \alpha$. Thus $\beta \subseteq \bigcup X$ as required.

Now that we have proved that $\bigcup X$ is an ordinal, it is meaningful to attempt (ii)(b). We have to show that

Least upper bounds were covered in Exercise 7.19 of Section 7.2.

(1) $\bigcup X$ is an upper bound of X, i.e. $\alpha \leq \bigcup X$ for all $\alpha \in X$, and

(2) $\bigcup X$ is the *least* such upper bound, i.e. if γ is an upper bound for X then $\bigcup X \leq \gamma$.

Equivalently we could show that: (2)$'$ if $\delta < \bigcup X$ then δ is *not* an upper bound of X.

For (1), suppose that $\alpha \in X$. Then $\alpha \subseteq \bigcup X$, so by Exercise 8.6 $\alpha \leq \bigcup X$ as required.

For (2), suppose that the ordinal γ is an upper bound for X. We shall show that $\bigcup X \subseteq \gamma$, so that by Exercise 8.6 we have $\bigcup X \leq \gamma$ as required. Take any $\beta \in \bigcup X$. Then $\beta \in \alpha$ for some $\alpha \in X$. As γ is an upper bound for X, we have $\alpha \leq \gamma$, which by Exercise 8.6 means that $\alpha \subseteq \gamma$. Thus $\beta \in \alpha$ and $\alpha \subseteq \gamma$, so that $\beta \in \gamma$. ■

In the theory of ordinals, it is quite common in arguments about order words, like 'upper bound', to switch between $<$ and $\in$.

Exercise 8.8

Prove Theorem 8.4(i), i.e. if α is an ordinal then α^+ is

(a) an ordinal;

(b) the successor of α.

That is, the least ordinal greater than α.

It will help here to introduce some streamlined terminology to describe different sorts of ordinals. We know from Exercise 7.12 that as the ordinals are linearly ordered, then every ordinal is either a minimum element of the order, or a successor, or a limit point of the order. The ordinals certainly have a minimum element, namely $\mathbf{0}$. We shall sometimes use the phrase *successor ordinal* for an ordinal which is a successor and *limit ordinal* for an ordinal which is a limit. As we now know that every ordinal α has a successor α^+, it will be a very straightforward for you to prove a simple but convenient reformulation of the definition of limit ordinal, given in the next exercise.

$\mathbf{0}$ is the set $\varnothing$.

Exercise 8.9

Let λ be a non-zero ordinal. Show that λ is a limit ordinal if and only if

for all ordinals α, if $\alpha < \lambda$ then $\alpha^+ < \lambda$.

Exercise 8.10

Identify the ordinal $\bigcup X$ for each of the following sets X:

(a) $X = \varnothing$;

(b) $X = \{\mathbf{2}, \mathbf{4}, \mathbf{6}\}$;

(c) $X = \{\mathbf{2} \cdot \mathbf{n} : \mathbf{n} \in \omega\}$.

For ordinals $\alpha < \beta$ means the same as $\alpha \in \beta$.

Exercise 8.11

Let λ be an ordinal.

(a) Suppose that λ is a limit ordinal. Show that $\bigcup \lambda = \lambda$.

(b) Suppose that $\bigcup \lambda = \lambda$. Show that λ is a limit ordinal.

Solution

(a) We have a choice of arguments: one using the terminology of $\in$ and $\bigcup \lambda$ as the union of a set; one using the terminology of $<$ and least upper bounds; and, when feeling cavalier, one mixing both lots of terminology! We shall give arguments of the first two sorts.

First let's use the language of $\in$ and unions. We need to show that $\bigcup \lambda \subseteq \lambda$ and $\lambda \subseteq \bigcup \lambda$.

Suppose that $\beta \in \bigcup \lambda$. Then $\beta \in \alpha$ for some $\alpha \in \lambda$. As λ is an ordinal we then have $\beta \in \lambda$. Thus $\bigcup \lambda \subseteq \lambda$.

Suppose that $\alpha \in \lambda$. Then as λ is a limit ordinal we have $\alpha^+ \in \lambda$. But $\alpha \in \alpha^+$, so that $\alpha \in \bigcup \lambda$. Thus $\lambda \subseteq \bigcup \lambda$.

$\alpha \in \{\alpha\} \subseteq \alpha^+$.

Now let's try the argument using the terminology of order. By Theorem 8.4 it is enough to show that the least upper bound of the set λ of ordinals is λ itself.

First we show that the set of ordinals λ has λ as an upper bound. If α is a member of λ, then $\alpha < \lambda$ (as $<$ is the same as $\in$). Thus the set λ has upper bound λ.

Now we must show that λ is the *least* upper bound. We will take the route of showing that no smaller ordinal γ is an upper bound. Take any $\gamma < \lambda$. As λ is a limit ordinal then $\gamma^+ < \lambda$, so that γ^+ is in the set λ of ordinals. Thus as $\gamma < \gamma^+$ it cannot be that γ is an upper bound of the set λ of ordinals. Hence λ is the least upper bound of the set λ, so must equal $\bigcup \lambda$ (by Theorem 8.4).

(b) Left for you!

Exercise 8.12

What is $\bigcup(\alpha^+)$ in terms of α, where α is an ordinal?

Exercise 8.13

Let X be a non-empty set of ordinals and let $\alpha = \bigcup X$. Show that if $\alpha \notin X$, then α is a limit ordinal.

Solution

Suppose that $\alpha \notin X$. Let $\beta \in \alpha$. We must show that there is an ordinal γ with $\beta \in \gamma \in \alpha$. As $\in$ is the same as $<$ for ordinals, we have $\beta < \alpha$, so that as $\alpha = \text{lub}\, X$, we know that β is *not* an upper bound for X. Thus there is some $\gamma \in X$ for which $\beta < \gamma$. As α is lub X, we have $\gamma \leq \alpha$. But $\alpha \notin X$, whereas $\gamma \in X$, so that $\gamma \neq \alpha$. Thus $\beta < \gamma < \alpha$, as required to show that α is a limit ordinal.

$\beta < \gamma < \alpha$ translates into $\beta \in \gamma \in \alpha$.

Now that we have some ways of constructing ordinals, we might be in a better position to show that each isomorphism class of well-ordered sets contains an ordinal. At the beginning of this section we showed that the ordinals included **0**, **1**, **2**, **3**, ..., ω and ω^+, but asked whether we had an ordinal order-isomorphic to the order sum $\omega + \omega$. From your work in Chapter 7, you might agree that $\omega + \omega$ is a pretty simple and straightforward well-ordered set! Recall its shape:

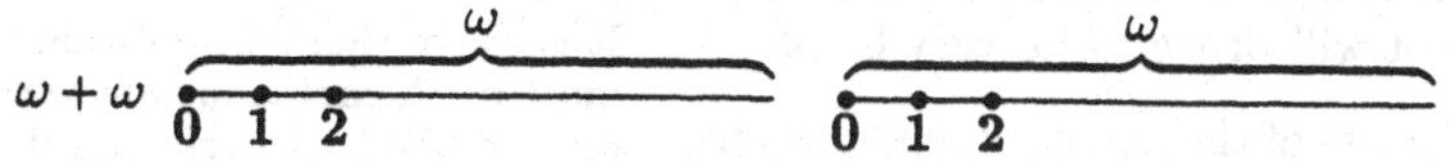

It should seem reasonable to say that $\omega + \omega$ is a 'smallest' well-ordered set which contains initial segments order-isomorphic to each of the order sums ω, $\omega + \mathbf{1}$, $\omega + \mathbf{2}$, ..., $\omega + \mathbf{n}$, ... for all $\mathbf{n} \in \omega$. Now apply the result of Theorem 8.4(i)(a) finitely many times to the ordinal ω to show that there is an ascending sequence of ordinals

Finite applications of such a result are legitimate in *ZF*. As ever, it's only infinite processes which cause a problem!

$$\omega, \omega^+, \omega^{++}, \omega^{+++}, \ldots, \omega^{\overbrace{+++\cdots+}^{n}}, \ldots,$$

We write ω^{++} for $(\omega^+)^+$ etc.

for each $\mathbf{n} \in \omega$. It is pretty plausible (and true!) that we can use Theorem 8.4(i)(b) to show that

$$\omega + \mathbf{n} \cong \omega^{\overbrace{+++\cdots+}^{n}}$$

for each $\mathbf{n}$. Then surely $\omega + \omega$ will be order-isomorphic to the least upper bound of all these $\omega^{\overbrace{+++\cdots+}^{n}}$s, which by Theorem 8.4(ii) would be the ordinal

$$\bigcup\{\omega, \omega^+, \omega^{++}, \omega^{+++}, \ldots\}.$$

This will all turn out to be correct, but note that there is a significant snag with what we have said so far: the application of Theorem 8.4(ii) requires not only that $\omega, \omega^+, \omega^{++}, \ldots$ are ordinals, which is unproblematic, but that there is a *set*

$$\{\omega, \omega^+, \omega^{++}, \omega^{+++}, \ldots\}$$

containing all of them. To show that this is a set we can exploit the recursion principle for $\mathbb{N}$ (in its class form, Theorem 4.5 of Section 4.5 of Chapter 4), defining a function f with domain $\mathbb{N}$ by

$$f(\mathbf{n}) = \begin{cases} \omega, & \text{if } \mathbf{n} = \mathbf{0}, \\ (f(\mathbf{k}))^+, & \text{if } \mathbf{n} = \mathbf{k}^+. \end{cases}$$

By the recursion principle, f is a function within *ZF*, so that Range(f), i.e. $\{\omega, \omega^+, \omega^{++}, \omega^{+++}, \ldots\}$, is indeed a set.

The recursion principle for $\mathbb{N}$, which we have just exploited, is a consequence of the axiom of replacement, *ZF8*. This axiom has a crucial role in the further development of the theory of ordinals, as we shall see not only in a moment, when we prove that every well-ordered set is order-isomorphic to some ordinal, but in the next section when we extend the recursion principle to define functions within *ZF* with domains much larger than $\mathbb{N}$. Recall that the axiom essentially says that if a well-formed formula $\phi(x, y)$ of set theory behaves like a function f with domain known to be a set, X, but no codomain is specified as a set, then the collection of images, Range(f), *is* a set. For a function f like the one we used above to show that $\{\omega, \omega^+, \omega^{++}, \omega^{+++}, \ldots\}$ is a set, we had a set ($\mathbb{N}$ or equivalently ω) as domain and sets as well-defined images of elements of the domain. But we could not specify in advance any set as codomain; and a function within *ZF* has to have a set as codomain. As we are trying to construct the ordinals within *ZF* and could not have otherwise shown that there are enough of them to provide a set as codomain for even this very modest f, the axiom of replacement will prove to be very helpful.

A function is a subset of $X \times Y$, where X and Y are sets.

It is the author's impression that most mathematicians are very relaxed about the crucial role of this axiom within set theory: arguably it is more natural and much less contentious than seemingly more basic axioms, like the power set axiom.

Now that we have identified the likely need for use of the axiom of replacement, we can prove a result which virtually achieves our aim of showing that there are 'enough' ordinals.

Theorem 8.5

Let X be a well-ordered set. Then for each $x \in X$ there is an ordinal α such that $\mathrm{Seg}_X(x)$ is order-isomorphic to α.

Proof

Let X be a well-ordered set. Let A be the set

$$\{x \in X : \text{ there is some ordinal } \alpha \text{ such that } \mathrm{Seg}_X(x) \cong \alpha\}.$$

We shall show by transfinite induction that $A = X$.

That is, using Theorem 7.6.

If X is empty then so is A, so that $A = X$. So now suppose that X is non-empty. We need to show that

(i) the least element of X is in A;

(ii) for all $x \in X$, if $\mathrm{Seg}_X(x) \subseteq A$ then $x \in A$.

The least element x_0 of X is in A because $\mathrm{Seg}_X(x_0)$ is just the empty set $\varnothing$, and $\varnothing$ is the ordinal $\mathbf{0}$.

For the inductive step, suppose that $x \in X$ is such that $\mathrm{Seg}_X(x) \subseteq A$, where x is not the least element of X. This means that for each x' with $x' <_X x$,

We are trying to show that $x \in A$, i.e. that there is some α such that $\mathrm{Seg}_X(x) \cong \alpha$.

there is some ordinal β for which $\mathrm{Seg}_X(x') \cong \beta$. By Theorem 8.2 the β for each such x' is unique. Thus the well-formed formula ϕ defined by

Take it on trust that ϕ can be built up within our formal language for sets.

if $x' < x$ then $\phi(x', \beta)$ if and only if $\mathrm{Seg}_X(x') \cong \beta$,
and $\beta = \mathbf{0}$ otherwise,

behaves like a function on the universe $\mathscr{V}$. This is just the sort of situation for which we included the axiom of replacement as one of the *ZF* axioms. By the axiom of replacement ϕ does define a function (i.e. a set) when its domain is restricted to the set

$$\{x' \in X : x' <_X x\}, \text{ i.e. } \mathrm{Seg}_X(x);$$

and the images of this function form a set, so that

$$\{\beta : \beta \text{ is an ordinal with } \beta \cong x' \text{ for some } x' <_X x\}$$

is a set, Y say. It should seem reasonable that the function g defined by ϕ is an order-isomorphism between $\mathrm{Seg}_X(x)$ and Y. So if we could prove that Y is in fact an ordinal, we would have shown that $x \in A$, as required.

Y is a set of ordinals, so is well-ordered by $\in$, by Theorem 8.3. The extra thing needed to show that Y is an ordinal is that if $\beta \in Y$ then $\beta \subseteq Y$. So suppose that $\beta \in Y$ and that $\gamma \in \beta$. As $\beta \in Y$ there is some $x' <_X x$ with $\mathrm{Seg}_X(x)' \cong \beta$. Let $f\colon \mathrm{Seg}_X(x') \longrightarrow \beta$ be the corresponding isomorphism. As $\gamma \in \beta$, there must be some $x'' \in \mathrm{Seg}_X(x')$ such that $f(x'') = \gamma$. Then by Exercise 7.55, we have

We shall show that $\gamma \in Y$.

$$\mathrm{Seg}_{\mathrm{Seg}_X(x')}(x'') \cong \mathrm{Seg}_\beta(\gamma).$$

By Exercise 7.54(c),

$$\mathrm{Seg}_{\mathrm{Seg}_X(x')}(x'') = \mathrm{Seg}_X(x''),$$

while by Exercise 8.4(b),

$$\mathrm{Seg}_\beta(\gamma) = \gamma.$$

Thus

$$\mathrm{Seg}_X(x'') \cong \gamma,$$

so as $x'' <_X x'$, and hence $x'' <_X x$, the ordinal γ must be $g(x'')$ and in the set Y.

This shows that if $\beta \in Y$ then $\beta \subseteq Y$, completing the argument that Y is an ordinal. This in turn completes the proof of the inductive step, i.e.

if $\mathrm{Seg}_X(x) \subseteq A$ then $x \in A$.

Thus by Theorem 7.6 $A = X$. ■

Exercise 8.14

Check that the function g in the proof above, i.e.

$$g\colon \mathrm{Seg}_X(x) \longrightarrow Y$$
$$x' \longmapsto \text{the unique ordinal } \beta \text{ such that } \mathrm{Seg}_X(x') \cong \beta$$

is an order-isomorphism.

We are now in a position to prove the promised theorem that there are enough ordinals to represent all well-ordered sets.

> ***Theorem 8.6***
>
> Let Y be a well-ordered set. Then there is a unique ordinal α which is isomorphic to Y.

Proof

The trick is to construct a well-ordered set X which contains Y, or an isomorphic copy of Y, as an initial segment $\text{Seg}_X(x)$ for some $x \in X$. Then Theorem 8.5 tells us that there is an ordinal α isomorphic to $\text{Seg}_X(x)$ and hence to Y. The uniqueness of α follows from Theorem 8.2.

There are several ways of constructing a suitable X. We shall define X to be the order-theoretic sum $Y + \mathbf{1}$, i.e. the set $(Y \times \{\mathbf{0}\}) \cup (\mathbf{1} \times \{\mathbf{1}\})$. By Exercise 7.47 $Y + \mathbf{1}$ is well-ordered. Our result follows because

$$\begin{aligned} Y &\cong Y \times \{\mathbf{0}\} \\ &= \text{Seg}_X((\mathbf{0}, \mathbf{1})) \end{aligned}$$

$(\mathbf{0}, \mathbf{1})$ is in $\mathbf{1} \times \{\mathbf{1}\}$ and is greater than any $(y, \mathbf{0})$.

■

We have finally justified that there are enough ordinals to represent, up to isomorphism, all possible well-ordered sets. The next question is: how big can ordinals get? We already know that there is no set of all ordinals, which usually suggests that there are too many of them to form a set! All our examples so far of well-ordered sets, and hence, up to isomorphism, of ordinals, just happen to be countable sets. And there are indeed too many countable well-ordered sets for there to be a *set* of all of them.

In the next section we shall exploit the structure of ordinals to simplify and extend the arithmetic of ordinals, to recapture Cantor's original vision.

Exercise 8.15

Explain why there is no set

$$\{X : X \text{ is a countable well-ordered set}\}.$$

Solution

If this was a set, then it would have as a subset

$$\{\{x\} : x \text{ is a set}\},$$

which would thus be a set in its own right. But we know that this is not a set, so that there is no set of all countable ordinals.

See Exercise 4.36 of Section 4.4.

Our solution to this exercise suggests that we did not get a set because we had included too many copies of each order-isomorphism class of countable well-ordered sets. If we look at only the countable ordinals, taking just one member of each order-isomorphism class, might we get a set? That is, is

$$\omega_1 = \{\alpha : \alpha \text{ is a countable ordinal}\}$$

a set? If so, then as there is no set of all ordinals, there must be some ordinals which are not countable, i.e. are uncountable. We shall now show not only that ω_1 is a set, but is also an uncountable ordinal.

> ***Theorem 8.7***
>
> Let ω_1 be defined by
>
> $$\omega_1 = \{\alpha : \alpha \text{ is a countable ordinal}\}.$$
>
> Then ω_1 is the smallest uncountable ordinal.

We shall later use the notation ω_0 for ω; and there'll also be an ω_α for each ordinal α.

Proof

We must show that

(i) ω_1 is a set;

(ii) that it is an ordinal;

(iii) that it is uncountable;

(iv) that it is the least uncountable ordinal.

Within the context of this section it won't be surprising that the axiom of replacement will help show that ω_1 is a set; and that some of the other parts of the proof will be similar to earlier arguments involving ordinals.

(i) Let α be a countable ordinal. By virtue of being countable, there must be a bijection between α and a subset of $\mathbb{N}$ (finite in the case that α is finite). The ordering on α induces an ordering on this subset of $\mathbb{N}$. And this last ordering is given as a set by some subset of $\mathbb{N} \times \mathbb{N}$. In fact there will in general be many subsets of $\mathbb{N} \times \mathbb{N}$ arising in this way from α, as there can be several different bijections from α to subsets of $\mathbb{N}$.

See Exercise 8.17 below.

Turning the above process on its head, by starting with a subset of $\mathbb{N} \times \mathbb{N}$ representing a well-order, there must be a countable ordinal α to which this well-order (on $\mathbb{N}$ or a subset) is order-isomorphic. Of course by Theorem 8.2 this ordinal α is unique. So that we can (plausibly!) define a formula $\phi(R, y)$ within set theory such that

> for all subsets R of $\mathbb{N} \times \mathbb{N}$ representing a well-order on a subset of $\mathbb{N}$ and all y,
>
> $\phi(R, y)$ if and only if y is the ordinal order-isomorphic to the subset of $\mathbb{N}$ well-ordered by R.
>
> and for any other R, take $y = \varnothing$ (or some other fixed set).

Now put

$$W = \{R \subseteq \mathbb{N} \times \mathbb{N} : R \text{ is a well-order on a subset of } \mathbb{N}\}.$$

W is a subset of $\mathscr{P}(\mathbb{N} \times \mathbb{N})$, so is a set. By the axiom of replacement, the formula ϕ above defines a function with domain W. It should be clear from the way we arrived at ϕ that the range of this function is ω_1. Thus ω_1 is a set.

(ii) ω_1 is a set of ordinals, so is well-ordered by $\in$. Thus to show that ω_1 is an ordinal, it remains to show that if $\alpha \in \omega_1$ then $\alpha \subseteq \omega_1$. So suppose that

See Theorem 8.3.

$\alpha \in \omega_1$ and that $\beta \in \alpha$. Then α is a countable ordinal, while β is an ordinal which is also a subset of α. Thus β is also countable and must be in ω_1. Hence $\alpha \subseteq \omega_1$. Thus ω_1 is an ordinal.

(iii) If ω_1 was countable then, as it is now known to be an ordinal, we would have $\omega_1 \in \omega_1$, which is impossible. Thus ω_1 is uncountable.

(iv) We shall leave it to you to show that $\bigcup \omega_1 = \omega_1$. Then by Theorem 8.4 ω_1 is the least upper bound of the set of all countable ordinals, so that any smaller ordinal than ω_1 must be countable. Thus as ω_1 is uncountable, it must therefore be the smallest uncountable ordinal. ■

Exercise 8.16

Prove that $\bigcup \omega_1 = \omega_1$.

Exercise 8.17

Give two different bijections from the ordinal ω^+ to $\mathbb{N}$ and give a bijection from ω^+ to a proper subset of $\mathbb{N}$.

We shall show in a later section that there are ordinals even larger, in the sense of cardinality, than ω_1. You might reasonably ask why we did not give examples earlier than now of uncountable well-ordered sets. Surely, for instance, we could have given an example of a well-ordering on the set $\mathbb{R}$ of all reals. As we saw earlier, the usual $<$ on $\mathbb{R}$ is not a well-order; but $\mathbb{R}$ has such rich structure – think of all that real analysis! – that one might think that some sort of well-order on it could be defined. This was what Cantor thought very early on in the development of his theory of infinite sets, but he came to realize that such a definition was difficult, so formulated the *well-ordering problem*: can $\mathbb{R}$ be well-ordered? As we shall see in the final chapter of the book, this major problem has a non-trivial resolution! You might also ask how the size of ω_1, which is the smallest uncountable ordinal, compares with that of other uncountable sets, in particular $\mathbb{R}$ or equivalently, in terms of cardinality, $\mathscr{P}(\mathbb{N})$ or $2^{\mathbb{N}}$. This, the other of Cantor's major problems, will likewise be discussed in the final chapter.

See the solution to Exercise 7.42(c) in Section 7.4.

See also Exercise 8.22 below.

Further exercises

Exercise 8.18

Which of the following assertions is true? Explain your answers.

(a) If α is an ordinal, then $\bigcup \alpha$ is an ordinal.

(b) If X is a set of ordinals, then X is an ordinal.

(c) If X is a set of ordinals and $\bigcup X = X$, then X is a limit ordinal.

(d) If X is a set such that $\bigcup X = X$, then X is an ordinal.

Exercise 8.19

(a) Let α and β be ordinals. Show that if $\alpha^+ = \beta^+$, then $\alpha = \beta$.

(b) Is the above result true for sets which are not necessarily ordinals? That is, if x and y are sets with $x^+ = y^+$, then $x = y$?

So that the successor (class) function is one–one on ordinals.

In general, $z^+ = z \cup \{z\}$.

Exercise 8.20

(a) Let α be an ordinal. Show that there is a limit ordinal μ with $\alpha < \mu$. [Hint: consider the set $\{\alpha, \alpha^+, \alpha^{++}, \dots\}$.]

(b) Show that there is no set of all limit ordinals, i.e. $\{\lambda : \lambda \text{ is a limit ordinal}\}$ is not a set.

Exercise 8.21

The theory of ordinals could have been developed without using results about general well-ordered sets like Theorem 7.5. Try proving the following results straight from the basic definition of an ordinal. Recall that the set α is an ordinal if

Our approach in this section has been to make heavy use of results of Chapter 7.

(i) α is well-ordered by $\in$, and

(ii) if $\beta \in \alpha$ then $\beta \subseteq \alpha$.

(a) Show that for any ordinal α, it is not the case that $\alpha \in \alpha$.

(b) Show that for any ordinals α and β,

$$\alpha \in \beta \text{ if and only if } \alpha \subset \beta.$$

$\alpha \subset \beta$ means that α is a *proper* subset of β.

[Hints: one way round follows from the definition. For the other, suppose that $\alpha \subset \beta$, put $\gamma = \min(\beta \setminus \alpha)$ and show that $\gamma = \alpha$.]

(c) Show that for any ordinals α and β, exactly one of the following holds:

$$\alpha \in \beta \text{ or } \alpha = \beta \text{ or } \beta \in \alpha.$$

[Hints: suppose that $\alpha \neq \beta$, so that without loss of generality $\alpha \setminus \beta \neq \varnothing$. Put $\gamma = \min(\alpha \setminus \beta)$ and show that $\gamma = \beta$. The result of the previous part should be of use.]

Exercise 8.22

(a) Without using AC show that

$$\omega_1 \preceq \mathscr{P}(\mathscr{P}(\mathbb{N})).$$

[Hints: if α is an infinite countable ordinal, then as in Exercise 7.25 of Section 7.2 there is at least one bijection between $\mathbb{N}$ and α. Using such a bijection, one can construct an order on $\mathbb{N}$ so that with this order $\mathbb{N}$ is order-isomorphic to α. This order on $\mathbb{N}$ is completely represented by a set of ordered pairs of natural numbers, i.e. by a subset of $\mathbb{N} \times \mathbb{N}$. In general there will be several such bijections, each giving a corresponding subset of $\mathbb{N} \times \mathbb{N}$. Use these ideas to construct a one–one function from ω_1 to $\mathscr{P}(\mathscr{P}(\mathbb{N}))$.]

(b) Use AC to show that $\omega_1 \preceq \mathscr{P}(\mathbb{N})$. [Hint: if you managed to follow the line of attack suggested by the hint for the previous part, you might be able to identify a stage at which use of AC can give the result for this part.]

See also Exercise 9.16 in Section 9.3.

8.3 Beginning ordinal arithmetic

In the previous section, when discussing whether there is an ordinal order-isomorphic to $\omega + \omega$, we mentioned that it would be easy to use Theorem 8.4(i)(b) to show that

$$\omega + \mathbf{n} \cong \omega^{\overbrace{+++\cdots+}^{n}}$$

for each **n**. This suggests the more general question of identifying the ordinals to which sums and products of well-ordered sets are isomorphic. For this we cannot simply take order sums and products of ordinals, as these do not in general give ordinals. Take, for instance, the sum of ω and the ordinal **1**: their sum $\omega + 1$ is the set

$$(\omega \times \{\mathbf{0}\}) \cup (\mathbf{1} \times \{\mathbf{1}\});$$

and this is not an ordinal, e.g. it isn't well-ordered by $\in$. However, $\omega + 1$ is order-isomorphic to the ordinal ω^+, which suggests that by repeatedly taking successors of ordinals, we might define an operation, $\oplus$ say, on ordinals, rather than general well-ordered sets, corresponding to order addition. How might such a definition go?

For a start, note that it is not enough to use repeated successors to obtain ordinal sums. While this works for adding ω and **n**, giving the ordinal

$$\omega^{\overbrace{+++\cdots+}^{n}},$$

it doesn't do for $\omega + \omega$, which in the previous section was given as order-isomorphic to the ordinal

$$\bigcup\{\omega, \omega^+, \omega^{++}, \omega^{+++}, \ldots\},$$

which involved taking the union of a set of ordinals.

There are other problems: how do we express the taking of n '+'s in

$$\omega^{\overbrace{+++\cdots+}^{n}}$$

within ZF; and for $\omega + \omega$ we need to know about all the ordinals corresponding to the smaller sums $\omega + \mathbf{n}$, for $\mathbf{n} \in \omega$.

The way to solve these problems is to adapt the method for defining $+$ on the set of natural numbers, namely recursion; i.e. define the sum of two ordinals in terms of the sum (or sums) of smaller ordinals. As for $+$ on $\mathbb{N}$, we shall define $\alpha \oplus \beta$ for a fixed ordinal α and all ordinals β. As with natural numbers, we distinguish between adding **0** and adding other ordinals β, setting

$$\alpha \oplus \mathbf{0} = \alpha.$$

To ensure that our definition gives the sensible result for finite ordinals, we then set

$$\alpha \oplus \gamma^+ = (\alpha \oplus \gamma)^+.$$

So far we have defined $\alpha \oplus \beta$ when β is **0** or a successor γ^+. What about when

The sum and product of two well-ordered sets are both well-ordered, by Exercise 7.47 in Section 7.4, so must both be order-isomorphic to an ordinal.

For this section, we shall use the symbol $\oplus$ for our new operation which gives an ordinal as the 'sum' of two ordinals, to distinguish it from the $+$ which has so far in this chapter meant order sum. And we shall call the new operation *ordinal sum*.

We defined $+$ on $\mathbb{N}$ by
$$\mathbf{m} + \mathbf{0} = \mathbf{m},$$
$$\mathbf{m} + \mathbf{n}^+ = (\mathbf{m} + \mathbf{n})^+,$$
for fixed **m** and all $\mathbf{n} \in \mathbb{N}$.

β is a limit ordinal? Following the example of $\omega + \omega$ above, we define, for a limit ordinal λ,

$$\omega + \omega \cong \bigcup\{\omega, \omega^+, \omega^{++}, \ldots\} = \bigcup\{\omega \oplus \mathbf{n} : \mathbf{n} \in \omega\}$$

$$\alpha \oplus \lambda = \bigcup\{\alpha \oplus \gamma : \gamma \in \lambda\}.$$

Putting these together, we appear to have defined a function, for a fixed ordinal α and all ordinals β, as follows:

$$\alpha \oplus \mathbf{0} = \alpha,$$
$$\alpha \oplus \gamma^+ = (\alpha \oplus \gamma)^+,$$
$$\alpha \oplus \lambda = \bigcup\{\alpha \oplus \gamma : \gamma \in \lambda\} \quad \text{for a limit ordinal } \lambda.$$

It's common to reserve the letter λ for a limit ordinal.

This tells us, by means of a finite description, how to add α to any ordinal, as it covers all possible cases: $\mathbf{0}$, a successor ordinal β^+, a limit ordinal λ. In each case (except when adding $\mathbf{0}$) the sum depends on the sum of α with smaller ordinals. Some questions arise:

1. Surely, as there is no set of all ordinals, we cannot have defined a function within *ZF*, i.e. a set?
2. Is $\alpha \oplus \beta$, as defined above, always an ordinal?
3. Does $\alpha \oplus \beta$, as defined above, meet our requirement of being order-isomorphic to the order sum of α and β?
4. Is definition by such a recursive process legitimate within *ZF*?

We can dodge the first question by fixing on some (large!) ordinal δ and using the definition for all $\beta \in \delta$ – this will turn out to give a function within *ZF*. Or we can face it more directly by saying that it defines a class function, on the class $\mathscr{O}$ of all ordinals, just as the operation of adding two arbitrary ordered sets is a class function – in these cases, the result of applying the function to sets gives a set.

We shall adopt the dodge above to show that the ordinal sum $\alpha \oplus \beta$ is always an ordinal. Given any β, take any ordinal δ greater than β, and show that $\alpha \oplus \beta$ is an ordinal for all $\beta \in \delta$.

$\delta = \beta^+$ would do!

Exercise 8.23

Let α and $\delta(> \mathbf{0})$ be ordinals. Use the principle of transfinite induction for ordinals (Theorem 8.1) to show that $\alpha \oplus \beta$ is an ordinal for all $\beta \in \delta$.

Solution

We want to show that the subset $A = \{\beta \in \delta : \alpha \oplus \beta \text{ is an ordinal}\}$ of δ is all of δ.

As $\delta > \mathbf{0}$, the least element of δ is $\mathbf{0}$. By definition $\alpha \oplus \mathbf{0}$ equals α, so that $\mathbf{0} \in A$.

We must now show that for $\beta \in \delta$, if $\beta \subseteq A$ then $\beta \in A$. The definition of $\alpha \oplus \beta$ deals with three cases for β, so our argument should deal with each case in turn.

The three cases are: β is $\mathbf{0}$, a successor or a limit.

The case when $\beta = \mathbf{0}$ has in fact already been done, as $\mathbf{0} = \varnothing \subseteq A$ and we have just shown that $\mathbf{0} \in A$. The remaining cases are when β is a successor and when β is a limit.

Suppose first that $\beta \subseteq A$ where β is a successor, γ^+ say. As $\gamma \in \gamma^+$ this means that $\gamma \in A$, so that $\alpha \oplus \gamma$ is an ordinal, by definition of A. Then

$$\begin{aligned}\alpha \oplus \beta &= \alpha \oplus \gamma^+ \\ &= (\alpha \oplus \gamma)^+\end{aligned}$$

which is the successor of an ordinal and is hence an ordinal, by Theorem 8.4(i). Thus $\beta \in A$.

Now suppose that β is a limit ordinal λ and that $\lambda \subseteq A$. This means that $\alpha \oplus \gamma \in A$ for all $\gamma \in \lambda$. Then

$$\alpha \oplus \lambda = \bigcup\{\alpha \oplus \gamma : \gamma \in \lambda\}.$$

As λ is a set and $\gamma \longmapsto \alpha \oplus \gamma$ behaves like a function on λ, the axiom of replacement ensures that it *is* a function and that its images $\{\alpha \oplus \gamma : \gamma \in \lambda\}$ form a set. Thus $\alpha \oplus \lambda$ is the union of a set of ordinals, which is an ordinal by Theorem 8.4(ii). So that $\beta \in A$, as required.

This is probably the trickiest part to justify properly, even though it hopefully seems very natural!

In all cases, we have shown that if $\beta \subseteq A$ then $\beta \in A$, so that by transfinite induction $A = \delta$.

Exercise 8.24

Why did we argue by induction on β rather than α in the exercise above?

Solution

One obvious answer is that the definition of $\alpha \oplus \beta$ is given for a fixed α and all β. However, the definition does give us a recipe for dealing with all α, so that we might expect to be able prove the result by induction on α for fixed β. A proof along these lines is going to run into complications quite quickly, as for instance it is not obvious from the definition how to relate $\alpha^+ \oplus \beta$ to $\alpha \oplus \beta$. Indeed as we are hoping to show that ordinal addition corresponds to the standard addition of ordered sets, in which one sometimes finds α and β for which $\alpha^+ + \beta$ equals $\alpha + \beta$, the method seems particularly messy.

Next let us show, at least in outline, that the ordinal sum of α and β is order-isomorphic to their order sum.

Their order sum is well-ordered, by Exercise 7.47 in Section 7.4, so is order-isomorphic to an ordinal.

Theorem 8.8

Let α and β be ordinals. Then the order sum $\alpha + \beta$ is order-isomorphic to the ordinal sum $\alpha \oplus \beta$.

Proof

As before, we shall prove this for a fixed α by transfinite induction on all β in some suitably large ordinal δ. Strictly speaking, we should define a subset A of all ordinals $\beta \in \delta$ for which the result holds, i.e.

$$A = \{\beta \in \delta : \alpha + \beta \cong \alpha \oplus \beta\},$$

and use induction to show that $A = \delta$. But instead we shall structure the argument, and similar ones to come, a bit more informally, as follows.

For $\beta = \mathbf{0}$, i.e. $\beta = \varnothing$, both $\alpha + \beta$ and $\alpha \oplus \beta$ equal α. So the result holds trivially for $\beta = \mathbf{0}$.

That is, $\mathbf{0} \in A$.

Next suppose that the result holds for all ordinals up to, and including, γ, and show that it holds for γ^+, i.e. $\alpha + \gamma^+ \cong \alpha \oplus \gamma^+$. On the one hand

This is essentially the argument to show that if $\beta \subseteq A$ then $\beta \in A$, in the case that β is a successor, γ^+.

$$\begin{aligned}\alpha + \gamma^+ &= (\alpha \times \{\mathbf{0}\}) \cup (\gamma^+ \times \{\mathbf{1}\})\\ &= (\alpha \times \{\mathbf{0}\}) \cup ((\gamma \cup \{\gamma\}) \times \{\mathbf{1}\})\\ &= ((\alpha \times \{\mathbf{0}\}) \cup (\gamma \times \{\mathbf{1}\})) \cup (\{\gamma\} \times \{\mathbf{1}\})\\ &= (\alpha + \gamma) \cup (\{\gamma\} \times \{\mathbf{1}\}),\end{aligned}$$

which is the ordered set $\alpha + \gamma$ with the extra element $(\gamma, \mathbf{1})$ at the top. On the other hand

$\{\gamma\} \times \{\mathbf{1}\} = \{(\gamma, \mathbf{1})\}$

$$\begin{aligned}\alpha \oplus \gamma^+ &= (\alpha \oplus \gamma)^+\\ &= (\alpha \oplus \gamma) \cup \{\alpha \oplus \gamma\},\end{aligned}$$

which consists of the ordinal $\alpha \oplus \gamma$, i.e. all the members of this ordinal, with the extra element $\alpha \oplus \gamma$ at the top. By hypothesis, there is an order-isomorphism from $\alpha + \gamma$ to $\alpha \oplus \gamma$. It should be reasonably clear that this order-isomorphism can be extended, by mapping $(\gamma, \mathbf{1})$ to the element $\alpha \oplus \gamma$ of $(\alpha \oplus \gamma)^+$, to give an order-isomorphism from $\alpha + \gamma^+$ to $\alpha \oplus \gamma^+$. Thus $\alpha + \gamma^+ \cong \alpha \oplus \gamma^+$ as required.

So that $\gamma^+ \in A$.

Now suppose that $\alpha + \gamma \cong \alpha \oplus \gamma$ for all $\gamma \in \lambda$, where λ is a limit ordinal. As λ is a limit ordinal,

This is essentially the argument to show that if $\beta \subseteq A$ then $\beta \in A$, in the case that β is a limit, λ.

$$\begin{aligned}\lambda &= \bigcup \lambda\\ &= \bigcup \{\gamma : \gamma \in \lambda\},\end{aligned}$$

so that

$$\begin{aligned}\alpha + \lambda &= (\alpha \times \{\mathbf{0}\}) \cup (\lambda \times \{\mathbf{1}\})\\ &= (\alpha \times \{\mathbf{0}\}) \cup (\bigcup \{\gamma : \gamma \in \lambda\} \times \{\mathbf{1}\})\\ &= (\alpha \times \{\mathbf{0}\}) \cup \bigcup \{\gamma \times \{\mathbf{1}\} : \gamma \in \lambda\}\\ &= \bigcup \{(\alpha \times \{\mathbf{0}\}) \cup (\gamma \times \{\mathbf{1}\}) : \gamma \in \lambda\}\\ &= \bigcup \{\alpha + \gamma : \gamma \in \lambda\}.\end{aligned}$$

For each $\gamma \in \lambda$, let f_γ be the order-isomorphism from $\alpha + \gamma$ to $\alpha \oplus \gamma$: regard f_γ as a set of ordered pairs. Now put

f_γ is unique by Exercise 7.58 of Section 7.4.

$$f = \bigcup \{f_\gamma : \gamma \in \lambda\}.$$

f is a set by the axiom of replacement.

If $\gamma, \gamma' \in \lambda$ with $\gamma < \gamma'$, then γ is an initial segment of γ', so that by Exercise 7.55 we have $f_\gamma \subseteq f_{\gamma'}$. This means that f, as a set of pairs, is a function. The domain of f is

$$\bigcup \{\mathrm{Dom}(f_\gamma) : \gamma \in \lambda\} = \bigcup \{\alpha + \gamma : \gamma \in \lambda\},$$

which from the argument above is $\alpha + \lambda$. And its image set is

$$\bigcup \{\alpha \oplus \gamma : \gamma \in \lambda\},$$

which is by definition $\alpha \oplus \lambda$. It is straightforward to show that f is an order-isomorphism, so that $\alpha + \lambda \cong \alpha \oplus \lambda$ as required.

So that $\lambda \in A$.

Thus by transfinite induction $\alpha + \beta \cong \alpha \oplus \beta$ for all $\beta \in \delta$. ■

Much of what we have done above hinges on the legitimacy of our use of recursion to define $\alpha \oplus \beta$. This legitimacy is guaranteed by the following, very important, theorem.

Subsequent induction proofs will be structured in the more informal style of the above proof.

Theorem 8.9 The recursion principle for ordinals

Let $\phi(x, y, z)$ be a formula of set theory such that for each ordinal β and set y there is a unique set z such that $\phi(\beta, y, z)$ holds. Thus ϕ essentially defines a class function

$$h\colon \mathscr{O} \times \mathscr{V} \longrightarrow \mathscr{V},$$

where $\mathscr{O}$ is the class of all ordinals and $\mathscr{V}$ is the universe of sets. Let δ be an ordinal. Then there is a unique function f with domain δ such that for every $\beta \in \delta$,

$$f(\beta) = h(\beta, f|_\beta).$$

f will be a genuine function within ZF, i.e. a set.

The function f is usually described as being defined by *transfinite recursion*. As $f|_\beta$ consists of the pairs $\langle \alpha, f(\alpha) \rangle$ for all $\alpha < \beta$, the definition of $f(\beta)$ as $h(\beta, f|_\beta)$ means that $f(\beta)$ depends potentially on all the values of $f(\alpha)$ for $\alpha < \beta$. We shall give the proof of this theorem at the end of this section. Right now it is more important to see how this formulation of the recursion principle relates to the way in which we defined $\alpha \oplus \beta$.

First, note that we defined $\alpha \oplus \beta$ for fixed α and variable β, so that the function f which we are defining has rule

$$f(\beta) = \alpha \oplus \beta.$$

Rather than defining straightaway the appropriate formula ϕ, it's easier to give first the corresponding function h. We defined $\alpha \oplus \beta$ in different ways, depending on whether β is $\mathbf{0}$, a successor ordinal γ^+, or a limit ordinal λ: so we would expect that the function h might likewise have different rules for these different cases.

For $\beta = \mathbf{0}$, we need to define $h(\mathbf{0}, f|_\mathbf{0})$ to give us the required value of $f(\mathbf{0})$, that is to say $\alpha \oplus \mathbf{0}$, which is just α. Of course $f|_\mathbf{0}$, the restriction of f to the empty set $\varnothing$, is just $\varnothing$. So we must define

$$h(\mathbf{0}, \varnothing) = \alpha.$$

Of course $\mathbf{0}$ equals $\varnothing$!

Actually, we need to define $h(\mathbf{0}, y)$ for any set y, even though we will only want its value when $y = \varnothing$. We can define, for all sets y,

$$h(\mathbf{0}, y) = \alpha.$$

For β a successor ordinal, say $\beta = \gamma^+$, we want $h(\beta, f|_\beta)$ to give $f(\beta)$, which is $(f(\gamma))^+$. How can we obtain $f(\gamma)$ using some formula of set theory involving the arguments of $h(\beta, f|_\beta)$, namely β and $f|\beta$? First, we can identify γ by the process which identifies that β is a successor, using a formula like

$$\begin{aligned} f(\beta) &= f(\gamma^+) \\ &= \alpha \oplus \gamma^+ \\ &= (\alpha \oplus \gamma)^+ \\ &= (f(\gamma))^+. \end{aligned}$$

$$\exists\gamma((\gamma \text{ is an ordinal}) \wedge \beta = \gamma^+).$$

Then we can obtain $f(\gamma)$ from the function $f|_\beta$, because as $\gamma \in \gamma^+ = \beta$, we have that γ is in the domain of this function and

$$f|_\beta(\gamma) = f(\gamma).$$

Of course, once we have recovered the value of $f(\gamma)$ in this way, we simply take its successor to obtain $f(\beta)$.

Lastly, we must deal with the case when β is a limit ordinal, say λ. For a limit ordinal λ we want

$$\begin{aligned} f(\lambda) &= \alpha \oplus \lambda \\ &= \bigcup\{\alpha \oplus \gamma : \gamma \in \lambda\} \\ &= \bigcup\{f(\gamma) : \gamma \in \lambda\} \\ &= \bigcup \mathrm{Range}(f|_\lambda). \end{aligned}$$

Thus we need $h(\lambda, f|_\lambda)$ to equal $\bigcup \mathrm{Range}(f|_\lambda)$, which we can achieve by defining, for β a limit ordinal,

$$h(\beta, y) = \begin{cases} \bigcup \mathrm{Range}(y|_\beta), & \text{if } y \text{ is a function with } \beta \subseteq \mathrm{Dom}(y), \\ \varnothing, & \text{otherwise.} \end{cases}$$

When y is $f|_\lambda$, we have
$$\begin{aligned} y|_\lambda &= (f|_\lambda)|_\lambda \\ &= f|_\lambda. \end{aligned}$$

We now have to define the formula $\phi(x, y, z)$ to capture the definition of $h(\beta, y)$ as above, coping with the three different sorts of ordinal β. For instance, $\phi(x, y, z)$ could be the disjunction of three formulas ϕ_0, $\phi_{\text{successor}}$ and ϕ_{limit}, where $\phi_0(x, y, z)$ is

The case where $\beta = \mathbf{0}$.

$$x = \mathbf{0} \wedge z = \alpha,$$

$\phi_{\text{successor}}(x, y, z)$ is

The case where β is a successor.

$$\exists\gamma((\gamma \text{ is an ordinal}) \wedge x = \gamma^+ \wedge (y \text{ is a function}) \\ \wedge \gamma \in \mathrm{Dom}(y) \wedge z = (y(\gamma))^+),$$

and $\phi_{\text{limit}}(x, y, z)$ is

The case where β is a limit.

$$(x \text{ is an ordinal}) \wedge (x \text{ is a limit}) \wedge (y \text{ is a function}) \\ \wedge x \subseteq \mathrm{Dom}(y) \wedge z = \bigcup \mathrm{Range}(y|_x).$$

Of course, strictly speaking, many of the informally expressed subformulas within the above formulas, like (y is a function), still need to be expressed formally within set theory – in fact, we saw how to do this in Chapter 4.

Exercise 8.25

Express the following formally within the language of set theory.

(a) R is a subset of the Cartesian product $x \times x$.

(b) R is a strict linear order on x.

(c) x is well-ordered by $\in$.

(d) x is an ordinal.

(e) x is a limit ordinal.

To summarize the above, we have given an outline of how to show that $\alpha \oplus \beta$ is well-defined as a set. It is pretty usual for books at this level to present the definition of a function f on an ordinal using transfinite recursion fairly

We needed a separate argument to show that $\alpha \oplus \beta$ is actually an ordinal.

informally, without giving any detail of the formula ϕ involved, or the corresponding class function h (in the notation of Theorem 8.9). But for further studies in set theory, it is often important to look at the structure, e.g. the logical complexity, of the ϕ – as well as to be sure that such a ϕ really exists!

In our arguments above about $\alpha \oplus \beta$ for fixed α and variable β, we took an appropriate ordinal δ and proved results about all $\beta \in \delta$. As this δ could be chosen to be any ordinal, we have effectively dealt with $\alpha \oplus \beta$ for all ordinals β. This means that we can regard $\beta \longmapsto \alpha \oplus \beta$ as a class function with 'domain' the class $\mathscr{O}$ of all ordinals. Whenever we restrict the domain of this class function to an ordinal δ, we return to the firm ground of sets. If we were to develop further the terminology of classes, we could then restate the principles of both transfinite recursion and induction for ordinals in somewhat more succinct terms, as follows.

Strictly speaking, as functions within ZF are sets, domains ought also to be sets. But it's clear what is meant here!

Theorem 8.9 The recursion principle, class form

Let $\phi(x, y, z)$ be a formula of set theory such that for each ordinal β and set y there is a unique set z such that $\phi(\beta, y, z)$ holds. Thus ϕ essentially defines a class function

$$h: \mathscr{O} \times \mathscr{V} \longrightarrow \mathscr{V},$$

where $\mathscr{O}$ is the class of all ordinals and $\mathscr{V}$ is the universe of sets. Then there is a unique class function f with 'domain' $\mathscr{O}$ such that for every ordinal β,

$$f(\beta) = h(\beta, f|_{\beta}).$$

Theorem 8.1 Transfinite induction for ordinals, class form

Let A be a 'subset' of the class $\mathscr{O}$ of all ordinals such that

(i) $\mathbf{0} \in A$;

(ii) for all ordinals β, if $\beta \subseteq A$ then $\beta \in A$.

Then $A = \mathscr{O}$.

As A turns out to be the class $\mathscr{O}$, it won't be a set, so that a word like 'subclass' would have to be invented to describe it, rather than 'subset'.

Another example of an important function defined by transfinite recursion is multiplication of ordinals. The definition will exploit the sum, namely $\oplus$, that we have previously defined – typically, we define functions on ordinals in some sort of hierarchy: for instance, we use successor to define sum, and then use sum to define product, just as we did for $\mathbb{N}$. We shall, for the moment, use the notation $\alpha \otimes \beta$ for the *ordinal product* of ordinals α and β, and define it, for a fixed α and variable β, as follows.

Remember that we want ordinal multiplication both to give an ordinal as its output and to correspond to our previous product of ordered sets.

$$\alpha \otimes \mathbf{0} = \mathbf{0},$$
$$\alpha \otimes \gamma^{+} = (\alpha \otimes \gamma) \oplus \alpha,$$
$$\alpha \otimes \lambda = \bigcup\{\alpha \otimes \gamma : \gamma \in \lambda\} \quad \text{for a limit ordinal } \lambda.$$

We defined $\cdot$ on $\mathbb{N}$ by

$$\mathbf{m} \cdot \mathbf{0} = \mathbf{0},$$
$$\mathbf{m} \cdot \mathbf{n}^{+} = (\mathbf{m} \cdot \mathbf{n}) + \mathbf{m},$$

for fixed $\mathbf{m}$ and all $\mathbf{n} \in \mathbb{N}$.

As for addition, the definition at least corresponds to that for the natural numbers within ZF, which are precisely the finite ordinals. For the rest of this book, we shall rely on it being 'obvious' to you and us that a definition like the above does really match the recursion principle, so that $\alpha \otimes \beta$ is a well-defined set, and that it is also an ordinal. The 'obviousness' is meant to follow from the transparency of what appears on the right-hand side of the $=$ signs in the definition. In the following exercises we ask you to check, just the once, that you can justify the 'obvious'!

Note that we shall often define functions of ordinals by transfinite recursion whose images are not ordinals.

Exercise 8.26

Show that there is a function $\alpha \otimes \beta$ defined as above, for a fixed ordinal α and all ordinals $\beta \in \delta$, where δ is some fixed ordinal, by identifying the f, h and ϕ in the statement of Theorem 8.9.

Exercise 8.27

Show that $\alpha \otimes \beta$ is an ordinal, for all ordinals α and β.

We shall now ask you to justify that ordinal multiplication of ordinals α and β does give an ordinal order-isomorphic to their order product, i.e. the Cartesian product $\alpha \times \beta$ with the anti-lexicographic order. (Although the latter is well-ordered, by Exercise 7.47, it is not in general an ordinal.)

Theorem 8.10

Let α and β be ordinals. Show that the order product $\alpha \times \beta$ is order-isomorphic to the ordinal product $\alpha \otimes \beta$.

Exercise 8.28

Prove Theorem 8.10. [Hints: use transfinite induction, as in the corresponding result for addition, Theorem 8.8. You will need this result for addition for part of the inductive step, as the definition of $\otimes$ exploits the function $\oplus$.]

We can now explain why we defined order product using the anti-lexicographic order, rather than the lexicographic order. The latter would have given us the aesthetically less pleasing result $\alpha \times \beta \cong \beta \otimes \alpha$!

In the next section we shall investigate the properties of ordinal addition and multiplication, to build up a full arithmetic of ordinals.

As another application of the recursion principle, let's look again at the cumulative hierarchy of sets. We first mentioned this in Section 4.5 of Chapter 4, but lacked the necessary machinery, namely ordinals, with which to give its full definition.

Definition

The *cumulative hierarchy of sets*, $\mathscr{V}_\alpha$ for all ordinals α, is defined as follows:

$$\mathscr{V}_0 = \varnothing,$$
$$\mathscr{V}_{\gamma^+} = \mathscr{P}(\mathscr{V}_\gamma),$$
$$\mathscr{V}_\lambda = \bigcup\{\mathscr{V}_\gamma : \gamma \in \lambda\} \quad \text{(for a limit ordinal } \lambda\text{).}$$

So $x \in \mathscr{V}_{\gamma^+}$ if and only if $x \subseteq \mathscr{V}_\gamma$.

The recursion principle ensures that there is a class function with rule $\alpha \longmapsto \mathscr{V}_\alpha$ for all ordinals α, with each $\mathscr{V}_\alpha$ being a set. We shall now justify our remarks in Chapter 4 that as a consequence of the axiom of foundation (along with other axioms of *ZF*) every set appears in some member of this cumulative hierarchy. Introducing the temporary notation $\mathscr{V}_\infty$ for the class of all sets which are in at least one of the $\mathscr{V}_\alpha$s, i.e.

$$\mathscr{V}_\infty = \bigcup\{\mathscr{V}_\alpha : \alpha \in \mathscr{O}\},$$

The union will give $\mathscr{V}_\infty$ as a proper class.

we shall show that $\mathscr{V}_\infty$ equals $\mathscr{V}$, the universe of sets.

Note first that for any set x in $\mathscr{V}_\infty$, so that $x \in \mathscr{V}_\alpha$ for some ordinal α, there must be a *least* ordinal for which this is true. What can we say about the least ordinal β for which $x \in \mathscr{V}_\beta$? First of all, we cannot have $\beta = \mathbf{0}$ as $\mathscr{V}_\mathbf{0}$ is the empty set.

Exercise 8.29

Suppose that β is the least ordinal such that $x \in \mathscr{V}_\beta$. Show that β cannot be a limit ordinal.

Solution

If β was a limit ordinal, then by definition $\mathscr{V}_\beta$ would equal $\bigcup\{\mathscr{V}_\gamma : \gamma \in \beta\}$. Thus if $x \in \mathscr{V}_\beta$, x would be in $\mathscr{V}_\gamma$ for some $\gamma \in \beta$, so that β couldn't be the least ordinal for which $x \in \mathscr{V}_\beta$.

Thus the least ordinal β for which $x \in \mathscr{V}_\beta$ must be a successor, α^+, for some α. This allows us to associate a unique ordinal with the set x as in the following definition.

Definition

For any set x in $\mathscr{V}_\infty$, the *rank* of x, written as $\rho(x)$, is the least ordinal α for which $x \in \mathscr{V}_{\alpha^+}$.

ρ is the Greek letter 'rho' corresponding to 'r'.

This ρ can be regarded as a class function:

$$\begin{aligned} \rho \colon \mathscr{V}_\infty &\longrightarrow \mathscr{O} \\ x &\longmapsto \rho(x). \end{aligned}$$

And of course we have $x \in \mathscr{V}_{\rho(x)^+}$, whenever $\rho(x)$ is defined.

Exercise 8.30

Show that for any set $x \in \mathscr{V}_\infty$, $\rho(x)$ is the least ordinal such that $x \subseteq \mathscr{V}_{\rho(x)}$.

The idea of rank will be used in the proof that $\mathscr{V} = \mathscr{V}_\infty$; and with this theorem, so that every set has a rank, the idea is correspondingly more powerful.

Theorem 8.11

$$\mathscr{V} = \bigcup\{\mathscr{V}_\alpha : \alpha \in \mathscr{O}\} \ (= \mathscr{V}_\infty).$$

So that every set has a rank.

Proof

We shall show that every set is an element of $\mathscr{V}_\alpha$ for some ordinal α. As each $\mathscr{V}_\alpha$ is a set (so that all elements of $\mathscr{V}_\alpha$ are sets), the result that the universe $\mathscr{V}$ equals the class $\mathscr{V}_\infty$ follows immediately. We need the following useful lemma.

Lemma: Suppose that every element of a set X is in $\mathscr{V}_\infty$, i.e. $X \subseteq \mathscr{V}_\infty$. Then X is also in $\mathscr{V}_\infty$.

Proof of Lemma: As X is a set and $\rho(x)$ is defined for all $x \in X$, the axiom of replacement gives that $\{\rho(x) : x \in X\}$ is a set (of ordinals). Thus $\bigcup\{\rho(x) : x \in X\}$ is an ordinal, say β; and there is a limit ordinal λ greater than this β. Note that as λ is a limit and $\beta < \lambda$, we also have $\beta^+ < \lambda$ – a detail we'll use below. For any $x \in X$ we then have

Such a limit λ exists by Exercise 8.20.

$$\begin{aligned} x &\in \mathscr{V}_{\rho(x)^+} \quad \text{(by definition of } \rho(x)) \\ &\subseteq \bigcup\{\mathscr{V}_\gamma : \gamma \in \lambda\} \quad (\text{as } \rho(x) \leq \beta, \text{ so that } \rho(x)^+ < \beta^+ < \lambda) \\ &= \mathscr{V}_\lambda \quad \text{(by definition)}, \end{aligned}$$

so that

$$X \subseteq \mathscr{V}_\lambda.$$

But then the definition of $\mathscr{V}_{\lambda^+}$ as $\mathscr{P}(\mathscr{V}_\lambda)$ gives

$$X \in \mathscr{V}_{\lambda^+},$$

so that $X \in \mathscr{V}_\infty$, proving the lemma. □

Now we prove the theorem. Let Y be any set. Note that if every element of Y is in $\mathscr{V}_\infty$, then the lemma gives that Y is in $\mathscr{V}_\infty$, as desired.

Now suppose that not every element of Y is in $\mathscr{V}_\infty$. We shall show that this leads to a contradiction. First we form the transitive closure $T(Y)$ of Y, i.e. the set

$T(Y)$ is $\in$-transitive. Transitive closure and some of its properties were the subject of Exercise 4.53 of Section 4.5.

$$T(Y) = Y \cup \bigcup Y \cup \bigcup\bigcup Y \cup \bigcup\bigcup\bigcup Y \cup \ldots,$$

with the property that if $x \in y \in T(Y)$ then $x \in T(Y)$.

Let Y' be the complement, $T(Y) \setminus \mathscr{V}_\infty$, of $\mathscr{V}_\infty$ in $T(Y)$ – taking the complement of a class in a set doesn't sound legal, but all is well, as we can express Y' as a set using the axiom of separation by

$$Y' = \{y \in T(Y) : \neg\exists\alpha((\alpha \text{ is an ordinal}) \wedge y \in \mathscr{V}_\alpha)\},$$

where 'α is an ordinal' and '$y \in \mathscr{V}_\alpha$' are both representable by formulas within our formal language.

As $Y \subseteq T(Y)$ and there are elements of Y not in $\mathscr{V}_\infty$, we have $Y' \neq \varnothing$. By the axiom of foundation Y' contains an $\in$-minimal element y_0, meaning that $y_0 \in Y'$ (so that also $y_0 \in T(Y)$) and $y_0 \cap Y' = \varnothing$. It is consideration of this set y_0 which is going to generate the contradiction, as follows.

This is the critical use of the axiom of foundation, *ZF9*. In fact it can be shown, using the system for set theory consisting of all the axioms of *ZF* except foundation, that *ZF9* is equivalent to $\mathscr{V} = \mathscr{V}_\infty$.

As $y_0 \in Y' \subseteq T(Y)$ and $T(Y)$ is $\in$-transitive, we have that if $x \in y_0$ then $x \in T(Y)$, so that $y_0 \subseteq T(Y)$. As $y_0 \cap Y' = \varnothing$ we thus have

$$\begin{aligned} y_0 &\subseteq T(Y) \setminus Y' \\ &= T(Y) \cap \mathscr{V}_\infty, \end{aligned}$$

so that every element of y_0 is in $\mathscr{V}_\infty$. Using the lemma above again, this gives $y_0 \in \mathscr{V}_\infty$, contradicting that $y_0 \in Y'$.

This contradiction means that every element of our original set Y was in $\mathscr{V}_\infty$, which we have already shown entails that Y itself is in $\mathscr{V}_\infty$, proving the theorem. ■

As a consequence of this theorem, every set x has a rank $\rho(x)$. We leave some of the properties of rank and working out the ranks of some famous sets, like $\mathbb{R}$, as further exercises at the end of the section.

Most of the interesting sets in everyday maths have quite small rank, less than $\omega + \omega$.

We shall end this section with a proof of the recursion principle. The proof has similarities with the proof of the recursion principle for $\mathbb{N}$ which we gave in Section 4.5 as Theorem 4.4. It makes vital use of transfinite induction and has a key step where the axiom of replacement is needed to guarantee that we are dealing with sets in ZF.

Theorem 8.9 The recursion principle for ordinals

Let $\phi(x, y, z)$ be a formula of set theory such that for each ordinal β and set y there is a unique set z such that $\phi(\beta, y, z)$ holds. Thus ϕ essentially defines a class function

$$h: \mathscr{O} \times \mathscr{V} \longrightarrow \mathscr{V},$$

where $\mathscr{O}$ is the class of all ordinals and $\mathscr{V}$ is the universe of sets. Let δ be an ordinal. Then there is a unique function f with domain δ such that for every $\beta \in \delta$,

$$f(\beta) = h(\beta, f|_\beta).$$

Proof

Our proof will investigate approximations to the desired function f which we shall call, just for the purposes of this proof, *γ-functions*. We shall say that g is a *γ-function*, where γ is an ordinal, if the following hold:

(i) g is a function;

(ii) the domain of g is γ;

(iii) for all ordinals $\beta \in \gamma$, g satisfies the rule $g(\beta) = h(\beta, g|_\beta)$.

That is, g is a genuine function represented by a set!

We shall use $h(\beta, g|_\beta)$ as a shorthand for the unique z such that $\phi(\beta, g|_\beta, z)$ holds.

So g is just like the required function f except that its domain is γ rather than δ. (The required f would be a δ-function.) We shall make heavy use of the following lemma.

Lemma: Suppose that g is a γ-function and g' is a γ'-function. Then

$$g|_{\gamma\cap\gamma'} = g'|_{\gamma\cap\gamma'}.$$

$\gamma \cap \gamma'$ is just the ordinal $\min\{\gamma, \gamma'\}$.

Proof of Lemma: We shall prove the lemma by transfinite induction on β in $\gamma \cap \gamma'$. We shall show that $g(0) = g'(0)$ and that if $g(\beta) = g'(\beta)$ for all $\beta \in \alpha$, where $\alpha \in \gamma \cap \gamma'$, then $g(\alpha) = g(\alpha')$: we can then deduce that $g(\beta) = g'(\beta)$ for all $\beta \in \gamma \cap \gamma'$, using transfinite induction.

In the notation of Theorem 8.1, we are taking A to be the set $\{\beta \in \gamma \cap \gamma' : g(\beta) = g(\beta')\}$.

As g is a γ-function, we have

$$\begin{aligned} g(\mathbf{0}) &= h(\mathbf{0}, g|_{\mathbf{0}}) \\ &= h(\mathbf{0}, \varnothing) \quad (\text{as } \mathbf{0} = \varnothing \text{ and } g|_{\varnothing} = \varnothing \text{ for any function } g). \end{aligned}$$

Similarly, we have $g'(\mathbf{0}) = h(\mathbf{0}, \varnothing)$, so that $g(\mathbf{0}) = g'(\mathbf{0})$ as required.

To have $\mathbf{0} \in \gamma \cap \gamma'$, we are assuming that $\gamma \cap \gamma' \neq \varnothing$. In the case that $\gamma \cap \gamma' = \varnothing$, the lemma is vacuously true as both $g|_{\gamma\cap\gamma'}$ and $g|_{\gamma\cap\gamma'}$ equal $\varnothing$.

For the inductive step, we assume that $g(\beta) = g'(\beta)$ for all $\beta \in \alpha$, where $\alpha \in \gamma \cap \gamma'$, and shall show that $g(\alpha) = g'(\alpha)$. Our assumption that $g(\beta) = g'(\beta)$ for all $\beta \in \alpha$ can be restated as $g|_\alpha = g'|_\alpha$. This means that

$$\begin{aligned} g(\alpha) &= h(\alpha, g|_\alpha) \quad (\text{as } g \text{ is a } \gamma\text{-function and } \alpha \in \gamma) \\ &= h(\alpha, g'|_\alpha) \quad (\text{by our assumption}) \\ &= g'(\alpha) \quad (\text{as } g' \text{ is a } \gamma'\text{-function and } \alpha \in \gamma'), \end{aligned}$$

as required. Our lemma now follows using transfinite induction. □

For this particular argument, we do not need to separate the inductive step into the cases when α is a successor ordinal or a limit ordinal.

An immediate consequence of this lemma is that if the desired function f exists on δ, it must be unique. (Any two such functions would both be δ-functions, so would agree on $\delta \cap \delta$, i.e. on δ, meaning that they are the same function.) Similarly, if for some ordinal γ there is a γ-function g, then g is unique for this γ, and in this way the lemma also plays a part in showing that the function f exists.

We shall use transfinite induction to show that a γ-function, which we shall write as g_γ, exists for all $\gamma \in \delta^+$. As we have just said, our lemma guarantees that each g_γ will be unique. And g_δ will be our required function f.

For this induction the set A in Theorem 8.1 is $\{\gamma \in \delta^+ : \text{a } \gamma\text{-function exists}\}$.

For $\gamma = \mathbf{0}$, we take the function $g_\mathbf{0}$ to be $\varnothing$, which is vacuously a $\mathbf{0}$-function.

If δ itself is $\mathbf{0}$, then we take f to be $\varnothing$. So we shall assume that $\delta > \mathbf{0}$.

Next we assume that g_γ exists and show that g_{γ^+} exists. Define a set of pairs F by

$$F = g_\gamma \cup \{\langle \gamma, h(\gamma, g_\gamma)\rangle\}.$$

Recall that a function is represented by a set of pairs.

It is straightforward to show that F is a γ^+-function and we leave this to you as Exercise 8.31 below. We can thus set g_{γ^+} to be this F.

Lastly, suppose that g_γ exists for all $\gamma \in \lambda$, where λ is a limit ordinal. We shall use these g_γs to construct a λ-function. Observe that, by our lemma above, if $\gamma \in \gamma' \in \lambda$, then $g_\gamma = g_{\gamma'}|_\gamma$, so that $g_\gamma \subseteq g_{\gamma'}$. Thus the g_γs for $\gamma \in \lambda$ can be arranged in a $\subseteq$-chain

For this argument, we have split the inductive step into the separate cases of a successor ordinal and limit ordinal.

$$\ldots \subseteq g_\gamma \subseteq g_{\gamma^+} \subseteq g_{\gamma^{++}} \subseteq \ldots.$$

Taking the union of this chain should give good prospects of giving a function with domain λ with the extra property needed to be a λ-function.

First of all, we must check that we are dealing with sets within ZF. As $\gamma \longmapsto g_\gamma$ behaves like a function on λ, the axiom of replacement guarantees that it is a function, so that its images form a set $\{g_\gamma : \gamma \in \lambda\}$. We can thus define a set F within ZF by

$$F = \bigcup\{g_\gamma : \gamma \in \lambda\}.$$

Is F a function? Any element of F is an element of some g_γ, so is an ordered pair. Suppose that $\langle \beta, x\rangle$ and $\langle \beta, x'\rangle$ are both in F. Then there must be $\gamma, \gamma' \in \lambda$ such that $\langle \beta, x\rangle \in g_\gamma$ and $\langle \beta, x'\rangle \in g_{\gamma'}$. Without loss of generality

$\gamma \leq \gamma'$, so that $g_\gamma \subseteq g'_{\gamma}$. This means that both ordered pairs belong to $g_{\gamma'}$, which is a function, forcing $x = x'$. Thus F is a function.

Recall that $\gamma \leq \gamma'$ means that $\gamma \in \gamma'$ or $\gamma = \gamma'$.

What is the domain of the function F? As F is a union of functions, its domain is just the union of their domains, i.e.

$$\begin{aligned} \text{Dom}(F) &= \bigcup\{\text{Dom}(g_\gamma) : \gamma \in \lambda\} \\ &= \bigcup\{\gamma : \gamma \in \lambda\} \quad \text{(as each } g_\gamma \text{ is a } \gamma\text{-function)} \\ &= \lambda \quad \text{(as } \lambda \text{ is a limit ordinal).} \end{aligned}$$

That $\{\text{Dom}(g_\gamma) : \gamma \in \lambda\}$ is a set also follows from the axiom of replacement.

Thus F has domain λ. For F to be a λ-function, all that remains is to show that $F(\beta) = h(\beta, F|_\beta)$ for all $\beta \in \lambda$. For any $\beta \in \lambda$, we also have $\beta^+ \in \lambda$, as λ is a limit ordinal. By the definition of F, the value of $F(\beta)$ is the same as that of $g_{\beta^+}(\beta)$; and the restriction $F|_\beta$ is the same as $g_{\beta^+}|_\beta$. Thus our required property for $F(\beta)$ is the same as $g_{\beta^+}(\beta) = h(\beta, g_{\beta^+}|_\beta)$, which holds because g_{β^+} is a β^+-function. This completes the proof that F is a λ-function.

It follows by transfinite induction that there is a γ-function for all $\gamma \in \delta^+$. In particular there is a δ-function which we can take as the required f. ■

Exercise 8.31

Show that if g_γ is a γ-function, then F, where $F = g_\gamma \cup \{(\gamma, h(\gamma, g_\gamma))\}$, is a γ^+-function.

Solution

We must show that F is a function, that its domain is γ^+, and that $F(\beta) = h(\beta, F|_\beta)$, for all $\beta \in \gamma^+$.

As g_γ consists of ordered pairs, then so does F. As the domain of g_γ is γ, adding on a pair with first coordinate γ itself does not break down the unique image property of functions (as $\gamma \notin \gamma = \text{Dom}(g_\gamma)$). Thus F is a function. This argument also shows that the domain of F is $\text{Dom}(g_\gamma) \cup \{\gamma\} = \gamma \cup \{\gamma\} = \gamma^+$, as required.

For $\beta \in \gamma$ we automatically have $F(\beta) = h(\beta, F|_\beta)$ because $F(\beta)$ and $F|_\beta$ are, by definition, respectively $g_\gamma(\beta)$ and $g_\gamma|_\beta$, and g_γ has the required property for this β (as it is a γ-function). As $\gamma^+ = \gamma \cup \{\gamma\}$, it only remains to check that $F(\gamma) = h(\gamma, F|_\gamma)$. By the definition of F, $F(\gamma) = h(\gamma, g_\gamma)$. And as $F|_\gamma$ is just the same as $\mathbf{g}_\gamma$, we indeed have that $F(\gamma) = h(\gamma, F|_\gamma)$.

Thus F is a γ^+-function.

Further exercises

These exercises concern the cumulative hierarchy $\mathscr{V}_\alpha$ for all ordinals α and the rank function ρ.

Exercise 8.32

Use transfinite induction to show the following.

(a) Show that each $\mathscr{V}_\alpha$ is $\in$-transitive, i.e. if $x \in y \in \mathscr{V}_\alpha$ then $x \in \mathscr{V}_\alpha$ (or, equivalently, if $y \in \mathscr{V}_\alpha$ then $y \subseteq \mathscr{V}_\alpha$).

(b) Show that if α, β are ordinals with $\alpha < \beta$ then $\mathscr{V}_\alpha \subseteq \mathscr{V}_\beta$.

Exercise 8.33

Using results from Exercise 8.32 as appropriate, show the following.

(a) $\mathscr{V}_\alpha = \{x : \rho(x) < \alpha\}$

(b) For any set y, if $x \in y$ then $\rho(x) < \rho(y)$.

(c) For any set y, $\rho(y) = \bigcup\{\rho(x) + 1 : x \in y\}$.

Exercise 8.34

(a) Use transfinite induction to show that $\rho(\alpha) = \alpha$ for all ordinals α. [Hint: the results of Exercise 8.33 might help.]

(b) Deduce that the set of ordinals in $\mathscr{V}_\alpha$ equals α, i.e. $\mathscr{V}_\alpha \cap \mathscr{O} = \alpha$.

Exercise 8.35

(a) Show that $\mathscr{V}_\mathbf{n}$ is finite for all $\mathbf{n} \in \mathbb{N}$.

(b) Show that $\mathscr{V}_\omega \approx \omega$ without using AC. [Hint: it is possible to define an explicit well-order on each $\mathscr{V}_\mathbf{n}$ by induction.]

(c) Show that $\mathscr{V}_{\omega+\mathbf{1}} \approx \mathbf{2}^\mathbb{N}$.

With AC, this is an immediate consequence of Theorem 6.6 of Section 6.4. To avoid AC, we need to be able to specify bijections between each $\mathscr{V}_\mathbf{n}$ and the natural number with which it is equinumerous. Specifying a well-ordering of $\mathscr{V}_\mathbf{n}$ effectively specifies such a bijection.

Exercise 8.36

(a) Let x be any set. Show that $\rho(\mathscr{P}(x)) = \rho(\{x\}) = \rho(x) + \mathbf{1}$.

(b) Let x, y be sets and let $\alpha = \max\{\rho(x), \rho(y)\}$. Show that $\rho(x \cup y) = \alpha$, $\rho(\{x, y\}) = \alpha + \mathbf{1}$ and that for the ordered pair $\langle x, y\rangle$, $\rho(\langle x, y\rangle) = \alpha + \mathbf{2}$.

(c) Take the sets of numbers $\mathbb{Z}$, $\mathbb{Q}$ and $\mathbb{R}$ as constructed in Chapter 2, with $\mathbb{R}$ constructed using Dedekind cuts. [The result of Exercise 4.35 of Section 4.4 will be useful for this part of the exercise.]

(i) Use the fact that $\rho(\mathbb{N}) = \rho(\omega) = \omega$ to show that $\rho(\mathbb{N} \times \mathbb{N}) = \omega$. [Hint: use the result above about $\rho(\langle x, y\rangle)$ and the result of part (c) in Exercise 8.33.]

(ii) Use the result of part (i) to deduce that $\rho(\mathbb{Z}) = \omega + \mathbf{1}$.

(iii) Show that $\rho(\mathbb{Z} \times \mathbb{Z}) = \omega + \mathbf{3}$. [Hints: every integer in $\mathbb{Z}$ has rank ω, as it is an infinite subset of $\mathbb{N} \times \mathbb{N}$. Now use the result above about $\rho(\langle x, y\rangle)$ and the result of part (c) in Exercise 8.33.]

(iv) Find $\rho(\mathbb{Q})$ and show that $\rho(\mathbb{R}) = \omega + \mathbf{5}$.

8.4 Ordinal arithmetic

In the previous section we defined the operations of ordinal addition and multiplication, and gave some thought as to the set-theoretic machinery needed to show that these were well-defined within *ZF*. We shall now investigate some of the properties of these operations, along with an extra operation of exponentiation (i.e. taking the power of one ordinal by another).

Virtually all the results of this section come from Cantor's 1897 paper in [12]; but the proofs are adjusted to match the definition of ordinal within *ZF*.

First, let us define these operations. Although you have already seen the definitions of addition and multiplication, $\oplus$ and $\otimes$ in the previous section, we shall give them again using the more customary notation of + and $\cdot$, which we shall use from now for the rest of the book. We have also used + for order sum: we hope that the context will make clear which + is being considered.

The order sum of two ordinals is order-isomorphic to their ordinal sum, by Theorem 8.8. But the order sum is not usually an ordinal.

We define ordinal addition, for a fixed ordinal α and all ordinals β, as follows:

$$\begin{aligned}\alpha + \mathbf{0} &= \alpha,\\ \alpha + \gamma^{+} &= (\alpha+\gamma)^{+},\\ \alpha + \lambda &= \bigcup\{\alpha+\gamma : \gamma \in \lambda\} \quad \text{for a limit ordinal } \lambda.\end{aligned}$$

We define ordinal multiplication, for a fixed ordinal α and all ordinals β, as follows.

$$\begin{aligned}\alpha \cdot \mathbf{0} &= \mathbf{0},\\ \alpha \cdot \gamma^{+} &= (\alpha\cdot\gamma)+\alpha,\\ \alpha \cdot \lambda &= \bigcup\{\alpha\cdot\gamma : \gamma \in \lambda\} \quad \text{for a limit ordinal } \lambda.\end{aligned}$$

The extra operation, *ordinal exponentiation*, is defined to coincide with the equivalent definition for natural numbers, i.e. the finite ordinals, and similarly exploits multiplication – this time of ordinals. For a fixed ordinal α and all ordinals β:

We defined $\mathbf{m}^{\mathbf{n}}$ on $\mathbb{N}$ by

$$\begin{aligned}\mathbf{m}^{\mathbf{0}} &= \mathbf{1},\\ \mathbf{m}^{\mathbf{n}^{+}} &= (\mathbf{m}^{\mathbf{n}})\cdot\mathbf{m},\end{aligned}$$

for fixed $\mathbf{m}$ and all $\mathbf{n} \in \mathbb{N}$.

$$\begin{aligned}\alpha^{\mathbf{0}} &= \mathbf{1}\ (= \mathbf{0}^{+}),\\ \alpha^{\gamma^{+}} &= (\alpha^{\gamma})\cdot\alpha,\\ \alpha^{\lambda} &= \bigcup\{\alpha^{\gamma} : \gamma \in \lambda\} \quad \text{for a limit ordinal } \lambda.\end{aligned}$$

You might have noticed that we often define $f(\lambda)$ for limit ordinals λ to be the union of all the $f(\gamma)$s for the previous γs.

These can all be shown to be well-defined functions, for fixed α and all $\beta \in \delta$, where δ is some fixed ordinal, by use of Theorem 8.9. And one can show that for all ordinals α and β, each of $\alpha+\beta$, $\alpha\cdot\beta$ and α^{β} is an ordinal. (Note also that in this chapter, when we write α^{β} we mean the ordinal obtained by ordinal exponentiation, *not* the set of all functions from the set β to the set α. These two notions do not coincide. Should we wish to look at the latter set of functions from one ordinal to another, we shall make this clear in the text.)

In theory, we should now be able to compute arithmetic expressions like α^{β} for some interesting ordinals α and β, along the lines of computing, say, $\mathbf{3}^{\mathbf{2}}$ for the natural numbers $\mathbf{3}$ $(= \mathbf{0}^{+++})$ and $\mathbf{2}$ $(= \mathbf{0}^{++})$ in ZF, with the aim of ending up with a 'simpler' answer, like $\mathbf{9}$. For instance, if our definition of ordinal exponentiation does capture some of our expectations for the same operation on natural numbers, we should be able to show that $\alpha^{\mathbf{2}} = \alpha\cdot\alpha$, for all ordinals α.

$$\begin{aligned}\mathbf{3}^{\mathbf{2}} &= \mathbf{3}^{\mathbf{1}^{+}}\\ &= (\mathbf{3}^{\mathbf{1}})\cdot\mathbf{3}\\ &= ((\mathbf{3}^{\mathbf{0}})\cdot\mathbf{3})\cdot\mathbf{3}\\ &= \ldots\end{aligned}$$

Exercise 8.37

Show that $\alpha^{\mathbf{2}} = \alpha\cdot\alpha$, for all ordinals α. (Don't be surprised if you hit a snag!)

$\mathbf{2} = \mathbf{1}^{+} = \mathbf{0}^{++}$.

Solution

It seems simple!

$$\begin{aligned}\alpha^{\mathbf{2}} &= \alpha^{\mathbf{1}^{+}}\\ &= (\alpha^{\mathbf{1}})\cdot\alpha\\ &= ((\alpha^{\mathbf{0}})\cdot\alpha)\cdot\alpha\\ &= ((\mathbf{1}\cdot\alpha)\cdot\alpha)\\ &= \ldots ?\end{aligned}$$

Oh dear! We cannot simply say, without some extra proof, that $\mathbf{1}\cdot\alpha = \alpha$, even though it is presumably true. The problem is that to use the definition of the ordinal product $\mathbf{1} \cdot \alpha$, we need to know more about the α, namely whether it is $\mathbf{0}$, or a successor, or a limit.

We know from Theorem 8.10 that the ordinal product $\alpha \cdot \beta$ is order-isomorphic to the order product $\alpha \times \beta$, and from Section 7.3 that order product is not commutative with respect to order-isomorphism. So we might have some ordinals α, β for which $\alpha \cdot \beta \neq \beta \cdot \alpha$. Indeed $\omega \cdot \mathbf{2} \neq \mathbf{2} \cdot \omega$.

Luckily, we have a method of proof tailor-made for coping with such a general α, namely the principle of transfinite induction for ordinals (Theorem 7.6). This is going to be our major proof technique, because it goes hand in hand with the definition of the arithmetic functions using transfinite recursion.

Exercise 8.38

Use the principle of transfinite induction for ordinals (Theorem 7.6) to show that $\mathbf{1} \cdot \alpha = \alpha$, for all ordinals α.

Solution

We shall present our solution in the more informal style of the proof of Theorem 8.8. This means proving the result for $\alpha = \mathbf{0}$ (the base step), and for α equal to a successor or limit ordinal on the basis that the result holds for all smaller ordinals (the inductive step). As ever, we shall prove the result for all α in a suitably large ordinal δ.

We shall increasingly omit mention of the 'suitably large δ'. We shall behave as though we have justified the class form of the principle of transfinite induction for ordinals mentioned in the previous section.

For $\alpha = \mathbf{0}$,

$$\begin{aligned} \mathbf{1} \cdot \alpha &= \mathbf{1} \cdot \mathbf{0} \\ &= \mathbf{0} \\ &= \alpha, \end{aligned}$$

as required.

Next suppose that the result holds up to, and including, γ and show that it holds for γ^+:

$$\begin{aligned} \mathbf{1} \cdot \gamma^+ &= (\mathbf{1} \cdot \gamma) + \mathbf{1} \\ &= \gamma + \mathbf{1} \quad \text{(as the result holds for } \gamma) \\ &= \gamma + \mathbf{0}^+ \\ &= (\gamma + \mathbf{0})^+ \quad \text{(using the definition of } +) \\ &= \gamma^+, \end{aligned}$$

Although we are assuming that the result holds for all ordinals $\beta \leq \gamma$, it is typical for the γ^+ case that we only use the result for γ.

again as required.

Lastly, suppose that the result holds for all $\gamma \in \lambda$, where λ is an initial ordinal. We must show that it holds for λ.

$$\begin{aligned} \mathbf{1} \cdot \lambda &= \bigcup \{\mathbf{1} \cdot \gamma : \gamma \in \lambda\} \quad \text{(using the definition of multiplication)} \\ &= \bigcup \{\gamma : \gamma \in \lambda\} \quad \text{(by the inductive hypothesis)} \\ &= \bigcup \lambda \\ &= \lambda \quad \text{(as } \lambda \text{ is a limit ordinal),} \end{aligned}$$

By Exercise 8.11.

as required.

As can often happen, there is an alternative method to that used above, exploiting Theorem 8.10, that an ordinal product is order-isomorphic to the corresponding order product. Here we have

$$\begin{aligned} \mathbf{1} \cdot \alpha &\cong \mathbf{1} \times \alpha \\ &= \{\mathbf{0}\} \times \alpha, \end{aligned}$$

which is pretty obviously order-isomorphic to α. As the ordinals $\mathbf{1} \cdot \alpha$ and α are order-isomorphic, they must be equal, by Theorem 8.2. This is a perfectly good proof, but many problems are not easily amenable to this sort of technique. *For the rest of this chapter, we shall use, and ask you to use, techniques exploiting the detailed structure of ordinals, rather than of order.* The ideas of order-theoretic sums and products will still be important, as they will help us form mental pictures of some of what is going on.

Not all results of ordinal arithmetic require proofs which consist entirely of transfinite induction. For instance, try the following exercise.

Exercise 8.39

Show that $\alpha \cdot \mathbf{1} = \alpha$, for all ordinals α. (Use the methods of ordinal arithmetic, rather than those of order arithmetic, even though these work well here.)

Solution

For any ordinal α,

$$\begin{aligned} \alpha \cdot \mathbf{1} &= \alpha \cdot \mathbf{0}^+ \\ &= (\alpha \cdot \mathbf{0}) + \alpha \\ &= \mathbf{0} + \alpha \quad \text{(from the definition of multiplication).} \end{aligned}$$

Obviously we need to show now that $\mathbf{0} + \alpha = \alpha$. As the expression $\mathbf{0} + \alpha$ has a general ordinal on the right, we will probably need to use transfinite induction: here goes!

For $\alpha = \mathbf{0}$,

$$\begin{aligned} \mathbf{0} + \alpha &= \mathbf{0} + \mathbf{0} \\ &= \mathbf{0} \quad \text{(by definition of +)} \\ &= \alpha \end{aligned}$$

If the result holds up to, and including, γ, then

$$\begin{aligned} \mathbf{0} + \gamma^+ &= (\mathbf{0} + \gamma)^+ \\ &= \gamma^+ \quad \text{(by the result for } \gamma\text{),} \end{aligned}$$

as required.

If the result holds for all $\gamma \in \lambda$, where λ is a limit ordinal,

$$\begin{aligned} \mathbf{0} + \lambda &= \bigcup\{\mathbf{0} + \gamma : \gamma \in \lambda\} \\ &= \bigcup\{\gamma : \gamma \in \lambda\} \quad \text{(by the induction hypothesis)} \\ &= \bigcup \lambda \\ &= \lambda \quad \text{(as } \lambda \text{ is a limit ordinal),} \end{aligned}$$

as required.

The result $\mathbf{0} + \alpha = \alpha$, for all ordinals α, follows by transfinite induction. Applying it to our original problem gives us

$$\begin{aligned} \alpha \cdot \mathbf{1} &= \mathbf{0} + \alpha \\ &= \alpha, \end{aligned}$$

for all α.

How can one judge whether to derive a result directly from the definition or by transfinite induction, or, as in the last exercise, by a mixture of methods? As a rule of thumb, if the result involves arithmetic expressions in which there's a general ordinal on the right, as in

$$\mathbf{1} \cdot \alpha, \quad \alpha + \beta, \quad (\alpha^{\beta})^{\gamma},$$

In $\alpha \cdot \mathbf{1}$, the general ordinal α is on the left-hand side of the expression.

you will probably need to use induction on the right-most variable, respectively

$$\alpha, \quad \beta, \quad \gamma,$$

keeping the other ordinals fixed. But, as in Exercise 8.39, you might not know that you will need induction until you are some way into the argument. Try the following exercises, in which you could need any of these methods.

Exercise 8.40

Show that each of the following holds for all ordinals α.

Use ordinal arithmetic, to get the practice!

(a) $\alpha \cdot \mathbf{2} = \alpha + \alpha$

(b) $\mathbf{0} \cdot \alpha = \mathbf{0}$

(c) $\alpha^{\mathbf{1}} = \alpha$

(d) $\alpha^{\mathbf{2}} = \alpha \cdot \alpha$

(e) $\mathbf{1}^{\alpha} = \mathbf{1}$

Let us try something more ambitious. Within order arithmetic, order addition is associative with respect to order-isomorphism, so that ordinal addition has to be associative. Thus we expect to have the following theorem.

Theorem 8.12

For all ordinals α, β and γ,

$$(\alpha + \beta) + \gamma = \alpha + (\beta + \gamma).$$

Exercise 8.41

Try to prove Theorem 8.12 by ordinal methods, by using transfinite induction on γ, for fixed α and β.

You may hit another rock!

Solution

For $\gamma = \mathbf{0}$, we have

$$\begin{aligned} (\alpha + \beta) + \gamma &= (\alpha + \beta) + \mathbf{0} \\ &= \alpha + \beta \quad \text{(by definition of } +\text{)}, \end{aligned}$$

and

$$\alpha + (\beta + \gamma) = \alpha + (\beta + \mathbf{0})$$
$$= \alpha + \beta,$$

which is the same. Thus the result holds for $\gamma = \mathbf{0}$.

Suppose the result holds up to, and including, γ. Then

$$\begin{aligned}(\alpha + \beta) + \gamma^+ &= ((\alpha + \beta) + \gamma)^+ \\ &= (\alpha + (\beta + \gamma))^+ \quad \text{(from the result for } \gamma) \\ &= \alpha + (\beta + \gamma)^+ \quad \text{(by definition of } +) \\ &= \alpha + (\beta + \gamma^+) \quad \text{(by definition of } +),\end{aligned}$$

so that the result holds for γ^+.

We hit the rock with the limit ordinal case! Suppose that the result holds for all ordinals $\gamma \in \lambda$, where λ is a limit ordinal. Then

$$\begin{aligned}(\alpha + \beta) + \lambda &= \bigcup\{(\alpha + \beta) + \gamma : \gamma \in \lambda\} \quad \text{(by definition of } +) \\ &= \bigcup\{\alpha + (\beta + \gamma) : \gamma \in \lambda\} \quad \text{(by the inductive hypothesis)},\end{aligned}$$

which looks promising! But how do we show that this last expression equals

We shall complete this argument in Exercise 8.47.

$$\alpha + \bigcup\{\beta + \gamma : \gamma \in \lambda\},$$

which equals $\alpha + (\beta + \lambda)$?

We need some extra machinery, which will come from deriving some very useful inequalities involving the three standard ordinal arithmetic functions. These basically show that for fixed α, all of $\alpha + \beta$, $\alpha \cdot \beta$ and α^β strictly increase with increasing β. For inequalities, it seems more natural to use the notation $<$, rather than $\in$, for the order relation on ordinals. But when reading or writing arguments, remember that $<$ and $\in$ mean the same thing for ordinals.

> ***Theorem 8.13***
>
> Let α, β, γ be ordinals. Then
>
> (i) if $\beta < \gamma$, then $\alpha + \beta < \alpha + \gamma$;
>
> (ii) if $\alpha > \mathbf{0}$ and $\beta < \gamma$, then $\alpha \cdot \beta < \alpha \cdot \gamma$;
>
> (iii) if $\alpha > \mathbf{1}$ and $\beta < \gamma$, then $\alpha^\beta < \alpha^\gamma$.

The need for the conditions on α in (ii) and (iii) should be obvious.

Proof

We shall prove (ii) and leave the other parts for you as an exercise. Our method is transfinite induction on γ for fixed α and β.

Induction on γ is a sensible choice as it is on the right of one of the ordinal products involved.

Suppose that $\alpha > \mathbf{0}$. We wish to show that

$$\text{if } \beta < \gamma \text{ then } \alpha \cdot \beta < \alpha \cdot \gamma$$

by induction on γ for fixed α, β.

It is pointless to put much effort into the case when $\gamma = \mathbf{0}$, as we cannot then have $\beta < \gamma$. This means that the result is vacuously true for $\gamma = \mathbf{0}$. Similarly the result is vacuously true for any γ with $\gamma \leq \beta$. The first interesting case is when γ is the least ordinal greater than β, namely β^+.

When $\gamma \leq \beta$, the statement $\beta < \gamma$ is false, so that 'if $\beta < \gamma$ then ...' is (vacuously) true.

Thus the base case is when $\gamma = \beta^+$. For this γ we have

$$\begin{aligned}\alpha \cdot \gamma &= \alpha \cdot \beta^+ \\ &= (\alpha \cdot \beta) + \alpha \quad \text{(by definition of multiplication)} \\ &> (\alpha \cdot \beta) + \mathbf{0} \quad \text{(by the result of (i), as } \alpha > \mathbf{0}) \\ &= \alpha \cdot \beta \quad \text{(by definition of +)},\end{aligned}$$

so that the result holds for $\gamma = \beta^+$.

Suppose that the result holds up to, and including, γ, where $\beta < \gamma$. We shall show that it also holds for γ^+. We have

We are only interested in ordinals γ greater than β.

$$\begin{aligned}\alpha \cdot \gamma^+ &= (\alpha \cdot \gamma) + \alpha \\ &> (\alpha \cdot \gamma) + \mathbf{0} \quad \text{(by the result of (i), as } \alpha > \mathbf{0}) \\ &= \alpha \cdot \gamma \\ &> \alpha \cdot \beta \quad \text{(as the result holds for } \gamma),\end{aligned}$$

as required.

Lastly, suppose that the result holds for all $\gamma \in \lambda$, where λ is a limit ordinal, with $\beta < \lambda$. We shall show that the result also holds for λ. By definition,

$$\alpha \cdot \lambda = \bigcup\{\alpha \cdot \gamma : \gamma \in \lambda\}.$$

As λ is a limit ordinal and $\beta < \lambda$, i.e. $\beta \in \lambda$, we have

$$\beta^+ \in \lambda,$$

so that

$$\alpha \cdot \beta^+ \subseteq \bigcup\{\alpha \cdot \gamma : \gamma \in \lambda\}.$$

We have already shown that $\alpha \cdot \beta < \alpha \cdot \beta^+$, i.e.

$$\alpha \cdot \beta \in \alpha \cdot \beta^+,$$

so that

$$\alpha \cdot \beta \in \bigcup\{\alpha \cdot \gamma : \gamma \in \lambda\},$$

which translates into

$$\alpha \cdot \beta < \alpha \cdot \lambda,$$

$<$ and $\in$ are the same!

as required. Result (ii) follows by the principle of transfinite induction. ■

Of course, the above argument relies on part (i) of the theorem, for addition, as we'd expect, given that the definition of multiplication depends on addition.

Exercise 8.42

Prove parts (i) and (iii) of Theorem 8.13, namely, for ordinals α, β, γ,

(i) if $\beta < \gamma$, then $\alpha + \beta < \alpha + \gamma$;

(iii) if $\alpha > \mathbf{1}$ and $\beta < \gamma$, then $\alpha^\beta < \alpha^\gamma$.

Solution

We give a solution only for (i). We shall use transfinite induction on γ, for fixed α and β.

The result is vacuously true for all γ with $\gamma \leq \beta$. The first interesting case is when $\gamma = \beta^+$. Then

$$\begin{aligned} \alpha + \beta^+ &= (\alpha + \beta)^+ \\ &> \alpha + \beta \quad (\text{as } \delta \in \delta^+, \text{ i.e. } \delta < \delta^+, \text{ for all } \delta), \end{aligned}$$

so that the result holds for β^+.

Suppose that the result holds up to, and including, γ, where $\beta < \gamma$. Then

$$\begin{aligned} \alpha + \gamma^+ &= (\alpha + \gamma)^+ \\ &> \alpha + \gamma \\ &> \alpha + \beta \quad (\text{using the result for } \gamma), \end{aligned}$$

so that the result holds for γ^+.

Finally, suppose that the result holds for all $\gamma \in \lambda$, where λ is a limit ordinal with $\beta < \lambda$. Then

$$\begin{aligned} \alpha + \beta &\in \alpha + \beta^+ \quad (\text{just shown above}) \\ &\subseteq \bigcup\{\alpha + \gamma : \gamma \in \lambda\} \quad (\text{as } \lambda \text{ is a limit and } \beta \in \lambda, \text{ so that } \beta^+ \in \lambda) \\ &= \alpha + \lambda \quad (\text{by definition of } +), \end{aligned}$$

which translates into the result for λ:

$$\alpha + \beta < \alpha + \lambda.$$

The result thus holds for all γ, by transfinite induction.

Theorem 8.13, with earlier results, gives us some idea of the order structure of the ordinals. For instance

$$\begin{aligned} &\omega < \omega + \mathbf{1} < \ldots < \omega + \mathbf{n} < \omega + \omega = \omega \cdot \mathbf{2} < \omega \cdot \mathbf{2} + \mathbf{1} < \ldots \\ &\quad \ldots < \omega \cdot \mathbf{2} + \omega = \omega \cdot \mathbf{3} < \ldots < \omega \cdot \omega = \omega^2 < \ldots < \omega^2 + \omega < \ldots \\ &\quad \ldots < \omega^2 + \omega^2 = \omega^2 \cdot \mathbf{2} < \ldots < \omega^2 \cdot \omega = \omega^3 < \ldots < \omega^\omega. \end{aligned}$$

We are now in a position where we can use ordinal, rather than order, arguments to show that ordinal addition and multiplication are not commutative. For instance, we can show that

$$\mathbf{1} + \omega = \omega,$$

whereas, by Theorem 8.13,

$$\begin{aligned} \omega + \mathbf{1} &> \omega + \mathbf{0} \\ &= \omega. \end{aligned}$$

Exercise 8.43

Show that $\mathbf{1} + \omega = \omega$.

Solution

By definition of $+$,

$$1 + \omega = \bigcup\{1 + n : n \in \omega\}.$$

As the finite ordinals are simply the natural numbers within ZF, we know that $1 + n = n + 1$, for all $n \in \omega$, so that

$$1 + \omega = \bigcup\{n + 1 : n \in \omega\}.$$

There are several straightforward ways of showing that $\bigcup\{n + 1 : n \in \omega\} = \omega$. One way is by the standard method of showing sets A and B are equal, namely proving that they have the same members. Another way is by exploiting the order-theoretic properties of ordinals, which we do below as an illustration of this attractive method.

By Theorem 8.4, $\bigcup\{n + 1 : n \in \omega\}$ is the ordinal which is the least upper bound of the set $\{n + 1 : n \in \omega\}$. So we shall show that ω is this least upper bound. First we must show that ω is an upper bound, which follows easily as $n + 1 \leq \omega$ for each $n \in \omega$. And we must also show that no smaller ordinal is an upper bound. This is also easy, as for a smaller ordinal m, we have

$$m < m + 1 \in \{n + 1 : n \in \omega\},$$

so that m is *not* an upper bound for the set. Thus ω is the least upper bound of the set, as required.

Exercise 8.44

(a) Show that $2 \cdot \omega = \omega$.

(b) Show that ordinal multiplication is not commutative.

Exercise 8.45

Give examples of non-zero ordinals α, β, γ for which $\alpha < \beta$, but $\alpha^\gamma \not< \beta^\gamma$. Justify your answer with computations of α^γ and β^γ for your α, β, γ.

Exercise 8.46

Let α, β, γ be ordinals. Show that if $\alpha \leq \beta$ then each of the following holds:

(a) $\alpha + \gamma \leq \beta + \gamma$;

(b) $\alpha \cdot \gamma \leq \beta \cdot \gamma$;

(c) $\alpha^\gamma \leq \beta^\gamma$.

Let us now fill in the hole in Exercise 8.12, where we asked you to prove that ordinal addition is associative. We had shown that

$$(\alpha + \beta) + \lambda = \bigcup\{\alpha + (\beta + \gamma) : \gamma \in \lambda\},$$

but had not shown that this equalled

$$\alpha + \bigcup\{\beta + \gamma : \gamma \in \lambda\},$$

which equals $\alpha + (\beta + \lambda)$. This requires a result which is worth proving for more general ordinal operations than just addition, and which needs a couple of technical definitions.

Definitions

Let δ, δ' be ordinals and $f: \delta \longrightarrow \delta'$ a function. Let $\lambda \in \delta$ be a limit ordinal. Then f is said to be *continuous* at λ if

$$f(\lambda) = \bigcup\{f(\gamma) : \gamma \in \lambda\}.$$

If f is continuous at all limit ordinals λ in its domain, then f is said to be a *continuous* function.

f might as well be given as a class function $f: \mathscr{O} \longrightarrow \mathscr{O}$.

Thus each of addition, multiplication and exponentiation is continuous, in the sense that the functions

$$\beta \longmapsto \alpha + \beta, \quad \beta \longmapsto \alpha \cdot \beta, \quad \beta \longmapsto \alpha^{\beta},$$

for a fixed α, are all continuous because of the way they are defined for limit ordinals.

Definition

Let δ, δ' be ordinals and $f: \delta \longrightarrow \delta'$ a function. Then f is said to be a *normal* function if it is strictly increasing and continuous.

Again, f might as well be given as a class function $f: \mathscr{O} \longrightarrow \mathscr{O}$.

By Theorem 8.13 each of addition, multiplication and exponentiation is normal, in the sense that the functions

$$\beta \longmapsto \alpha + \beta, \quad \beta \longmapsto \alpha \cdot \beta \text{ (for } \alpha > 0), \quad \beta \longmapsto \alpha^{\beta} \text{ (for } \alpha > 1),$$

for a fixed α, are all normal.

We need the conditions on α for the functions to be strictly increasing.

Theorem 8.14

Let δ, δ' be ordinals and $f: \delta \longrightarrow \delta'$ a normal function. Then for any non-empty set X of ordinals,

$$f(\bigcup X) = \bigcup\{f(\gamma) : \gamma \in X\}.$$

Proof

By Theorem 8.4 $\bigcup X$ is an ordinal β and is $\operatorname{lub} X$, the least upper bound of X. The proof considers separately two possibilities for β: that it is in X, and that it isn't.

Consider the case when $\beta \in X$. Then for any $\gamma \in X$,

$$\gamma \leq \beta \quad (\text{as } \beta \text{ is } \operatorname{lub} X),$$

so that, as f is increasing (because it is normal),

$$f(\gamma) \leq f(\beta).$$

As $\beta \in X$, this means that $f(\beta)$ is the least upper bound of $\{f(\gamma) : \gamma \in X\}$, so that by Theorem 8.4

$$f(\beta) = \bigcup\{f(\gamma) : \gamma \in X\}.$$

Thus

$$\begin{aligned} f(\bigcup X) &= f(\beta) \\ &= \bigcup\{f(\gamma) : \gamma \in X\}, \end{aligned}$$

as required.

Now consider the case when $\beta \notin X$. Then by the result of Exercise 8.13, β is in fact a limit ordinal. As f is continuous (as it is normal), we then have

$$f(\beta) = \bigcup\{f(\gamma) : \gamma \in \beta\}.$$

For our theorem we are interested in $\bigcup\{f(\gamma) : \gamma \in X\}$. As $\beta = \text{lub}\, X$, we have

$$\{f(\gamma) : \gamma \in X\} \subseteq \{f(\gamma) : \gamma \in \beta\},$$

so that

$$\begin{aligned} \bigcup\{f(\gamma) : \gamma \in X\} &= \text{lub}\{f(\gamma) : \gamma \in X\} \\ &\leq \text{lub}\{f(\gamma) : \gamma \in \beta\} \\ &= f(\beta), \end{aligned}$$

Thus $f(\beta)$ is an upper bound of $\{f(\gamma) : \gamma \in X\}$. Is $f(\beta) = \text{lub}\{f(\gamma) : \gamma \in X\}$? Given any α with

$$\alpha < f(\beta) = \text{lub}\{f(\gamma) : \gamma \in \beta\},$$

α can't be an upper bound for $\{f(\gamma) : \gamma \in \beta\}$, so that there is some $\gamma \in \beta$ with $\alpha < f(\gamma)$. As $\beta = \text{lub}\, X$, γ is not an upper bound of X, so that there is some $\gamma' \in X$ with $\gamma < \gamma'$. As f is increasing, $f(\gamma) < f(\gamma')$, so that

We want to show that α is not an upper bound of $\{f(\gamma) : \gamma \in X\}$.

$$\alpha < f(\gamma) < f(\gamma'),$$

Although the γ might not have been in X, the γ' is.

so that α is not an upper bound for $\{f(\gamma) : \gamma \in X\}$. Thus $f(\beta)$ is the least upper bound of this set, so that

$$\begin{aligned} f(\bigcup X) &= f(\beta) \\ &= \text{lub}\{f(\gamma) : \gamma \in X\} \\ &= \bigcup\{f(\gamma) : \gamma \in X\}, \end{aligned}$$

as required. ■

We are now in a position to complete the proof that ordinal addition is associative, and to prove a variety of other identities of ordinal arithmetic.

Exercise 8.47

Suppose that α, β are ordinals and that λ is a limit ordinal. Show that

$$\bigcup\{\alpha + (\beta + \gamma) : \gamma \in \lambda\} = \alpha + (\beta + \lambda).$$

This completes the argument of Exercise 8.41, to show that ordinal addition is associative.

Solution

We shall exploit Theorem 8.14. Take f to be the function defined by

$$f(\eta) = \alpha + \eta,$$

for all ordinals η (in some arbitrarily chosen ordinal δ). As we have already observed, f is a normal function. And let X be the set

$$X = \{\beta + \gamma : \gamma \in \lambda\}.$$

Then by Theorem 8.14,

$$\bigcup\{f(\eta) : \eta \in X\} = f(\bigcup X).$$

The left-hand side of this equation is

$$\begin{aligned}\bigcup\{\alpha + \eta : \eta \in X\} &= \bigcup\{\alpha + (\beta + \gamma) : (\beta + \gamma) \in X\} \\ &= \bigcup\{\alpha + (\beta + \gamma) : \gamma \in \lambda\},\end{aligned}$$

As $(\beta + \gamma) \in X$ if and only if $\gamma \in \lambda$.

and the right-hand side is

$$\alpha + \bigcup\{\beta + \gamma : \gamma \in \lambda\},$$

which, by definition of $\beta + \lambda$ for a limit ordinal λ, equals

$$\alpha + (\beta + \lambda).$$

Thus the equation becomes

$$\bigcup\{\alpha + (\beta + \gamma) : \gamma \in \lambda\} = \alpha + (\beta + \lambda),$$

as required.

Armed with the normality of the three standard arithmetic operations and methods like proof by transfinite induction, we can prove several more, rather familiar-looking, identities for ordinal arithmetic.

Theorem 8.15

For all ordinals α, β and γ,

(i) $\alpha \cdot (\beta + \gamma) = (\alpha \cdot \beta) + (\alpha \cdot \gamma)$;

(ii) $(\alpha \cdot \beta) \cdot \gamma = \alpha \cdot (\beta \cdot \gamma)$;

(iii) $\alpha^{\beta+\gamma} = \alpha^{\beta} \cdot \alpha^{\gamma}$;

(iv) $(\alpha^{\beta})^{\gamma} = \alpha^{\beta \cdot \gamma}$.

The distributive law

Associativity of multiplication.

So e.g. $\omega \cdot \omega^{n} = \omega^{1+n} = \omega^{n+1}$.

Proof

We shall prove (i), the distributive law, and leave the rest as exercises. We shall use transfinite induction on γ, for fixed α and β.

For $\gamma = \mathbf{0}$ we have

$$\alpha \cdot (\beta + \mathbf{0}) = \alpha \cdot \beta,$$

and

$$\begin{aligned}(\alpha \cdot \beta) + (\alpha \cdot \mathbf{0}) &= (\alpha \cdot \beta) + \mathbf{0} \\ &= \alpha \cdot \beta,\end{aligned}$$

so that

$$\alpha \cdot (\beta + \mathbf{0}) = (\alpha \cdot \beta) + (\alpha \cdot \mathbf{0}),$$

i.e. the result holds for $\gamma = \mathbf{0}$.

Now suppose that the result holds up to, and including γ, and show that it also holds for γ^+:

$$\begin{aligned}\alpha\cdot(\beta+\gamma^+) &= \alpha\cdot(\beta+\gamma)^+ \quad \text{(by definition of +)}\\ &= (\alpha\cdot(\beta+\gamma))+\alpha \quad \text{(by definition of multiplication)}\\ &= ((\alpha\cdot\beta)+(\alpha\cdot\gamma))+\alpha \quad \text{(using the result for } \gamma)\\ &= (\alpha\cdot\beta)+((\alpha\cdot\gamma)+\alpha) \quad \text{(by the associativity of +)}\\ &= (\alpha\cdot\beta)+(\alpha\cdot\gamma^+) \quad \text{(by definition of +),}\end{aligned}$$

so that the result also holds for γ^+.

Lastly, suppose that the result holds for all $\gamma \in \lambda$, where λ is a limit ordinal, and show that it also holds for λ:

$$\begin{aligned}\alpha\cdot(\beta+\lambda) &= \alpha\cdot\bigcup\{\beta+\gamma:\gamma\in\lambda\} \quad \text{(by definition of + at a limit)}\\ &= \bigcup\{\alpha\cdot(\beta+\gamma):\gamma\in\lambda\} \quad \text{(by the normality of multiplication)}\\ &= \bigcup\{(\alpha\cdot\beta)+(\alpha\cdot\gamma):\gamma\in\lambda\} \quad \text{(using the result for all } \gamma\in\lambda)\\ &= (\alpha\cdot\beta)+\bigcup\{\alpha\cdot\gamma:\gamma\in\lambda\} \quad \text{(by the normality of +)}\\ &= (\alpha\cdot\beta)+(\alpha\cdot\lambda) \quad \text{(by definition of multiplication),}\end{aligned}$$

showing that the result holds for λ.

The result follows by transfinite induction. ■

Exercise 8.48

Prove the rest of Theorem 8.15, i.e. for all ordinals α, β and γ:

(a) $(\alpha\cdot\beta)\cdot\gamma = \alpha\cdot(\beta\cdot\gamma)$;

(b) $\alpha^{\beta+\gamma} = \alpha^\beta\cdot\alpha^\gamma$;

(c) $(\alpha^\beta)^\gamma = \alpha^{\beta\cdot\gamma}$.

Exercise 8.49

Find counterexamples to each of the following statements.

(a) $(\alpha+\beta)\cdot\gamma = (\alpha\cdot\gamma)+(\beta\cdot\gamma)$, for all ordinals α, β, γ.

(b) $(\alpha+\beta)^2 = \alpha^2+(\alpha\cdot\beta)+(\beta\cdot\alpha)+\beta^2$, for all ordinals α, β.

Just as in Exercises 8.43 and 8.44, we can prove that $\mathbf{m}+\omega=\omega$ and $\mathbf{m}\cdot\omega=\omega$ for any $\mathbf{m}\in\omega$. Taken with the distributive law, this produces some curiosities of ordinal arithmetic, for instance that if $\mathbf{j}, \mathbf{m}\in\omega$ then

$$(\omega^{\mathbf{j}}\cdot\mathbf{m})+\omega^{\mathbf{j+1}} = \omega^{\mathbf{j+1}}.$$

This follows because

$$\begin{aligned}(\omega^{\mathbf{j}}\cdot\mathbf{m})+\omega^{\mathbf{j+1}} &= (\omega^{\mathbf{j}}\cdot\mathbf{m})+(\omega^{\mathbf{j}}\cdot\omega)\\ &= \omega^{\mathbf{j}}\cdot(\mathbf{m}+\omega) \quad \text{(by the distributive law)}\\ &= \omega^{\mathbf{j}}\cdot\omega\\ &= \omega^{\mathbf{j+1}}.\end{aligned}$$

What has happened here is that addition of a smaller ordinal, $\omega^{\mathbf{j}}\cdot\mathbf{m}$, to the *left* of the larger ordinal, $\omega^{\mathbf{j+1}}$, has simply produced this larger ordinal. In the next exercise we ask you to investigate this phenomenon further.

As $\mathbf{m}<\omega$ we have
$$\omega^{\mathbf{j}}\cdot\mathbf{m} < \omega^{\mathbf{j}}\cdot\omega = \omega^{\mathbf{j+1}}.$$

Exercise 8.50

(a) Suppose that $\mathbf{i}, \mathbf{m} \in \omega$ with $\mathbf{i} > \mathbf{1}$. Show that $\mathbf{m} + \omega^{\mathbf{i}} = \omega^{\mathbf{i}}$. [Hint: use the result of Exercise 8.46 to show that $\omega^{\mathbf{i}} \leq \mathbf{m} + \omega^{\mathbf{i}} \leq \omega^{\mathbf{i-1}} + \omega^{\mathbf{i}}$.]

(b) Suppose that $\mathbf{i}, \mathbf{j}, \mathbf{m} \in \omega$ with $\mathbf{i} > \mathbf{j}$. Show that $(\omega^{\mathbf{j}} \cdot \mathbf{m}) + \omega^{\mathbf{i}} = \omega^{\mathbf{i}}$.

(c) Show that $\mathbf{1} + \omega^{\mathbf{3}} + (\omega^{\mathbf{2}} \cdot \mathbf{2}) + \omega + \omega^{\mathbf{2}} + \mathbf{3} = \omega^{\mathbf{3}} + (\omega^{\mathbf{2}} \cdot \mathbf{3}) + \mathbf{3}$.

As addition is associative, we can write the sum without brackets.

Solution

(a) Taking the hint, we have

$$\begin{aligned}
\omega^{\mathbf{i}} &= \mathbf{0} + \omega^{\mathbf{i}} \\
&\leq \mathbf{m} + \omega^{\mathbf{i}} \quad \text{(by Exercise 8.46, as } \mathbf{0} \leq \mathbf{m}) \\
&\leq \omega^{\mathbf{i-1}} + \omega^{\mathbf{i}} \quad \text{(by Exercise 8.46, as } \mathbf{m} \leq \omega^{\mathbf{i-1}}) \\
&= \omega^{\mathbf{i-1}} \cdot (\mathbf{1} + \omega) \quad \text{(by the distributive law)} \\
&= \omega^{\mathbf{i-1}} \cdot \omega \\
&= \omega^{\mathbf{i}}.
\end{aligned}$$

This forces $\mathbf{m} + \omega^{\mathbf{i}} = \omega^{\mathbf{i}}$, as required.

(b) We have already shown the result in the case that $\mathbf{i} = \mathbf{j} + \mathbf{1}$. If $\mathbf{i} > \mathbf{j} + \mathbf{1}$, so that $\mathbf{i} - \mathbf{j} > \mathbf{1}$ and the result of part (a) can be used, we have

$$\begin{aligned}
(\omega^{\mathbf{j}} \cdot \mathbf{m}) + \omega^{\mathbf{i}} &= \omega^{\mathbf{j}} \cdot (\mathbf{m} + \omega^{\mathbf{i-j}}) \\
&= \omega^{\mathbf{j}} \cdot \omega^{\mathbf{i-j}} \\
&= \omega^{\mathbf{i}}.
\end{aligned}$$

(c)

$$\begin{aligned}
\mathbf{1} + \omega^{\mathbf{3}} + (\omega^{\mathbf{2}} \cdot \mathbf{2}) + \omega + \omega^{\mathbf{2}} + \mathbf{3} &= (\mathbf{1} + \omega^{\mathbf{3}}) + (\omega^{\mathbf{2}} \cdot \mathbf{2}) + (\omega + \omega^{\mathbf{2}}) + \mathbf{3} \\
&= \omega^{\mathbf{3}} + (\omega^{\mathbf{2}} \cdot \mathbf{2}) + \omega^{\mathbf{2}} + \mathbf{3} \\
&= \omega^{\mathbf{3}} + ((\omega^{\mathbf{2}} \cdot \mathbf{2}) + \omega^{\mathbf{2}}) + \mathbf{3} \\
&= \omega^{\mathbf{3}} + (\omega^{\mathbf{2}} \cdot \mathbf{3}) + \mathbf{3}.
\end{aligned}$$

This exercise shows that in certain circumstances, adding an ordinal α smaller than β to the *left* of β (to form $\alpha + \beta$) simply gives β. (Beware! This doesn't always hold.) And part (c) suggests that complicated ordinal sums can be simplified into sums consisting of terms which decrease as one moves from left to right. This is indeed the case, as we shall see later (in e.g. Exercise 8.56).

Exercise 8.51

Give an example of infinite ordinals α, β with $\alpha < \beta$ for which $\alpha + \beta \neq \beta$.

An example involving finite ordinals α, β would be bit too easy!

The arithmetic of ordinals has great elegance, but one might well ask what other value it has. It's worth comparing ordinal arithmetic with that of $\mathbb{N}$ in everyday mathematics. The latter has manifold practical applications, but also provides important insights into the structure of $\mathbb{N}$. Cantor wanted to exploit the ordinals as a way of counting beyond the finite, e.g. when iterating a process more than finitely many times, and needed insights into the structure of infinite ordinals. For instance, the classification of non-zero ordinals as either successors or limits gives some insight into this structure. One benefit of ordinal arithmetic is that it gives further insights. Another example of such an insight is given by the next theorem. It deals with the following question.

If α is an ordinal, how near is it to a limit ordinal?

Of course, if α is finite, the nearest limit ordinal is ω. What happens if α is infinite, i.e. if $\alpha \geq \omega$?

> ***Theorem 8.16***
>
> Suppose that $\alpha \geq \omega$. Then there is a limit ordinal λ and an $\mathbf{n} \in \omega$ such that
>
> $$\alpha = \lambda + \mathbf{n}.$$
>
> Furthermore the λ and $\mathbf{n}$ are unique.

Proof

We shall deal with the uniqueness first. Suppose that

$$\alpha = \lambda + \mathbf{n} = \lambda' + \mathbf{n}',$$

where λ, λ' are limit ordinals and $\mathbf{n}, \mathbf{n}' \in \omega$. Without loss of generality we can suppose that $\lambda \leq \lambda'$, and we shall show first that the two limit ordinals must be equal.

As a general strategy, one usually tackles the big ordinals first.

Suppose that $\lambda < \lambda'$ and try to derive a contradiction. As λ' is a limit ordinal and $\lambda < \lambda'$, we have $\lambda^+ < \lambda'$, so that similarly $\lambda^{++} < \lambda'$, and so on. Continuing this process, we would expect that

$$\lambda + \mathbf{n} = \lambda^{\overbrace{++\cdots+}^{n}} < \lambda' \leq \lambda' + \mathbf{n}',$$

contradicting that $\lambda + \mathbf{n} = \lambda' + \mathbf{n}'$. Of course the '...' in the expression above for $\lambda + \mathbf{n}$ tells us that we need a general proof that

$$\lambda + \mathbf{n} < \lambda',$$

for all $\mathbf{n} \in \omega$. This proof is a very straightforward induction on all $\mathbf{n} \in \omega$, exploiting the facts that λ' is a limit ordinal and that $\lambda < \lambda'$ (which gives the base step, that $\lambda + \mathbf{0} < \lambda'$).

This is left as a straightforward exercise.

The upshot of the above is that $\lambda = \lambda'$, so that

$$\alpha = \lambda + \mathbf{n} = \lambda + \mathbf{n}',$$

where, without loss of generality, $\mathbf{n} \leq \mathbf{n}'$. Suppose, to obtain a contradiction, that $\mathbf{n} < \mathbf{n}'$. Then by Theorem 8.13,

$$\lambda + \mathbf{n} < \lambda + \mathbf{n}',$$

giving us a contradiction.

This completes the proof of the uniqueness of the λ and $\mathbf{n}$. We must now prove that these ordinals exist. We shall discuss several different methods, to emphasize the variety of proof techniques that one can exploit: transfinite induction is one such technique, but there are others which can be useful, especially when one gets stuck on an induction!

Method 1: Transfinite induction

We shall prove that λ and $\mathbf{n}$ exist by transfinite induction on α, where $\alpha \geq \omega$.

For $\alpha = \omega$ we have

The base case.

$$\alpha = \omega + \mathbf{0},$$

which is in the required form.

Next suppose that the result holds up to, and including, γ and show that it holds for γ^+. As the result holds for γ, there is a limit λ and an $\mathbf{n} \in \omega$ such that $\gamma = \lambda + \mathbf{n}$. Then

$$\begin{aligned} \gamma^+ &= (\lambda + \mathbf{n})^+ \\ &= \lambda + \mathbf{n}^+, \end{aligned}$$

which is in the required form (as $\mathbf{n}^+ \in \omega$).

Lastly, suppose that the result holds for all $\gamma \in \alpha$, where α is a limit ordinal, and show that it holds for α. We are in luck, as the result is immediate! α is a limit, so that

$$\alpha = \alpha + \mathbf{0},$$

which is in the required form, and our result follows by transfinite induction.

One reason for showing you alternative proofs of this result is that we were *very* lucky with the limit ordinal case above! We didn't actually have to make use of the inductive hypothesis, namely that for each $\gamma < \alpha$ there were a limit λ and $\mathbf{n}$ such that $\alpha = \lambda + \mathbf{n}$. If we had had to do so, we might well have found it a problem that both λ and $\mathbf{n}$ depend on the γ: there's no guarantee that the λs, or the $\mathbf{n}$s, are nicely related to each other for all the γs less than α.

Method 2: Minimal counterexample

The method is to suppose that the result does not hold for all ordinals and to let α be the least for which it fails. Then try to show that this leads to a contradiction. We shall not go through the details of such a proof, as they are essentially the same as those for a proof by transfinite induction, namely dealing with three different cases for α: the least possible value of α, normally $\mathbf{0}$, but for this theorem ω; the successor case, when $\alpha = \gamma^+$; and α a limit ordinal.

This should not be surprising, given that our proof of the principle of transfinite induction made essential use of the method of the minimal counterexample.

Method 3: Direct construction

This method involves describing how to construct the required λ and $\mathbf{n}$. Here our first step is to define a set X of ordinals by

$$X = \{\mu \leq \alpha : \mu \text{ is a limit ordinal}\}.$$

Note that X is a set, because it is a subset of α^+. Also X is non-empty, as $\alpha \geq \omega$, so that $\omega \in X$.

Let $\lambda = \operatorname{lub} X$. We claim that λ is a limit ordinal. There are two cases to consider, depending on whether or not $\lambda \in X$. If $\lambda \in X$, then λ is a limit, by definition of X. If $\lambda \notin X$, then λ is a limit by the result of Exercise 8.13. Whichever is the case, λ is a limit.

As α is an upper bound of X, we must have that $\lambda \leq \alpha$. This means that λ is the greatest limit ordinal with $\lambda \leq \alpha$.

If μ is a limit with $\mu \leq \alpha$, then $\mu \in X$, so that $\mu \leq \text{lub}\, X = \lambda$.

Now define a subset Y of ω by

$$Y = \{\mathbf{n} \in \omega : \lambda + \mathbf{n} \leq \alpha\}.$$

Y is non-empty, as $\lambda + \mathbf{0} = \lambda \leq \alpha$, so that $\mathbf{0} \in Y$. Put $\beta = \text{lub}\, Y$, so that $\beta \leq \omega$, as ω is an upper bound for Y.

As α is an upper bound for $\{\lambda + \mathbf{n} : \mathbf{n} \in Y\}$, we have

$$\begin{aligned} \alpha &\geq \text{lub}\{\lambda + \mathbf{n} : \mathbf{n} \in Y\} \\ &= \bigcup\{\lambda + \mathbf{n} : \mathbf{n} \in Y\} \\ &= \lambda + \bigcup\{\mathbf{n} : \mathbf{n} \in Y\} \quad \text{(as + is normal)} \\ &= \lambda + \beta \quad (\text{as } \beta = \text{lub}\, Y = \bigcup Y). \end{aligned}$$

If we had $\beta = \omega$, this would mean that the limit ordinal $\lambda + \omega$, which is greater than λ, would satisfy $\lambda + \omega \leq \alpha$, contradicting the maximality of λ. Thus $\beta < \omega$, so that $\beta \in Y$.

As $\mathbf{0} < \omega$, we have $\lambda = \lambda + \mathbf{0} < \lambda + \omega$.

We have $\lambda + \beta \leq \alpha$. Could it be that $\lambda + \beta < \alpha$? If so, then

We would really like to show that $\lambda + \beta = \alpha$ for this β.

$$\begin{aligned} \lambda + \beta^+ &= (\lambda + \beta)^+ \\ &\leq \alpha, \end{aligned}$$

so that $\beta^+ \in Y$, contradicting that β is $\text{lub}\, Y$. Thus $\lambda + \beta$ must equal α, and as λ is a limit and $\beta \in \omega$, we have proved the theorem.

This method can be very useful in the circumstances mentioned above, where transfinite induction doesn't seem to work well.

Method 4: Arithmetic of order

Using the arithmetic of order, rather than ordinal arithmetic, plus the results that order sums/products of ordinals are order-isomorphic to their ordinal sums/products, sometimes works well – not here though! ■

Exercise 8.52

Fill in a minor detail arising from one of the proofs of the previous theorem. Suppose that λ' is a limit ordinal and that β is any ordinal with $\beta < \lambda$. Show that $\beta + \mathbf{n} < \lambda'$, for all $\mathbf{n} \in \omega$. Does this mean that $\beta + \omega < \lambda'$?

Exercise 8.53

Theorem 8.16 deals with the nearest limit λ to an infinite ordinal α which satisfies $\lambda \leq \alpha$. Which limit λ, if there is one at all, is the smallest satisfying $\alpha \leq \lambda$?

You should be able to derive the following further results about ordinals and their arithmetic. They are given as exercises rather than theorems, but many of them have a status important enough to be classed as theorems! Some of them lead towards what are called *normal forms*, namely standard ways of writing ordinals as sums of certain simple ordinals. (So, for instance, to test if two ordinals are equal, one might try to compute their equivalents in a particular normal form and then see whether these are the same.)

Exercise 8.54

Suppose that α and β are ordinals with $\alpha \geq \beta$. In this exercise we would like you to use two different methods to show that there is an ordinal δ such that $\alpha = \beta + \delta$. In fact the δ is unique, as we shall also ask you to show.

So δ is $\alpha - \beta$.

(a) Let X be the set $\{\gamma \in \alpha^+ : \beta + \gamma \leq \alpha\}$.

(i) Show that X is an ordinal.

(ii) Show that $X \neq \mathbf{0}$.

(iii) Show that X is not a limit ordinal.

(iv) As X is neither $\mathbf{0}$ nor a limit ordinal, it must be a successor, δ^+ say. Show that $\alpha = \beta + \delta$. [Hints: why is $\beta + \delta \leq \alpha$? And why can we not have $\beta + \delta^+ \leq \alpha$?]

(b) Consider the complement $\alpha \setminus \beta$ of β in α. This is a well-ordered set so is order-isomorphic to an ordinal δ. Why does $\alpha = \beta + \delta$? [Hint: use the arithmetic of order.]

(c) Show that the δ above is unique, i.e. if $\alpha = \beta + \delta = \beta + \delta'$, then $\delta = \delta'$.

(d) Is there a unique ordinal δ such that $\alpha = \delta + \beta$?

Exercise 8.55

Let α be an ordinal with $\alpha < \omega^2$.

(a) Show that if $\omega \cdot \mathbf{n} \leq \alpha < \omega \cdot (\mathbf{n} + \mathbf{1})$ for some $\mathbf{n} \in \omega$, then there is a unique $\mathbf{m} \in \omega$ such that $\alpha = (\omega \cdot \mathbf{n}) + \mathbf{m}$.

(b) Show that there is indeed some $\mathbf{n} \in \omega$ such that $\omega \cdot \mathbf{n} \leq \alpha < \omega \cdot (\mathbf{n} + \mathbf{1})$.

So that if $\alpha < \omega^2$, then α is of the form $\omega \cdot \mathbf{n} + \mathbf{m}$, for some $\mathbf{n}, \mathbf{m} \in \omega$.

Exercise 8.56

Let α be an ordinal with $\alpha < \omega^\omega$

(a) Show that there is some $\mathbf{n} \in \omega$ for which $\alpha < \omega^{\mathbf{n}+\mathbf{1}}$.

(b) Let $\mathbf{n}$ be as above. Show that there is some $\mathbf{m} \in \omega$ such that $\omega^{\mathbf{n}} \cdot \mathbf{m} \leq \alpha < \omega^{\mathbf{n}} \cdot (\mathbf{m} + \mathbf{1})$.

(c) Let $\mathbf{n}$ and $\mathbf{m}$ be as above. Show that there is some $\beta < \omega^{\mathbf{n}}$ such that $\alpha = (\omega^{\mathbf{n}} \cdot \mathbf{m}) + \beta$.

(d) With $\mathbf{n}$ as above, show that there are $\mathbf{m}_0, \mathbf{m}_1, \ldots, \mathbf{m}_n \in \omega$ for which

$$\alpha = (\omega^{\mathbf{n}} \cdot \mathbf{m_n}) + (\omega^{\mathbf{n}-\mathbf{1}} \cdot \mathbf{m}_{n-1}) + \ldots + (\omega \cdot \mathbf{m}_1) + \mathbf{m}_0.$$

This is a normal form for ordinals less than ω^ω.

(e) Suppose that α, β are ordinals less than ω^ω. Show that both $\alpha + \beta$ and $\alpha \cdot \beta$ are less than ω^ω. [Hints: there are $\mathbf{m}, \mathbf{n} \in \omega$ with $\alpha < \omega^{\mathbf{m}}$ and $\beta < \omega^{\mathbf{n}}$. Use the results of Theorem 8.13 and Exercises 8.46 and 8.50 to get suitable upper bounds.]

Taken with results of the sort of Exercise 8.50, we can easily computes sums and products of ordinals less than ω^ω and give answers in the normal form.

Exercise 8.57

Let α, β be ordinals with $\beta > \mathbf{0}$. Show that there are unique ordinals γ, δ such that $\alpha = (\beta \cdot \gamma) + \delta$, where $\mathbf{0} \leq \delta < \beta$, as follows.

This is the quotient–remainder theorem for ordinals.

(a) Show that $\alpha \leq \beta \cdot \alpha < \beta \cdot \alpha^+$.

(b) Let X be the set $\{\gamma \in \alpha^+ : \beta \cdot \gamma \leq \alpha\}$. Show that X is a successor ordinal, γ^+ say. [Hint: show first that X is an ordinal, and then show it is neither $\mathbf{0}$ nor a limit.]

(c) With γ as in the last part, explain why

$$\beta \cdot \gamma \leq \alpha < \beta \cdot \gamma^+.$$

(d) Let Y be the set $\{\delta \in \beta : (\beta \cdot \gamma) + \delta \leq \alpha\}$. Show that Y is a successor ordinal, δ^+ say.

(e) With δ as in the last part, show that

$$(\beta \cdot \gamma) + \delta \leq \alpha < (\beta \cdot \gamma) + \delta^+,$$

and deduce that $\alpha = (\beta \cdot \gamma) + \delta$.

Let us now look more deeply at the results of a couple of earlier exercises, in which you were asked to show that

Exercises 8.43 and 8.44.

$$\mathbf{1} + \omega = \omega \text{ and } \mathbf{2} \cdot \omega = \omega.$$

We can describe the first of these results by saying that the function f, where

$$f(\alpha) = \mathbf{1} + \alpha \text{ for all } \alpha,$$

has a fixed point at ω, i.e. $f(\omega) = \omega$. Similarly the function g defined by

In general, a function f has a *fixed point* at α if $f(\alpha) = \alpha$.

$$g(\alpha) = \mathbf{2} \cdot \alpha \text{ for all } \alpha,$$

has a fixed point at ω, as $\mathbf{2} \cdot \omega = \omega$. Do these functions have other fixed points? For f the answer is 'Yes', as the result of Exercise 8.50 with $\mathbf{m} = \mathbf{1}$ gives

$$f(\omega^{\mathbf{i}}) = \mathbf{1} + \omega^{\mathbf{i}} = \omega^{\mathbf{i}},$$

for all $\mathbf{i} \in \omega$. How big can fixed points of f be? The answer is that f has arbitrarily large fixed points. In general, any normal function – this particular f is normal – has arbitrarily large fixed points, as we shall now show.

First, we shall ask you to prove a helpful lemma as an exercise.

Exercise 8.58

Let f be a normal function. Show that $\alpha \leq f(\alpha)$, for all ordinals α.

Solution

We shall argue by transfinite induction on α.

For $\alpha = \mathbf{0}$, we have that $f(\mathbf{0})$ is some ordinal, so that $f(\mathbf{0}) \geq \mathbf{0}$.

Next, suppose that the result holds for all ordinals up to, and including, γ. We shall show that it holds for γ^+. We have

$$\begin{aligned} f(\gamma^+) &> f(\gamma) \quad \text{(as } \gamma^+ > \gamma \text{ and } f \text{ is strictly increasing)} \\ &\geq \gamma \quad \text{(by the result for } \gamma\text{).} \end{aligned}$$

As γ^+ is the least ordinal greater than γ, and $f(\gamma^+)$ is some ordinal greater than γ, this means that $f(\gamma^+) \geq \gamma^+$. Hence the result also holds for γ^+.

Lastly, suppose that the result holds for all $\gamma < \lambda$, where λ is a limit ordinal, and show that it also holds for λ. We have

$$\begin{aligned} f(\lambda) &= \bigcup\{f(\gamma) : \gamma \in \lambda\} \quad \text{(as } f \text{ is continuous at all limit ordinals)} \\ &\geq f(\gamma), \text{ for all } \gamma \in \lambda, \\ &\geq \gamma, \text{ for all } \gamma \in \lambda \quad \text{(as the result holds for all } \gamma \in \lambda\text{).} \end{aligned}$$

Thus $f(\lambda)$ is an upper bound of the set $\{\gamma : \gamma \in \lambda\}$, i.e. the set λ, and is thus greater than equal to the least upper bound of this set, which is also λ as λ is a limit ordinal. Hence $f(\lambda) \geq \lambda$, i.e. the result holds for λ.

Thus the result holds for all ordinals by transfinite induction.

We can now show that any normal function has arbitrarily large fixed points.

> ***Theorem 8.17***
>
> Let f be a normal function and let α be any ordinal. Then there is an ordinal β with $\alpha \leq \beta$ for which $f(\beta) = \beta$.

Proof

If $f(\alpha)$ already equals α, then we are done. So suppose that $f(\alpha) \neq \alpha$, which, by the result of Exercise 8.58 above, means that $\alpha < f(\alpha)$. The general idea is to construct the set

$$X = \{\alpha, f(\alpha), f^2(\alpha), \ldots, f^n(\alpha), \ldots\},$$

$f^2(\alpha)$ meaning $f(f(\alpha))$ etc.

and put $\beta = \bigcup X$. Then

$$\begin{aligned} f(\beta) &= f(\textstyle\bigcup X) \\ &= f(\textstyle\bigcup\{f^n(\alpha) : n \in \mathbb{N}\}) \\ &= \textstyle\bigcup\{f(f^n(\alpha)) : n \in \mathbb{N}\} \quad \text{(as } f \text{ is normal)} \\ &= \textstyle\bigcup\{f^{n+1}(\alpha) : n \in \mathbb{N}\} \\ &= \beta. \end{aligned}$$

Is it the case that $\alpha \leq \beta$? As $\alpha < f(\alpha)$ and $f(\alpha) \subseteq \bigcup\{f^n(\alpha) : n \in \mathbb{N}\}$, we in fact have $\alpha < \beta$, so that β is as required.

The only detail that needs tidying up is the construction of the set X. All this needs is the definition by recursion on ω of an ordinal-valued function g by

As ever, g will be a function in ZF because of the axiom of replacement.

$$\begin{aligned} g(\mathbf{0}) &= \alpha \\ g(\mathbf{n}^+) &= f(g(\mathbf{n})), \text{ for all } \mathbf{n} \in \omega. \end{aligned}$$

Then $X = \text{Range}(g)$. ∎

It is quite interesting to see what fixed points the construction in this proof identifies, for specific normal functions. For instance, with f defined by $f(\alpha) = \mathbf{1} + \alpha$, and starting with $\alpha = \mathbf{1}$, the fixed point β would be given by

$$\begin{aligned} \beta &= \textstyle\bigcup\{\mathbf{1}, \mathbf{1}+\mathbf{1}, \mathbf{1}+(\mathbf{1}+\mathbf{1}), \ldots\} \\ &= \omega. \end{aligned}$$

But starting with $\alpha = \omega + \mathbf{1}$, the construction would give

$$\begin{aligned} \beta &= \textstyle\bigcup\{\omega+\mathbf{1}, \mathbf{1}+(\omega+\mathbf{1}), \mathbf{1}+(\mathbf{1}+(\omega+\mathbf{1})), \ldots\} \\ &= \textstyle\bigcup\{\omega+\mathbf{1}, \omega+\mathbf{1}, \omega+\mathbf{1}, \ldots\} \\ &= \textstyle\bigcup\{\omega+\mathbf{1}\} \\ &= \omega, \quad \text{(by the result we hope you obtained in Exercise 8.12)} \end{aligned}$$

$\mathbf{1} + (\omega + \mathbf{1}) = (\mathbf{1} + \omega) + \mathbf{1}$
$= \omega + \mathbf{1}$ (as $\mathbf{1} + \omega = \omega$
and so on.

which is less than α. Given that the theorem guarantees us a fixed point β with $\beta \geq \alpha$, it must be that α was already a fixed point – it wasn't appropriate for us to use the construction!

Exercise 8.59

Show that $\mathbf{1} + \alpha = \alpha$ for all ordinals α with $\omega \leq \alpha$.

So that all infinite ordinals are fixed points of $f(\alpha) = \mathbf{1} + \alpha$.

Exercise 8.60

For each of the following (normal) functions f and ordinals α, find the corresponding fixed point β with $\alpha \leq \beta$ given by the proof of Theorem 8.17.

(a) $f(\alpha) = \omega + \alpha$ with $\alpha = \mathbf{0}$ and ω.
(b) $f(\alpha) = \mathbf{2} \cdot \alpha$ with $\alpha = \mathbf{1}$ and ω.
(c) $f(\alpha) = \omega \cdot \alpha$ with $\alpha = \mathbf{0}$ and $\mathbf{1}$.
(d) $f(\alpha) = \mathbf{2}^{\alpha}$ with $\alpha = \mathbf{0}$ and ω.

Solution

(a) For $\alpha = \mathbf{0}$, the fixed point β is

$$\bigcup\{\omega, \omega + \omega, \omega + \omega + \omega, \ldots\} = \omega \cdot \omega$$
$$= \omega^{\mathbf{2}}.$$

For $\alpha = \omega$, we get the same fixed point, $\omega^{\mathbf{2}}$.

(b) For $\alpha = \mathbf{1}$, we get $\beta = \omega$. And for $\alpha = \omega$, the β is also α, as α is already a fixed point.

(c) For $\alpha = \mathbf{0}$, the β is also $\mathbf{0}$. For $\alpha = \mathbf{1}$, the β is

$$\bigcup\{\omega, \omega \cdot \omega, \omega \cdot (\omega \cdot \omega), \ldots\} = \bigcup\{\omega, \omega^{\mathbf{2}}, \omega^{\mathbf{3}}, \ldots\}$$
$$= \omega^{\omega}.$$

(d) For $\alpha = \mathbf{0}$, the β is

$$\bigcup\{\mathbf{1}, \mathbf{2}, \mathbf{4}, \ldots, \mathbf{2}^{\mathbf{n}}, \ldots\} = \omega.$$

And for $\alpha = \omega$, the β is also ω.

Note that $\mathbf{2}^{\omega} = \omega$.

Exercise 8.61

The proof of Theorem 8.17 constructs, for a given α and a normal function f such that $\alpha < f(\alpha)$, a fixed point β of f with $\alpha < \beta$. Could there be a fixed point γ of f with $\alpha < \gamma < \beta$?

The interest of the fixed points increases with the complexity of the function f. For instance, take f defined by $f(\alpha) = \omega^{\alpha}$. Let us find the fixed point β of f given by the construction in the proof of Theorem 8.17 for $\alpha = \omega$. We obtain

$$\beta = \bigcup\{\omega, \omega^{\omega}, \omega^{\omega^{\omega}}, \omega^{\omega^{\omega^{\omega}}}, \ldots\},$$

which must be pretty huge! It's not easy to picture this as an ordered set – none of our previous illustrations of well-ordered sets have involved anything more complicated than ω^{ω}, and you'd have to have dug around in one of the further exercises at the end of Section 7.4 to find an example of an order as complex as even that! This monster ordinal β is called ε_0 and is often written as

$$\varepsilon_0 = \omega^{\omega^{\omega^{\cdot^{\cdot^{\cdot}}}}}.$$

ε_0, pronounced 'epsilon zero', is the least of what Cantor called the *epsilon-numbers*.

Just how big is ε_0? The obvious measure of 'big' is in terms of cardinalities, and for numbers like ε_0 we run into problems unless we assume some form of the axiom of choice. Let us assume a weak form of AC and get an insight into the size of ε_0, in the next, fairly long, exercise. We shall leave most of the work for you, as much of it is useful practice of what we have already seen.

We shall be looking at how cardinals and ordinals are related, once AC is assumed, in the next chapter.

Exercise 8.62

In this exercise, you may assume that the union of countably many countable sets is countable. Also recall that ω_1 is the least uncountable ordinal, so that $\alpha < \omega_1$ if and only if α is a countable ordinal.

This is a weak consequence of AC.

(a) Let X be the set

$$X = \{\alpha_n : n \in \mathbb{N}\},$$

where each α_n is a countable ordinal. Show that $\bigcup X$ is also a countable ordinal. [Hints: we know that $\bigcup X$ is an ordinal: why it is countable? Some form of AC is needed because for each of infinitely many α we may have to choose a 'counting' of α – there may not necessarily be a uniform rule for selecting such a bijection between α and $\mathbb{N}$.]

(b) Let α, β be ordinals with $\alpha, \beta < \omega_1$. Show that $\alpha + \beta$, $\alpha \cdot \beta$ and α^β are all less than ω_1. [Hints: fix α and use a well-known method of proof! And in which order should you prove the results? The result of the first part will be useful.]

(c) Show that

$$\omega + \omega_1 = \omega \cdot \omega_1 = \omega^{\omega_1} = \omega_1.$$

(d) Show that $\varepsilon_0 < \omega_1$.

(e) We shall say that the ordinal γ is an *ε-number* if $\omega^\gamma = \gamma$. By using the construction in the proof of Theorem 8.17, show that there is a countable ε-number greater than ε_0. [Hint: start with $\alpha = \varepsilon_0 + \mathbf{1}$.]

(f) Show that there are uncountably many ε-numbers less than ω_1. [Hints: argue by contradiction. Assume that there are countably many ε-numbers, show that there is a countable ordinal larger than all of them, and show that there is a countable ε-number greater than this, to obtain a contradiction.]

Let us end this section with another of Cantor's normal forms for ordinals, this time one which holds for all ordinals. The result is an exercise for you.

Exercise 8.63

Let α be an ordinal.

(a) Explain why $\alpha \leq \omega^\alpha$.

(b) Show that there is some ordinal β with

$$\omega^\beta \leq \alpha < \omega^{\beta^+}.$$

(c) With β as in the previous exercise, show that there are an $\mathbf{m} \in \omega$ and $\gamma < \omega^\beta$ such that

$$\alpha = (\omega^\beta \cdot \mathbf{m}) + \gamma.$$

(d) Show that for some $n \in \mathbb{N}$ there are ordinals $\beta_1, \beta_2, \ldots, \beta_n$ and finite ordinals $\mathbf{m}_0, \mathbf{m}_1, \mathbf{m}_2, \ldots, \mathbf{m}_n \in \omega$ such that

$$\alpha = (\omega^{\beta_n} \cdot \mathbf{m}_n) + (\omega^{\beta_{n-1}} \cdot \mathbf{m}_{n-1}) + \ldots + (\omega^{\beta_1} \cdot \mathbf{m}_1) + \mathbf{m}_0.$$

Further exercises

Exercise 8.64

Suppose that α, β are non-zero ordinals with $\alpha, \beta \in \omega^\omega$. Show that

$$\alpha + \beta = \beta + \alpha$$

if and only if there are $\mathbf{n}, a, b \in \omega$ and $\gamma < \omega^{\mathbf{n}}$ for which

$$\alpha = (\omega^{\mathbf{n}} \cdot a) + \gamma \text{ and } \beta = (\omega^{\mathbf{n}} \cdot b) + \gamma.$$

[Hint: use the normal form for ordinals in ω^ω given by the result of Exercise 8.56.]

If β, say, is $\mathbf{0}$, we know that $\alpha + \mathbf{0} = \mathbf{0} + \alpha$ $(= \alpha)$.

Exercise 8.65

Let α be an ordinal. We say that α is *closed under addition* if, for all $\beta, \gamma \in \alpha$, we have $\beta + \gamma \in \alpha$. Similarly α is *closed under multiplication* (respectively *exponentiation*) if $\beta \cdot \gamma \in \alpha$ (respectively $\beta^\gamma \in \alpha$) for all $\beta, \gamma \in \alpha$. In each of the following cases, find the smallest ordinal $\alpha > \omega$ with the given property:

(a) α is closed under addition;

(b) α is closed under multiplication;

(c) α is closed under exponentiation.

Of course ω is closed under addition, multiplication and exponentiation.

Exercise 8.66

Cantor's original work, and many modern texts, present some of the results of this section in terms of limits of sequences of ordinals, rather than unions (which are also least upper bounds) of sets of ordinals. For the sake of completeness, we give some of the definitions and corresponding results below.

A *sequence* of ordinals is a function $f\colon \alpha \longrightarrow \delta$, where α, δ are ordinals. We write β_γ for $f(\gamma)$ and use the notation

$$\langle \beta_\gamma \rangle_{\gamma < \alpha}$$

to represent the sequence.

Let λ be a limit ordinal and let $\langle \beta_\gamma \rangle_{\gamma < \lambda}$ be a sequence of ordinals. We say that the ordinal α is the *limit* of the sequence, and write

$$\alpha = \lim_{\gamma < \lambda} \beta_\gamma,$$

if for each $\delta < \alpha$, there is some $\gamma_\delta < \lambda$ such that for all γ with $\gamma_\delta < \gamma < \lambda$,

$$\delta < \beta_\gamma \leq \alpha.$$

(a) Show that if $\langle\beta_\gamma\rangle_{\gamma<\lambda}$ has a limit, then this limit is unique.

(b) Suppose that $\langle\beta_\gamma\rangle_{\gamma<\lambda}$ is an increasing sequence of ordinals, where λ is a limit ordinal. Then $\lim\limits_{\gamma<\lambda}\beta_\gamma$ exists and equals $\bigcup\{\beta_\gamma : \gamma\in\lambda\}$.

(c) Suppose that f is a normal function and let λ be a limit ordinal. Show that

$$f(\lambda) = \lim_{\gamma<\lambda} f(\gamma).$$

8.5 The $\aleph$s

Let us now look at the connection between the two aspects of Cantor's theory of infinite numbers, by asking about the size (cardinality) of the ordinals. We have already seen that there is an uncountable ordinal, namely ω_1, where

See Theorem 8.7.

$$\omega_1 = \{\alpha : \alpha \text{ is a countable ordinal}\}.$$

Is there an ordinal α larger than ω_1, in the sense that $\omega_1 \prec \alpha$? The answer (which is 'Yes') is one of the consequences of the following theorem.

> ***Theorem 8.18 (Hartogs' theorem)***
>
> Let X be a set. Then there is an ordinal α such that $\alpha \not\preceq X$, ie. it is not the case that $\alpha \preceq X$.

Surely $X \prec \alpha$! See Exercise 8.67 below to see why this might not be the case.

Proof

We shall actually construct the least such ordinal α. As every smaller ordinal β would have the property that $\beta \preceq X$, we will try to define α by

That is, 'smaller' in the $\in$-order of the ordinals.

$$\alpha = \{\beta : \beta \text{ is an ordinal and } \beta \preceq X\}.$$

Suppose that this does define an ordinal α. Is it the case that $\alpha \not\preceq X$? If not, i.e. if we had $\alpha \preceq X$, then by definition of α we would have $\alpha \in \alpha$, which is of course impossible (by Theorem 8.3). Thus $\alpha \not\preceq X$ as required, while the definition of α guarantees that it is the least ordinal with this property.

We have yet to show first that this defines a set and next that it is also an ordinal.

Let us show that α is indeed a set. First, note that for any ordinal β with $\beta \preceq X$, there is at least one subset Y of X and a well-ordering R on Y such that β is order-isomorphic to Y ordered by R. This is because $\beta \preceq X$, so that there is some one–one function $f\colon \beta \longrightarrow X$; and all we need do is take $Y = \mathrm{Im}(f)$ and define R on Y by

$$yRy' \text{ if and only if } f^{-1}(y) \in f^{-1}(y'),$$

which turns f into an order-isomorphism between β and Y.

Now define the set W of all well-orderings R on subsets of X, i.e.

$$W = \{R \subseteq X \times X : R \text{ is a well-ordering on } Y \text{ for some } Y \subseteq X\},$$

so that for all ordinals β with $\beta \preceq X$, there is some R in W representing a well-order on a subset of X order-isomorphic to β. Likewise, for every R in

W there is some ordinal β order-isomorphic to a $Y \subseteq X$ well-ordered by R, so that $\beta \approx Y \subseteq X$, giving $\beta \preceq X$. Note that W is a set, as W is a subset of $\mathscr{P}(X \times X)$. Now define a formula ϕ within set theory such that

$\phi(R, \beta)$ if and only if β is the ordinal order-isomorphic to the subset of X well-ordered by R.

The β for a given R is unique by Theorem 8.2, so that ϕ essentially defines a function on W. Then using the axiom of replacement, the image set of this function is a set; and clearly this image set is

$$\{\beta : \beta \text{ is an ordinal and } \beta \preceq X\}.$$

Thus α is a set.

Is α an ordinal? Certainly α is a set of ordinals, so is well-ordered by $\in$. So all we need to show is that for all β, if $\beta \in \alpha$ then $\beta \subseteq \alpha$. Take any $\beta \in \alpha$ and $\gamma \in \beta$. Then β is an ordinal (with $\beta \preceq X$), so that γ is also an ordinal with $\gamma \subseteq \beta$. But then as $\beta \preceq X$ we have $\gamma \subseteq \beta \preceq X$, so that $\gamma \preceq X$. Thus $\gamma \in \alpha$ as required.

And try to show $\gamma \in \alpha$!

Thus α is an ordinal, completing the proof of Hartogs' theorem. ■

This proof of Hartogs' theorem is very similar to that of Theorem 8.7, showing that $\omega_1 = \{\alpha : \alpha \text{ is a countable ordinal}\}$ is the least uncountable ordinal: the proof of the latter is effectively that of Hartogs' theorem with $X = \mathbb{N}$. For ω_1 we were able to conclude not only that $\omega_1 \not\preceq \mathbb{N}$, but also that $\mathbb{N} \prec \omega_1$. Can't we likewise conclude that in general the α in Hartogs' theorem satisfies $X \prec \alpha$?

Exercise 8.67

Explain why we can't conclude that $X \prec \alpha$.

Solution

Intuitively as the size (cardinality) of α is neither smaller than nor equal to that of X, one itches to say that its size is actually bigger! But to say that $X \prec \alpha$, we need to have a one–one function from X to α. In the special case where $X = \mathbb{N}$, we have $\mathbb{N} = \omega \subseteq \omega_1$, so that there is a natural one–one function (the identity map) from $\mathbb{N}$ to ω_1. But when X has no known helpful structure, how do we construct a one–one function?

If we add AC to our list of axioms, our problem is solved!

Exercise 8.68

Suggest a class of sets for which we *can* conclude that $X \prec \alpha$ for any X in the class.

Solution

How about the class of all ordinals! Suppose that X is an ordinal, β say, and that α is an ordinal as given by Hartogs' theorem, so that $\alpha \not\preceq \beta$. As α and β are ordinals, we must still have $\alpha \subseteq \beta$ or $\beta \subseteq \alpha$. It cannot be that $\alpha \subseteq \beta$, as then $\alpha \preceq \beta$. Then it must be that $\beta \subseteq \alpha$, so that $\beta \preceq \alpha$. As $\alpha \not\preceq \beta$, we cannot have $\alpha \approx \beta$, so that $\beta \prec \alpha$ as required.

Our solution to Exercise 8.68 proves the following result about the sizes of ordinals.

Theorem 8.19

There is no ordinal of largest cardinality, i.e. for any ordinal β, there is an ordinal α with $\beta \prec \alpha$.

We shall exploit this theorem, along with the principle of transfinite recursion, to construct special ordinals which will fill the role of cardinal numbers. These ordinals will have the property that any ordinal which is smaller in the ordinal sense is also smaller in 'size': formally they are defined as follows.

Definition

An infinite ordinal α is said to be an *initial ordinal* if for all $\beta < \alpha$, we also have $\beta \prec \alpha$.

From what we have already done, we know that ω and ω_1 are initial ordinals. And there are plenty of other initial ordinals. For any infinite ordinal β, Theorem 8.19 gives an ordinal α such that $\beta \prec \alpha$, so that

$$\{\gamma \leq \alpha : \beta \prec \gamma\}$$

is a non-empty set of ordinals, and thus has a least element γ_0. The minimality of γ_0 ensures that if $\delta < \gamma_0$, then

$$\begin{aligned} \delta &\preceq \beta \\ &\prec \gamma_0, \end{aligned}$$

so that γ_0 is an initial ordinal. $\beta < \gamma_0$. We have thus proved part (i) of the following result.

Clearly $\beta < \gamma_0$, as if $\gamma_0 \leq \beta$ then $\gamma_0 \subseteq \beta$, so that $\gamma_0 \preceq \beta$: this contradicts that $\beta \prec \gamma_0$.

Theorem 8.20

For any infinite ordinal β,

(i) there is a least initial ordinal γ_0 such that $\beta \prec \gamma_0$;

(ii) there is a unique initial ordinal β_0 such that $\beta \approx \beta_0$.

Note that $\beta_0 \leq \beta < \gamma_0$.

Exercise 8.69

Prove part (ii) of Theorem 8.20. [Hints: let $A = \{\gamma \leq \beta : \gamma \approx \beta\}$. Then $A \neq \varnothing$, as $\beta \in A$. Let β_0 be the least element of A. Show that β_0 has the required properties.]

Exercise 8.70

Let β and γ be initial ordinals. Show that $\beta < \gamma$ if and only if $\beta \prec \gamma$.

The initial ordinals clearly have special significance for describing the sizes of infinite ordinals. We shall attempt to tie down their structure a bit more by defining the following (class) function $\aleph$ on all ordinals by transfinite recursion:

$$\aleph(0) = \omega$$
$$\aleph(\gamma^+) = \text{the least initial ordinal } \alpha \text{ such that } \aleph(\gamma) \prec \alpha$$
$$\aleph(\lambda) = \bigcup\{\aleph(\gamma) : \gamma \in \lambda\}, \text{ for a limit ordinal } \lambda.$$

$\aleph$, written in English as 'aleph' and pronounced *alef*, is the first letter of the Hebrew alphabet.

This exists by Theorem 8.20.

From the definition, $\aleph(\alpha)$ is an initial ordinal for α equal to 0 or a successor. Is $\aleph(\lambda)$ an initial ordinal for each limit ordinal λ? The answer is yes, and you might like to prove this for yourself in the next exercise.

Exercise 8.71

Let λ be a limit ordinal. Show that $\aleph(\lambda)$ is an initial ordinal. [Hints: clearly $\aleph(\lambda)$ is infinite, as $0 < \lambda$, so that $\omega = \aleph(0) \leq \aleph(\lambda)$. So it remains to show that if $\beta < \aleph(\lambda)$ then $\beta \prec \aleph(\lambda)$.]

$\aleph(\lambda)$ is the lub of all the $\aleph(\gamma)$s for $\gamma < \lambda$. So that one way of interpreting this result is that $\aleph(\lambda)$ is the least initial ordinal greater than all the previous $\aleph(\gamma)$s.

Solution

Suppose that $\beta < \aleph(\lambda)$, i.e. $\beta \in \aleph(\lambda)$. As $\aleph(\lambda) = \bigcup\{\aleph(\gamma) : \gamma \in \lambda\}$, there must be some $\gamma \in \lambda$ for which $\beta \in \aleph(\gamma)$. We then have

$$\beta \subseteq \aleph(\gamma) \quad (\text{as } \beta \in \aleph(\gamma))$$
$$\prec \aleph(\gamma^+) \quad (\text{by definition of } \aleph(\gamma^+))$$
$$\subseteq \aleph(\lambda) \quad (\text{as } \gamma \in \lambda \text{ and } \lambda \text{ is a limit, so that } \gamma^+ \in \lambda),$$

so that $\beta \prec \aleph(\lambda)$, as required.

Thus every $\aleph(\alpha)$ is an initial ordinal. We hope that it seems likely to you that every initial ordinal is an $\aleph(\alpha)$ for some α. Let us introduce some new notation.

See Exercise 8.72 below.

> ***Notation***
>
> For each ordinal α, we shall write $\aleph(\alpha)$ as $\aleph_\alpha$.

Authors often use the notation ω_α when discussing the ordinal properties of the set. And $\aleph_\alpha$ is used when one is interested in cardinality properties.

$\aleph_0$ is called 'aleph-zero' or 'aleph-null'.

So rewriting the definition of the $\aleph(\alpha)$s in terms of this new notation, we have

$$\aleph_0 = \omega$$
$$\aleph_{\gamma^+} = \text{the least initial ordinal } \alpha \text{ such that } \aleph_\gamma \prec \alpha$$
$$\aleph_\lambda = \bigcup\{\aleph_\gamma : \gamma \in \lambda\} \text{ for a limit ordinal } \lambda.$$

Let us now show that every initial ordinal is an $\aleph_\gamma$ for some γ.

Exercise 8.72

(a) Let α be an ordinal. Explain why $\alpha \leq \aleph_\alpha$.

(b) Now let α be an initial ordinal. Define a set X of ordinals by

$$X = \{\beta \in \alpha^+ : \aleph_\beta \leq \alpha\}.$$

Show that X is a successor ordinal, say γ^+, and explain why α equals $\aleph_\gamma$.

Solution

(a) It is easy to show that the $\aleph$ function above is in fact normal. Thus by the result of Exercise 8.58 in the previous section, we have $\alpha \leq \aleph(\alpha) = \aleph_\alpha$.

(b) First we must show that X is an ordinal. X is a set of ordinals, so all we need to show is that if $\beta \in X$, then $\beta \subseteq X$. So suppose that $\beta \in X$ and take any $\gamma \in \beta$. Then

$$\begin{aligned} \aleph_\gamma &< \aleph_\beta \quad (\text{as } \gamma < \beta) \\ &\leq \alpha \quad (\text{as } \beta \in X), \end{aligned}$$

so that $\gamma \in X$. Thus $\beta \subseteq X$ as required, so that X is an ordinal.

Next we show that X is a successor ordinal by showing that it is neither 0 nor a limit. As α is an initial ordinal, α has to be infinite, so that $\omega \leq \alpha$, or, equivalently, $\aleph_0 \leq \alpha$. This means that $0 \in X$, so that X is non-empty, i.e. $X \neq 0$. Could X be a limit ordinal? If so, then as $\aleph_\gamma \leq \alpha$ for all $\gamma \in X$, we would have α as an upper bound for the set $\{\aleph_\gamma : \gamma \in X\}$, so that

$$\begin{aligned} \alpha &\geq \bigcup\{\aleph_\gamma : \gamma \in X\} \quad (= \operatorname{lub}\{\aleph_\gamma : \gamma \in X\}) \\ &= \aleph_X \quad (\text{as } X \text{ is a limit}); \end{aligned}$$

but this would mean that $X \in X$, which is impossible. Thus X cannot be a limit, which only leaves the possibility that X is a successor ordinal, say γ^+.

As $\gamma \in \gamma^+ = X$, we must have $\aleph_\gamma \leq \alpha$. But as $X \notin X$, we have $\aleph_{\gamma^+} \not\leq \alpha$, so that $\alpha < \aleph_{\gamma^+}$. Thus

$$\aleph_\gamma \leq \alpha < \aleph_{\gamma^+}.$$

All three of $\aleph_\gamma$, α and $\aleph_{\gamma^+}$ are initial ordinals, and by definition $\aleph_{\gamma^+}$ is the least initial ordinal greater than $\aleph_\gamma$. We must thus have

$$\aleph_\gamma = \alpha,$$

as required.

Thus the $\aleph_\gamma$s completely describe the different possible sizes (or cardinals) of the infinite ordinals, as by Theorem 8.20 every infinite ordinal is equinumerous with some initial ordinal, hence to some $\aleph_\gamma$. In Chapter 6 we avoided saying what we meant by the size or cardinal of an infinite set. For ordinals, we are now in a position to say what we mean!

Definition

Let α be an ordinal. The *cardinal* of α, written as $\operatorname{Card}(\alpha)$, is defined as follows:

For α finite, $\operatorname{Card}(\alpha)$ is simply α.

For α infinite, $\operatorname{Card}(\alpha)$ is the least $\aleph_\gamma$ equinumerous with α.

Furthermore, if an infinite set X can be well-ordered, we can find an $\aleph_\gamma$ with which it is equinumerous, so that this $\aleph_\gamma$ serves as a measure of the size of X. This is the content of the next exercise.

We could thus extend the definition of cardinal by setting $\text{Card}(X) = \aleph_\gamma$.

Exercise 8.73

Let X be an infinite set which can be well-ordered. Explain why there is an $\aleph_\gamma$ such that $X \approx \aleph_\gamma$.

Solution

As X can be well-ordered, by R say, there is some ordinal α order-isomorphic to X; and the order-isomorphism is a bijection between X and α. Thus $X \approx \alpha$. As X is infinite, so too is α, so that there is some $\aleph_\gamma$ with $\alpha \approx \aleph_\gamma$. Hence $X \approx \aleph_\gamma$.

By Theorem 8.6.

So if every set can be well-ordered (the well-ordering principle), the $\aleph_\gamma$s, along with the natural numbers, would completely describe the cardinals of all sets. In Cantor's original theory, the $\aleph_\gamma$s were indeed his cardinal numbers, but with hindsight this depended on heavy use of equivalents of the axiom of choice, including the well-ordering principle. In the next chapter we shall investigate what happens when one assumes AC.

For the moment, let's investigate the structure of the $\aleph_\gamma$s a bit more. Those that we have met before, namely $\aleph_0$ and $\aleph_1$ (i.e. ω and ω_1), are both limit ordinals. The next exercise shows that this is the case for all the $\aleph_\gamma$s.

Exercise 8.74

(a) Let β be an infinite ordinal, so that $\omega \leq \beta$. Construct a bijection to show that $\beta^+ \approx \beta$.

(b) Hence show that an $\aleph_\gamma$ is a limit ordinal.

Solution

(a) We can exploit the fact that ω is a subset of β and hence of β^+ $(= \beta \cup \{\beta\})$. Define a function f by

$$f: \beta^+ \longrightarrow \beta$$
$$\gamma \longmapsto \begin{cases} 0, & \text{if } \gamma = \beta \\ \gamma + 1, & \text{if } \gamma < \omega \\ \gamma, & \text{if } \omega \leq \gamma < \beta \end{cases}$$

It is easy to show that f is a bijection.

(b) If $\aleph_\gamma$ is a successor ordinal, say β^+, then $\aleph_\gamma$ is equinumerous with the ordinal β, which is smaller in the sense that $\beta < \aleph_\gamma$. Thus $\aleph_\gamma$ cannot be initial. Hence any initial ordinal can be neither a successor nor 0 (as an initial ordinal must be infinite), and must thus be a limit.

What happens, in terms of cardinality, when one adds $\aleph_\gamma$s to each other? It helps to investigate the ordinal sum $\mu + \mu$ for a limit ordinal μ.

First of all, this sum is order-isomorphic to the order sum $\mu + \mu$, so that this order-isomorphism is a bijection between $\mu + \mu$ and $(\mu \times \{0\}) \cup (\mu \times \{1\})$.

Next note that every element of μ is either of the form n for some $n \in \omega$ or of the form $\lambda + n$ for some limit ordinal λ and $n \in \omega$. Then we can define a function $f: \mu \times \{0\} \longrightarrow \mu$ by

By Theorem 8.16 every infinite ordinal is of the form $\lambda + n$, for some limit λ and $n \in \omega$.

$$f(n,0) = 2n, \text{ for } n \in \omega,$$
$$f(\lambda + n, 0) = \lambda + 2n, \text{ for each limit } \lambda \in \mu \text{ and } n \in \omega,$$

and similarly a function $g: \mu \times \{1\} \longrightarrow \mu$ by

$$g(n,1) = 2n+1, \text{ for } n \in \omega,$$
$$g(\lambda + n, 1) = \lambda + 2n + 1, \text{ for each limit } \lambda \in \mu \text{ and } n \in \omega.$$

You can check that as μ is a limit ordinal, f and g are well-defined and one–one, and that by gluing f and g together we obtain a bijection from $(\mu \times \{0\}) \cup (\mu \times \{1\})$ to μ. Thus $\mu + \mu \approx \mu$.

This means that for each γ,

$$\aleph_\gamma + \aleph_\gamma \approx \aleph_\gamma.$$

In the next chapter we shall use the symbol + to stand for a sum operation on cardinals, so could write this as $\aleph_\gamma + \aleph_\gamma = \aleph_\gamma$. Care will be needed to distinguish between cardinal + and ordinal +: for the latter $\aleph_\gamma + \aleph_\gamma > \aleph_\gamma$.

The next exercise looks at the cardinality of the sum of two different $\aleph_\gamma$s.

Exercise 8.75

Show that

$$\aleph_\alpha + \aleph_\beta \approx \max\{\aleph_\alpha, \aleph_\beta\} \ (= \aleph_{\max\{\alpha,\beta\}}).$$

[Hint: use Theorem 8.13 and Exercise 8.46 to obtain suitable upper and lower bounds for the sum in terms of the biggest of $\aleph_\alpha$ and $\aleph_\beta$.]

There is a corresponding result for the multiplication of two $\aleph_\gamma$s. Recall that the ordinal product $\alpha \cdot \beta$ of two ordinals is order-isomorphic to their Cartesian product $\alpha \times \beta$ with the anti-lexicographic order: this means that in terms of cardinality, $\alpha \cdot \beta \approx \alpha \times \beta$. The key result which follows makes much more use of the fact that an $\aleph_\gamma$ is an initial ordinal, not just a limit ordinal, than in our argument for addition.

By Theorem 8.10 of Section 8.3.

All we want out of the order-isomorphism here is that it is a bijection!

Theorem 8.21

Let κ be an initial ordinal. Then

$$\kappa \cdot \kappa \approx \kappa.$$

In terms of the $\aleph_\gamma$ notation, this becomes, for any ordinal γ,

$$\aleph_\gamma \cdot \aleph_\gamma \approx \aleph_\gamma.$$

It's easier to write κ than $\aleph_\gamma$!

Proof

We shall show by transfinite induction that for every infinite ordinal α,

$$\alpha \times \alpha \approx \alpha,$$

which, by Theorem 8.10, implies that $\alpha \cdot \alpha \approx \alpha$.

So we are looking at all infinite ordinals, not just those which are initial. The proof is not all that easy ...

The result holds for $\alpha = \omega$, the smallest infinite ordinal, by Theorem 6.4 of Section 6.4.

Suppose that the result holds for an infinite ordinal β. We shall show that it holds for β^+. By Exercise 8.74, $\beta^+ \approx \beta$. Thus

$$\begin{aligned}\beta^+ \times \beta^+ &\approx \beta \times \beta \quad (\text{as } \beta^+ \approx \beta)\\ &\approx \beta \cdot \beta\\ &\approx \beta \quad (\text{by induction hypothesis})\\ &\approx \beta^+,\end{aligned}$$

as required to show that the result holds for β^+.

Now suppose that the result holds for all infinite γ less than a limit ordinal λ, where $\lambda > \omega$. We shall show that it holds for λ. There are two cases. The easy case is when λ is *not* an initial ordinal, as then $\lambda \approx \gamma$ for some $\gamma < \lambda$, so that

$$\begin{aligned}\lambda \times \lambda &\approx \gamma \times \gamma\\ &\approx \gamma \quad (\text{by induction hypothesis})\\ &\approx \lambda.\end{aligned}$$

The hard case is when λ is an initial ordinal, in which case we have $\gamma \prec \lambda$ for all $\gamma < \lambda$. We shall show that $\lambda \times \lambda \approx \lambda$ by constructing a special ordering R on $\lambda \times \lambda$ which makes the set order-isomorphic to λ, so that there is a bijection between the sets, as is required to show that they are equinumerous. We define R by

This R is the same order as dealt with in Exercise 7.69 in Section 7.4.

$$\begin{aligned}(\alpha,\beta)R(\alpha',\beta') \text{ if (i)} \quad &\max\{\alpha,\beta\} < \max\{\alpha',\beta'\},\\ \text{or (ii)} \quad &\max\{\alpha,\beta\} = \max\{\alpha',\beta'\} \text{ and } \beta < \beta',\\ \text{or (iii)} \quad &\max\{\alpha,\beta\} = \max\{\alpha',\beta'\} \text{ and } \beta = \beta' \text{ and } \alpha < \alpha'.\end{aligned}$$

It can be shown (using the result of Exercise 7.69) that R is a well-order on $\lambda \times \lambda$. It is easy to show that

$$\lambda \times \lambda = \bigcup\{\gamma \times \gamma : \gamma \in \lambda\}.$$

This is left as an easy exercise for you.

Also from Exercise 7.69, for any $\gamma, \gamma' \in \lambda$ with $\gamma < \gamma'$, we can show that $\gamma \times \gamma$ is an initial segment of both $\gamma' \times \gamma'$ and $\lambda \times \lambda$ under the order R. Thus $\{\gamma \times \gamma : \gamma \in \lambda\}$ forms a $\subseteq$-chain with union order-isomorphic to $\lambda \times \lambda$, with all the sets ordered by R.

It is the initial segment $\mathrm{Seg}_{\lambda\times\lambda}((\gamma, 0))$.

For each $\gamma \in \lambda$ let $\theta(\gamma)$ be the ordinal number order-isomorphic to $\gamma \times \gamma$ ordered by R. Then $\{\theta(\gamma) : \gamma \in \lambda\}$ forms a $\subseteq$-chain, this time of ordinals, with each element order-isomorphic to the corresponding element of the earlier chain $\{\gamma \times \gamma : \gamma \in \lambda\}$, so that

$\theta(\gamma)$ is unique by Theorem 8.6 of Section 8.2.

$$\begin{aligned}\lambda \times \lambda &\cong \bigcup\{\gamma \times \gamma : \gamma \in \lambda\} \quad (\text{ordered by } R)\\ &\cong \bigcup\{\theta(\gamma) : \gamma \in \lambda\}.\end{aligned}$$

But for each $\gamma \in \lambda$ we have $\theta(\gamma) \cong \gamma \times \gamma$, so that $\theta(\gamma) \approx \gamma \times \gamma$, which by the induction hypothesis means that $\theta(\gamma) \approx \gamma$. As in this case λ is an initial ordinal and $\gamma \in \lambda$, so that $\gamma \prec \lambda$, this means that $\theta(\gamma) \prec \lambda$, for each $\gamma \in \lambda$. Thus λ is an upper bound for the set of ordinals $\{\theta(\gamma) : \gamma \in \lambda\}$, so that

$$\begin{aligned}\lambda \times \lambda &\cong \bigcup\{\theta(\gamma) : \gamma \in \lambda\}\\ &\leq \lambda \quad (\text{as } \lambda \text{ is an upper bound}).\end{aligned}$$

Thus $\lambda \times \lambda \preceq \lambda$. As trivially $\lambda \preceq \lambda \times \lambda$, this gives us $\lambda \times \lambda \approx \lambda$, as required. ■

Exercise 8.76

Fill in the following detail of the proof above. Use the fact that λ is a limit ordinal to show that

$$\lambda \times \lambda = \bigcup\{\gamma \times \gamma : \gamma \in \lambda\}.$$

Theorem 8.21 along with the result of Exercise 6.26 of Section 6.3 provides a simple alternative proof that $\aleph_\gamma + \aleph_\gamma \approx \aleph_\gamma$. We leave the details for you as an exercise. Likewise we shall leave the proof of the analogous result to Exercise 8.75 above for the multiplication of different $\aleph_\gamma$s as a straightforward exercise for you.

Exercise 8.77

Use Theorem 8.21 and the result of Exercise 6.26 to show that $\aleph_\gamma + \aleph_\gamma \approx \aleph_\gamma$.

Exercise 8.78

Show that

$$\aleph_\alpha \cdot \aleph_\beta \approx \max\{\aleph_\alpha, \aleph_\beta\} \ (= \aleph_{\max\{\alpha,\beta\}}).$$

So if the $\aleph_\gamma$s are going to be our infinite cardinal numbers, their addition and multiplication as cardinals is a bit dull! In the next chapter, we shall see how the axiom of choice resolves many of the issues surrounding cardinals, and connects them firmly with ordinals, in particular the $\aleph_\gamma$s, as in Cantor's original theory.

But exponentiation of the $\aleph_\gamma$s is *very* interesting!

Further exercises

Exercise 8.79

Let $\aleph_\alpha$ be any initial ordinal and $\mathbf{n} \in \mathbb{N}$. Show that

$$\aleph_\alpha + \mathbf{n} \approx \aleph_\alpha \cdot \mathbf{n} \approx \aleph_\alpha.$$

Exercise 8.80

Show that if κ is an initial ordinal and X, Y are sets such that $\kappa \preceq X \times Y$, then $\kappa \preceq X$ or $\kappa \preceq Y$. [Hints: suppose that $f \colon \kappa \longrightarrow X \times Y$ is one–one. Put

$$A = \{x \in X : (x, y) \in \text{Range}(f) \text{ for some } y \in Y\}$$

and

$$B = \{y \in Y : (x, y) \in \text{Range}(f) \text{ for some } x \in X\},$$

so that $\kappa \approx \text{Range}(f) \subseteq A \times B$. Show that both A and B can be well-ordered, so that there are initial ordinals $\aleph_\alpha$ and $\aleph_\beta$ such that $A \approx \aleph_\alpha$ and $B \approx \aleph_\beta$. Now exploit the fact that κ is initial and the result of Exercise 8.78.]

9 SET THEORY WITH THE AXIOM OF CHOICE

9.1 Introduction

We are now in a position to construct the remainder of Cantor's remarkable theory of infinite numbers. That we have taken so much space to get here, compared to Cantor's original work, is mainly a reflection of the very complex issues that this raised and the consequent attempts to place his work on a more rigorous footing. One of these issues is the role of the axiom of choice: as you will see in this chapter, AC is inextricably caught up with Cantor's work linking cardinals and ordinals.

First of all, let us remind ourselves of some of Cantor's results which can be obtained without using AC.

There are as many natural numbers as rationals. That is, $\mathbb{N} \approx \mathbb{Q}$.

There are more real numbers than rationals. That is, $\mathbb{Q} \prec \mathbb{R}$.

There are uncountably many transcendental real numbers.

There are as many points on the real line as in the plane. That is, $\mathbb{R} \approx \mathbb{R}^2$.

Along with these very impressive results, we have proved important technical results, like the Schröder–Bernstein theorem, telling us much about how to compare the 'sizes' of sets, without having used AC. But we have avoided the issue of saying what the size, Card(X), of an infinite set X actually is! Cantor's definition of Card(X) was effectively equivalent to saying that it is the class $\{Y : X \approx Y\}$, which is a proper class. If this class were to contain some well-ordered sets, then within the theory that we have developed so far, the class would contain a unique initial ordinal $\aleph_\alpha$ – we could then take Card(X) to be the corresponding $\aleph_\alpha$. But if X is equinumerous with some ordinal, then the bijection between them can be used to define a well-order on X. That's where AC comes into the theory of cardinals, as the principle from which Zermelo proved that every set can be well-ordered.

Cantor didn't actually attempt to represent Card(X) by a set. Instead he represented it by the property '... is equinumerous with X'; but the corresponding class is a proper class.

In the next section of this chapter we shall look at the relationship between AC and the *well-ordering principle*, (which we abbreviate as WO) namely that every set can be well-ordered. And in the third section we shall look at the impact of AC on cardinal arithmetic.

We started the book by looking at Cantor's and Dedekind's answers to the question:

What are the real numbers?

Much of this last chapter hinges on two questions which Cantor raised as a consequence of his work on infinite sets:

1. Is there a well-ordering of $\mathbb{R}$?
2. Is there an infinite subset of $\mathbb{R}$ equinumerous with neither $\mathbb{N}$ nor $\mathbb{R}$?

That is, is there an $X \subseteq \mathbb{R}$ with $\mathbb{N} \prec X \prec \mathbb{R}$?

In the final section of the chapter (and book) we shall discuss these questions and look at some of the ways in which set theory can be further developed.

9.2 The well-ordering principle

Zermelo's answer to Cantor's question of whether $\mathbb{R}$ can be well-ordered was effectively to prove that every set (so in particular $\mathbb{R}$) could be well-ordered, on the basis of his axiom of choice. In this section we shall look at a proof of not only this result, but also of the equivalence of the following three apparently unrelated mathematical principles.

> ***The axiom of choice (AC)***
> Suppose that M is a non-empty set. Then there is a function $h\colon \mathscr{P}(M) \setminus \{\varnothing\} \longrightarrow M$ such that for all non-empty subsets A of M, $h(A) \in A$.
>
> ***The well-ordering principle (WO)***
> Every set can be well-ordered.
>
> ***Zorn's lemma (ZL)***
> Let P be a non-empty set partially ordered by R with the property that every chain $\mathscr{C}$ in P has an upper bound in P. Then P contains at least one maximal element.

We have taken the power set form of AC for convenience.

We shall break down the proof that these principles are equivalent into separate stages. We shall give separate proofs that WO $\Rightarrow$ AC, AC $\Rightarrow$ WO, ZL $\Rightarrow$ AC, AC $\Rightarrow$ ZL.

There are many attractive alternative routes, e.g. proving WO $\Rightarrow$ AC $\Rightarrow$ ZL $\Rightarrow$ WO.

First, let's look at the proof that WO implies AC. This is so straightforward that you might like to try it as the next exercise.

Exercise 9.1

Assume WO, i.e. for every set X there is a well-order R on X; and suppose that M is a non-empty set. Show that there is a function $h\colon \mathscr{P}(M) \setminus \{\varnothing\} \longrightarrow M$ such that for all non-empty subsets A of M, $h(A) \in A$. [Hint: by WO there is a well-order R on M. Exploit this to define each $h(A)$.]

Solution

Let R be a well-order on M. Then each non-empty subset A of M contains a least element $\min A$ under this order. Thus we can define a choice function h by

$$\begin{aligned} h\colon \mathscr{P}(M) \setminus \{\varnothing\} &\longrightarrow M \\ A &\longmapsto \min A. \end{aligned}$$

We have thus proved the following theorem.

> ***Theorem 9.1***
>
> The well-ordering principle implies the axiom of choice.

The proof of the converse is more complicated, so we shall actually show it to you!

> ***Theorem 9.2***
>
> The axiom of choice implies the well-ordering principle.

Proof

Let X be any non-empty set. To define a well-order on X, we shall first construct a function f from an appropriately chosen ordinal *onto* X. It will then be very straightforward to define a well-order R on X, exploiting the well-order on the ordinal.

If $X = \varnothing$ the result is trivially true.

But how do we choose the ordinal? The answer comes from Hartogs' theorem (Theorem 8.18 in Section 8.5). This gives us an ordinal α such that $\alpha \not\preccurlyeq X$. Although we cannot assume that this means that $X \prec \alpha$, we have the feeling that in some sense α is bigger than X! This will show up in our construction below, in the following way. We shall try to define a one–one function f from α to X by transfinite recursion: as $\alpha \not\preccurlyeq X$, we must run out of elements of X as possible images under f *before* we have used up all of the domain α. This means that we shall need some set c not in X for the remaining elements of α to map to – a by now familiar trick!

Recall that $\alpha \not\preccurlyeq X$ means that there is no one–one function from α into X.

Where will AC fit in? Our method will be to define $f(\beta)$, for each $\beta \in \alpha$, to be one of the elements of X not already used up as $f(\gamma)$ for some $\gamma \in \beta$, i.e. an element of X not in $\mathrm{Range}(f|_\beta)$. To choose one of these elements we shall need a choice function on non-empty subsets of X. So, assuming AC, as is permitted for this theorem, take a choice function $h\colon \mathscr{P}(X) \setminus \{\varnothing\} \longrightarrow X$ such that for all non-empty subsets A of X, $h(A) \in A$.

We can now fit all the ingredients of our construction together. Define a function f on α by transfinite recursion as follows:

For this definition, we don't need to distinguish between β being a successor ordinal or a limit ordinal.

$$
\begin{aligned}
f\colon \alpha &\longrightarrow X \cup \{c\}\\
0 &\longmapsto h(X),\\
\beta &\longmapsto \begin{cases} h\left(X \setminus \mathrm{Range}(f|_\beta)\right), & \text{if } X \setminus \mathrm{Range}(f|_\beta) \neq \varnothing,\\ c, & \text{otherwise.}\end{cases}
\end{aligned}
$$

We shall leave you to exploit f to complete the proof as Exercise 9.2 below. ■

This construction is very reminiscent of others in this book, for instance that used in Theorem 6.2 of Section 6.4 showing that if $A \subseteq \mathbb{N}$, then A is finite or $A \approx \mathbb{N}$.

Exercise 9.2

Work through the details of the end of the proof of Theorem 9.2 as follows.

(a) Show that if neither $f(\gamma)$ nor $f(\beta)$ equals c, so that both are in X, with $\gamma < \beta$, then $f(\gamma) \neq f(\beta)$.

(b) If $c \notin \mathrm{Range}(f)$, then we can infer from part (a) that f is one–one, contradicting that $\alpha \not\preccurlyeq X$. Thus we must have $c \in \mathrm{Range}(f)$. Let δ be the least element of α such that $f(\delta) = c$.

Show that $f|_\delta$ is a bijection between δ and X.

(c) Exploit $f|_\delta$ to define a well-order on X.

Let us now turn to the equivalence of AC and ZL. We asked you to show that ZL implies AC as Exercise 5.24 in Section 5.4. So all we need to show here is that AC implies ZL.

One of the interests of this is that ZL implies the dichotomy principle for cardinals, namely that for any sets A, B, we have $A \preceq B$ or $B \preceq A$. This is Theorem 5.7 in Section 5.4.

> ***Theorem 9.3***
>
> The axiom of choice implies Zorn's lemma.

Proof

Let P be a non-empty set partially ordered by R – we shall take R to be a strict order – with the property that every chain $\mathscr{C}$ in P has an upper bound in P. We have to show that P contains at least one maximal element.

In terms of a strict order R, an upper bound for $\mathscr{C}$ is a $y \in P$ such that for all $x \in \mathscr{C}$, xRy or $x = y$.

Our method will be to take a suitably big ordinal α and define an order-embedding f by transfinite recursion from α into P. We shall attempt to define $f(\beta)$, for each $\beta \in \alpha$, so that $f(\gamma)Rf(\beta)$ for all $\gamma \in \beta$. As an ordinal is linearly ordered by $\in$, the images of f will thus form an R-chain. What we mean by a 'suitably big ordinal α' is one as given by Hartogs' theorem, so that $\alpha \npreceq P$. This will ensure that something has to go wrong with our construction, because we cannot end up with f providing an order-embedding, hence a one–one function, from α into P. What will turn out to have gone wrong is that for some β the element $f(\beta)$ of P is a maximal element of P.

It might be more appropriate to say "what has gone right" here, as we actually want to find such a maximal element of P!

Let us assemble the ingredients required for our attempt at an order-embedding. First, we use Hartogs' theorem to give us an ordinal α with $\alpha \npreceq P$. Next, let us have a compact notation for the subset of P consisting of all elements, if any, greater under the order R than a given x: we shall call this subset S_x, so that

$$S_x = \{y \in P : xRy\}.$$

Note that x is a maximal element of P if and only if S_x is empty. Next, as we are assuming AC, let h be a choice function on non-empty subsets of P, i.e. $h\colon \mathscr{P}(P) \setminus \{\varnothing\} \longrightarrow P$ such that for all non-empty subsets A of P, $h(A) \in A$. Lastly, as so often in these arguments, we shall need a set c not in P as a possible image for when we have run out of suitable images in P; and for the sake of convenience in the definition below, we set $S_c = \varnothing$. Now define a function f as follows:

$$f\colon \alpha \longrightarrow P \cup \{c\}$$
$$f(0) = h(P)$$
$$f(\beta^+) = \begin{cases} h(S_{f(\beta)}), & \text{if } S_{f(\beta)} \neq \varnothing, \\ c, & \text{otherwise,} \end{cases}$$
$$\text{for a limit } \lambda,\ f(\lambda) = \begin{cases} h\left(\bigcap\{S_{f(\gamma)} : \gamma \in \lambda\}\right), & \text{if } \bigcap\{S_{f(\gamma)} : \gamma \in \lambda\} \neq \varnothing, \\ c, & \text{otherwise.} \end{cases}$$

Just to make sure that you see the significance of the conditions determining the definition above, note that

$S_{f(\beta)} \neq \varnothing$ if and only if there is some y with $f(\beta)Ry$,

and

$\bigcap\{S_{f(\gamma)} : \gamma \in \lambda\} \neq \varnothing$ if and only if there is some y with $f(\gamma)Ry$ for all $\gamma \in \lambda$.

Furthermore, if $f(\gamma)$ equals c for some $\gamma \in \alpha$, then $f(\beta) = c$ for all $\beta > \gamma$. Also note that until c is obtained as an image, f will be strictly order-preserving and thus one–one (which we shall ask you to confirm in Exercise 9.3). As $\alpha \not\preceq P$ there can be no one–one function from α to P, so that we must have $f(\gamma) = c$ for some $\gamma \in \alpha$.

Put $\delta = \min\{\gamma \in \alpha : f(\gamma) = c\}$. As P is non-empty, δ cannot be $\mathbf{0}$, so that δ is either a successor, β^+, for some β, or a limit λ. If δ is of the form β^+, what can we say about $f(\beta)$? As $f(\beta)$ doesn't equal c, it must be some element of P. And as $f(\beta^+) = c$, it must be that the set $S_{f(\beta)}$, i.e. the set of all elements of P greater than $f(\beta)$ in the order R, is empty. This would mean that $f(\beta)$ is a maximal element of P, which is what we are looking for.

Recall that we are trying to show that P contains a maximal element with respect to its order by R.

What would happen if δ was a limit ordinal? This would mean that $f(\gamma)$ was defined for all $\gamma \in \delta$ and that $\bigcap\{S_{f(\gamma)} : \gamma \in \delta\} = \varnothing$, so that there was no y with $f(\gamma)Ry$ for all $\gamma \in \lambda$. But $f|_\delta$ is order-preserving and, of course, the ordinal δ is linearly ordered by $\in$, so that the subset $\mathscr{C}$ defined by

$$\mathscr{C} = \{f(\gamma) : \gamma \in \delta\}$$

of P is an R-chain. By the special condition on P (the only place in this proof where we use it) this chain has an upper bound, y say, in P. Can this y be one of the elements of the chain $\mathscr{C}$? If y did equal $f(\gamma)$ for some $\gamma \in \delta$, then as δ is a limit ordinal, we also have $\gamma^+ \in \delta$, so that $f(\gamma^+)$ is also an element of $\mathscr{C}$. But the construction of $f(\gamma^+)$ guarantees that $f(\gamma)Rf(\gamma^+)$, so that $f(\gamma)$ is not an upper bound of $\mathscr{C}$. It follows from this contradiction that y does not equal $f(\gamma)$ for any $\gamma \in \delta$, so that $\bigcap\{S_{f(\gamma)} : \gamma \in \delta\} \neq \varnothing$, giving us a further contradiction. Thus δ cannot be a limit ordinal.

Remember that R is a strict order.

We are left with the conclusion that δ is indeed of the form β^+ for some β, and that $F(\beta)$ is a maximal element of P, as required. ■

Exercise 9.3

Confirm the following detail of the proof of Theorem 9.3. Show that if neither $f(\gamma)$ nor $f(\beta)$ equals c, so that both are in P, with $\gamma < \beta$, then $f(\gamma)R(\beta)$. [Hint: use transfinite induction on β for fixed γ.]

9.3 Cardinal arithmetic and the axiom of choice

Given the results of the previous section, on the equivalence of AC to WO and ZL, we are now able to see the impact of AC on the theory of cardinals. What we shall really be doing is resurrecting much of Cantor's work, as in [12], within the more rigorous setting of *ZF*. Assuming the axiom of choice we can define the cardinal number Card(X) of an infinite set, can show that such numbers are linearly ordered by $\preceq$, and can perform arithmetic with them. We shall also see the extent to which Cantor's edifice of cardinals actually requires the axiom of choice, by showing that many of his results about cardinals, which follow from AC, are actually equivalent to AC.

By and large, we could not do these things in Chapter 6, where we were avoiding use of AC.

Let's look first at how to define the cardinal number of an infinite set X. Cantor effectively associated the cardinal of X with the property 'is equinumerous with X', which in terms of ZF would give rise to the class

$$\{Y : Y \approx X\}.$$

But this is a proper class, not a set, and we want the cardinal of X within ZF to be a set. What AC allows us to do is to select a special set out of this class to be the cardinal of X, namely an ordinal, as follows. Assuming AC, which is equivalent to WO, X can be well-ordered. There are then several ordinals to which possible well-orders on X are order-isomorphic, and we will specify one of them as the cardinal of X. The obvious candidate is the least such ordinal. This is essentially the content of Exercise 8.73 of Section 8.5, which says that there is an initial ordinal $\aleph_\gamma$ with $X \approx \aleph_\gamma$. We can restate this as a theorem.

Without AC, the best we can do is define Card(X) to be X itself and say that Card(X) = Card(Y) when $X \approx Y$. The resulting theory is as in Chapter 6.

Theorem 9.4 Every set has a cardinal

AC implies that for every infinite set X there is an $\aleph_\gamma$ such that $X \approx \aleph_\gamma$.

This allows us to define the cardinal of X as follows.

Definitions

Let X be an infinite set. Assuming AC, the *cardinal number* (or, more simply, *cardinal*) of X, written as Card(X), is the initial ordinal $\aleph_\gamma$ for which $X \approx \aleph_\gamma$.

The class $\mathscr{C}$ of *cardinal numbers* consists of the finite and initial ordinals, i.e.

$$\mathscr{C} = \{\alpha : \alpha \text{ is a finite or initial ordinal}\},$$

or equivalently

$$\mathscr{C} = \omega \cup \{\aleph_\gamma : \gamma \text{ is an ordinal}\}.$$

For two sets X and Y, we say that the sets *have the same size* if Card(X) = Card(Y)

For a set X which is finite, so that $X \approx \mathbf{n}$ for some $\mathbf{n} \in \mathbb{N}$, we have already defined Card(X) to be $\mathbf{n}$ (in Chapter 6).

An immediate consequence of these definitions is that

$$X \approx \text{Card}(X),$$

for all sets X.

In the next (we hope very straightforward) exercise we ask you to show that there is only one cardinal number of each 'size'.

Exercise 9.4

Show that for any sets X, Y, Card(X) = Card(Y) if and only if $X \approx Y$.

We are implicitly assuming AC in these exercises, so that Card(X) is always defined.

Exercise 9.5

Show that for any cardinal κ, Card(κ) = κ.

Exercise 9.6

Show that $\mathscr{C}$ is not a set.

Assuming that $\mathscr{C}$ is a set leads to a contradiction, produced by Cantor in 1899 (see [11]) and often called Cantor's paradox.

Ordinals have a very rich structure, which cardinals (assuming AC) then inherit. By Exercise 9.4 the relation of equality ($=$) on cardinals is much less cumbersome than that of equinumerosity ($\approx$) on sets, which is merely an equivalence relation. Let's look now at the order structure of cardinals. The class $\mathscr{O}$ of all ordinals is well-ordered by the usual $<$ (i.e. the $\in$ relation), and the subclass $\mathscr{C}$ must then also be well-ordered by this $<$. For cardinals κ and λ, i.e. finite and initial ordinals, the relationship $\kappa < \lambda$ has the desired special significance that $\kappa < \lambda$ if and only if $\kappa \prec \lambda$. So we shall take the usual $<$ as our order on the cardinals.

See e.g. Exercise 8.70 of Section 8.5.

> ***Definition***
>
> Let κ, λ be cardinals. Then we shall say that $\kappa < \lambda$ exactly when $\kappa \in \lambda$, i.e. when $\kappa < \lambda$ in the order of the ordinals.

Of course one of the properties of $<$ on ordinals is the linearity property which, in terms of the associated weak order $\leq$, can be written as

$$\text{for all } \kappa,\ \lambda,\ \kappa \leq \lambda \text{ or } \lambda \leq \kappa.$$

Linearity also means that for any sets X, Y, their cardinals have a maximum, written as $\max\{\mathrm{Card}(X), \mathrm{Card}(Y)\}$.

This means that AC resolves a deficiency of $\preceq$ left over from Chapter 6, as follows.

> ***Theorem 9.5 The dichotomy principle***
>
> AC implies the dichotomy principle, namely that for any sets A, B,
>
> $$A \preceq B \text{ or } B \preceq A.$$

Proof

One proof uses the fact that, assuming AC, every set X is equinumerous to its cardinal, namely the ordinal $\mathrm{Card}(X)$. By the linearity of the usual order on the ordinals, either $\mathrm{Card}(A) \leq \mathrm{Card}(B)$ or $\mathrm{Card}(B) \leq \mathrm{Card}(A)$. Without loss of generality, let's suppose that $\mathrm{Card}(A) \leq \mathrm{Card}(B)$, which implies that $\mathrm{Card}(A) \subseteq \mathrm{Card}(B)$. This gives

$$\begin{aligned} A &\approx \mathrm{Card}(A) \\ &\preceq \mathrm{Card}(B) \quad (\text{as } A \subseteq B) \\ &\approx B, \end{aligned}$$

so that $A \preceq B$. ∎

Another proof exploits the fact that AC implies Zorn's Lemma. The result then follows from Theorem 5.7 in Section 5.4.

Of course if $\mathrm{Card}(B) \leq \mathrm{Card}(A)$, we end up with $B \preceq A$.

Exercise 9.7

Let X be a non-empty set and assume AC. Show that there is an element $x_0 \in X$ such that $x_0 \preceq x$, for all $x \in X$. Does the result still hold if X is a proper class?

Using AC we can also define arithmetic operations on cardinals. With these operations, many of the results of Chapter 6 translate into very familiar and attractive properties of this arithmetic.

Definitions

Assume AC and let X, Y be any sets. Then the operations of *cardinal addition, multiplication and exponentiation* are defined on the cardinal numbers as follows:

$$\begin{aligned}\mathrm{Card}(X) +_{\mathscr{C}} \mathrm{Card}(Y) &= \mathrm{Card}((X \times \{\mathbf{0}\}) \cup (Y \times \{\mathbf{1}\}));\\ \mathrm{Card}(X) \cdot_{\mathscr{C}} \mathrm{Card}(Y) &= \mathrm{Card}(X \times Y);\\ \mathrm{Card}(X)^{\mathrm{Card}(Y)} &= \mathrm{Card}\left(X^Y\right).\end{aligned}$$

So, for instance,

$$\mathrm{Card}(\mathbb{N}) +_{\mathscr{C}} \mathrm{Card}(\mathbb{N}) = \mathrm{Card}(\mathbb{N}),$$

because $(\mathbb{N} \times \{\mathbf{0}\}) \cup (\mathbb{N} \times \{\mathbf{1}\}) \approx \mathbb{N}$. And

By Theorem 6.3 of Section 6.4.

$$\mathrm{Card}(\mathbb{N}) \cdot_{\mathscr{C}} \mathrm{Card}(\mathbb{R}) = \mathrm{Card}(\mathbb{R}),$$

because $\mathbb{N} \times \mathbb{R} \approx \mathbb{R}$. Also

By Exercise 6.54 of Section 6.5.

$$\mathrm{Card}(\mathbf{2})^{\mathrm{Card}(\mathbb{N})} = \mathrm{Card}(\mathbf{2}^{\mathbb{N}}).$$

The notation for cardinal exponentiation is potentially confusing. If κ and λ are cardinals (and thus sets, namely natural numbers or ordinals $\aleph_\gamma$), does κ^λ mean the set of functions from λ to κ or the cardinal of this set, which is how the cardinal exponential κ^λ is defined? We hope that it will be clear from the context which of these is meant. If κ^λ appears in an equation or inequality involving other cardinals and the symbols $=$, $\leq$ or $<$, we shall usually mean the cardinal exponential.

The results of Section 8.5 about adding and multiplying the ordinals $\aleph_\gamma$ mean that the addition and multiplication of cardinals is dull, perhaps surprisingly so.

Theorem 9.6 The absorption law

Assume AC and let X, Y be sets, at least one of which is infinite. Then

$$\begin{aligned}\mathrm{Card}(X) +_{\mathscr{C}} \mathrm{Card}(Y) &= \mathrm{Card}(X) \cdot_{\mathscr{C}} \mathrm{Card}(Y)\\ &= \max\{\mathrm{Card}(X), \mathrm{Card}(Y)\}.\end{aligned}$$

The bigger of Card(X) and Card(Y) 'absorbs' the smaller.

Proof

For any two ordinals γ, δ, their ordinal sum $\gamma + \delta$ is equinumerous with $(\gamma \times \{\mathbf{0}\}) \cup (\delta \times \{\mathbf{1}\})$. In particular, this result holds when γ and δ are cardinals. As at least one of X and Y is infinite, at least one of Card(X) and Card(Y) is an $\aleph_\alpha$, for some α; and the other is either an $\aleph_\beta$ or an $\mathbf{n} \in \mathbb{N}$. The

The sets are equinumerous because they are order-isomorphic.

results of Exercises 8.75 and 8.79 of Section 8.5 then translate into the result for cardinal addition:

$$\mathrm{Card}(X) +_{\mathscr{C}} \mathrm{Card}(Y) = \max\{\mathrm{Card}(X), \mathrm{Card}(Y)\};$$

and the results of Exercises 8.78 and 8.79 of the same section translate into the equivalent result for cardinal multiplication. ■

The result of this theorem can be written directly in terms of cardinals. Assuming AC, for any cardinals κ, λ, at least one of which is infinite,

$$\begin{aligned}\kappa +_{\mathscr{C}} \lambda &= \kappa \cdot_{\mathscr{C}} \lambda \\ &= \max\{\kappa, \lambda\}.\end{aligned}$$

Exercise 9.8

Show that $2^{\mathbb{R}} \times 2^{\mathbb{R}} \approx 2^{\mathbb{R}}$. You may assume AC.

Solution

Using AC we have

$$\begin{aligned}2^{\mathbb{R}} \times 2^{\mathbb{R}} &\approx \mathrm{Card}\left(2^{\mathbb{R}}\right) \times \mathrm{Card}\left(2^{\mathbb{R}}\right) \\ &\approx \mathrm{Card}\left(2^{\mathbb{R}}\right) \cdot_{\mathscr{C}} \mathrm{Card}\left(2^{\mathbb{R}}\right) \\ &= \max\left\{\mathrm{Card}\left(2^{\mathbb{R}}\right), \mathrm{Card}\left(2^{\mathbb{R}}\right)\right\} \quad \text{(by Theorem 9.6)} \\ &= \mathrm{Card}\left(2^{\mathbb{R}}\right) \\ &\approx 2^{\mathbb{R}}.\end{aligned}$$

An alternative proof exploiting AC uses that $\mathrm{Card}(2^{\mathbb{R}})$ is an $\aleph_\gamma$, along with Theorem 8.21 of Section 8.5 which says that $\aleph_\gamma \cdot \aleph_\gamma \approx \aleph_\gamma$. This latter theorem says that in general

$$\kappa \cdot \kappa = \kappa,$$

for all infinite cardinals κ.

This result can in fact be derived quite easily *without* using AC, by the methods of Section 6.5, which some might consider gives it a higher status! The use of AC does, however, make the derivation even easier.

You might like to try this yourself as a revision of the use of the Schröder–Bernstein theorem.

Exercise 9.9

Why does the result of Theorem 9.6 require at least one of X and Y to be infinite?

The result of Exercise 9.8 can be written in terms of cardinal arithmetic as $\mathrm{Card}\left(2^{\mathbb{R}}\right) \cdot_{\mathscr{C}} \mathrm{Card}\left(2^{\mathbb{R}}\right) = \mathrm{Card}\left(2^{\mathbb{R}}\right)$. There are many other results of cardinal arithmetic which, like this one, follow from earlier results obtained without using AC, especially results from Chapter 6. Take, for instance, the result that for all sets A, B, $A \times B \approx B \times A$. In terms of cardinals, this becomes

This is the result of Exercise 6.13 of Section 6.3.

$$\text{for all cardinals } \kappa, \lambda,\ \kappa \cdot_{\mathscr{C}} \lambda = \lambda \cdot_{\mathscr{C}} \kappa,$$

as follows. We have

$$\begin{aligned}\kappa \cdot_{\mathscr{C}} \lambda &\approx \kappa \times \lambda \\ &\approx \lambda \times \kappa \quad (\text{as } A \times B \approx B \times A \text{ for all } A, B) \\ &\approx \lambda \cdot_{\mathscr{C}} \kappa,\end{aligned}$$

and as two cardinals are equal if and only if they are equinumerous the result follows. We can similarly state and prove other attractive equalities involving cardinals, fully justifying Cantor's description of cardinals as possessing an arithmetic.

By Exercise 9.4 above.

Theorem 9.7

Assume AC and let κ, λ, μ be any cardinals. Then each of the following holds.

(i) $\kappa +_{\mathscr{C}} \lambda = \lambda +_{\mathscr{C}} \kappa$.

(ii) $\kappa +_{\mathscr{C}} (\lambda +_{\mathscr{C}} \mu) = (\kappa +_{\mathscr{C}} \lambda) +_{\mathscr{C}} \mu$.

(iii) $\kappa \cdot_{\mathscr{C}} \lambda = \lambda \cdot_{\mathscr{C}} \kappa$.

(iv) $\kappa \cdot_{\mathscr{C}} (\lambda \cdot_{\mathscr{C}} \mu) = (\kappa \cdot_{\mathscr{C}} \lambda) \cdot_{\mathscr{C}} \mu$.

(v) $\kappa \cdot_{\mathscr{C}} (\lambda +_{\mathscr{C}} \mu) = (\kappa \cdot_{\mathscr{C}} \lambda) +_{\mathscr{C}} (\kappa \cdot_{\mathscr{C}} \mu)$.

(vi) $\kappa^{\lambda +_{\mathscr{C}} \mu} = \kappa^{\lambda} \cdot_{\mathscr{C}} \kappa^{\mu}$.

(vii) $\kappa^{\lambda \cdot_{\mathscr{C}} \mu} = \left(\kappa^{\lambda}\right)^{\mu}$.

Here κ^{λ} etc. are cardinals, not sets of functions.

Exercise 9.10

Prove Theorem 9.7. [Hint: use Exercise 6.13 of Section 6.3.]

Likewise, there are very attractive results about cardinal inequalities, for instance those arising from Exercise 6.16 of Section 6.3, as follows.

Theorem 9.8

Assume AC and let κ, λ, μ be any cardinals with $\kappa \leq \lambda$. Then

(i) $\kappa +_{\mathscr{C}} \mu \leq \lambda +_{\mathscr{C}} \mu$

(ii) $\kappa \cdot_{\mathscr{C}} \mu \leq \lambda \cdot_{\mathscr{C}} \mu$

(iii) $\kappa^{\mu} \leq \lambda^{\mu}$

(iv) $\mu^{\kappa} \leq \mu^{\lambda}$

Exercise 9.11

Prove Theorem 9.8.

Exercise 9.12

Prove the following results of cardinal arithmetic assuming AC, where κ and λ are infinite cardinals.

(a) $\mathbf{2}^{\kappa} +_{\mathscr{C}} \kappa = \mathbf{2}^{\kappa}$

(b) $\mathbf{2}^{\kappa} \cdot_{\mathscr{C}} \mathbf{2}^{\lambda} = \mathbf{2}^{\max\{\kappa,\lambda\}}$

(c) If $\kappa \leq \lambda$ then $\kappa^{\lambda} = \mathbf{2}^{\lambda}$. [Hint: show that $\mathbf{2}^{\lambda} \leq \kappa^{\lambda}$ and $\kappa^{\lambda} \leq \mathbf{2}^{\lambda}$.]

As the context of this exercise is cardinal arithmetic, the $\mathbf{2}^{\kappa}$ in this exercise is the cardinal resulting from cardinal exponentiation, rather than the set of functions from κ to $\mathbf{2}$. Similarly with the $\mathbf{2}^{\lambda}$ and κ^{λ}.

We have already remarked that the addition and multiplication of infinite cardinals are essentially rather dull, thanks to Theorem 9.6. It is cardinal exponentiation which is exciting, in that it often produces bigger cardinals than those that it is operating on! This of course is a consequence of Cantor's theorem, which in terms of cardinal arithmetic is written as

Theorem 6.5 of Section 6.4.

$$\kappa < 2^{\kappa},$$

for all cardinals κ. Using AC we have that for any infinite cardinal κ, 2^{κ} is also a cardinal, so that for instance $2^{\aleph_0}$ is an $\aleph_\gamma$ for some γ. The natural question is 'What is γ?'. Of course $\gamma > 0$, as $2^{\aleph_0} > \aleph_0$; and it turns out that there are some other restrictions, such as $\gamma \neq \omega$. But otherwise the issue *cannot be resolved within* ZF *alone*! We shall return to this problem, the *continuum problem*, in the next section.

The problem is one of the major problems posed by Cantor. The cardinal Card $(2^{\aleph_0})$ is the size of $\mathbb{R}$, the continuum.

To end this section, let us look at how some of the results above about cardinals which stem from AC actually entail AC themselves. That such principles are then equivalent to AC illustrates the extent to which Cantor's original theory of cardinals is intimately bound up with Zermelo's axiom of choice.

First, let us look at the result of Theorem 9.4:

> AC implies that for every infinite set X there is an $\aleph_\gamma$ such that $X \approx \aleph_\gamma$.

Let's investigate the converse of this result. If X is an infinite set for which $X \approx \aleph_\gamma$, for some $\aleph_\gamma$, then the bijection between X and the (initial) ordinal $\aleph_\gamma$ can be used to define a well-order on X. Thus if this holds for all infinite sets X, then every infinite set can be well-ordered, so that the well-ordering principle holds. (What happens for finite sets? Look at Exercise 9.13 below.) As WO is equivalent to AC, we have proved the converse to Theorem 9.4:

If $f\colon X \longrightarrow \aleph_\gamma$ is the bijection, define $<_X$ by $x <_X x'$ if $f(x) < f(x')$ (with the usual ordinal $<$ on $\aleph_\gamma$). It is very easy to show that $<_X$ is a well-order on X.

Theorem 9.9

If for every infinite set X there is an $\aleph_\gamma$ such that $X \approx \aleph_\gamma$, then AC holds.

With Theorem 9.4, this means that this principle is equivalent to AC.

Exercise 9.13

Explain why every finite set X can be well-ordered, without needing to use AC or any equivalent.

Next, let us look at the dichotomy principle, which Theorem 9.5 shows to be a consequence of AC. We shall show that this principle actually implies AC.

Theorem 9.10

Suppose that the dichotomy principle holds, i.e. for any sets A, B

$$A \preceq B \text{ or } B \preceq A.$$

Then AC holds.

Thus the dichotomy principle is equivalent to AC.

Proof

As with the previous theorem, we shall actually show that dichotomy implies WO. We take any set X and show that X can be well-ordered. This time, our argument will use some high-powered machinery, namely Hartogs' theorem.

Theorem 8.18 of Chapter 8.

Assume the dichotomy principle and let X be a set. Then by Hartogs' theorem there is an ordinal α such that $\alpha \not\preceq X$. By the dichotomy principle we have that either $\alpha \preceq X$ or $X \preceq \alpha$, so as $\alpha \not\preceq X$ we must have $X \preceq \alpha$. This means that there is a one–one function f from X into α. This f can be exploited to define a well-order $<_X$ on X (which you can show as the next exercise).

Actually $X \prec \alpha$, but this fact isn't required for the argument.

Thus every set can be well-ordered, and as WO is equivalent to AC, this means that AC holds. ■

Exercise 9.14

Complete the proof of Theorem 9.10 by showing how to exploit the one–one function $f\colon X \longrightarrow \alpha$ (with codomain the ordinal α) to define a well-order $<_X$ on X.

Let's look now at the absorption law in Theorem 9.6. If we remove the terminology of cardinals and cardinal arithmetic, this theorem can be rewritten as follows: assuming AC, then for any sets X, Y, at least one of which is infinite,

$$\begin{aligned}(X \times \{\mathbf{0}\}) \cup (Y \times \{\mathbf{1}\}) &\approx X \times Y \\ &\approx \text{one of } X \text{ and } Y.\end{aligned}$$

We do this because the terminology of cardinals, like Card(X), requires use of AC to make sense: and we are going to show that the absorption law implies AC.

Theorem 9.11

Suppose that for all sets X, Y, at least one of which is infinite,

$$(X \times \{\mathbf{0}\}) \cup (Y \times \{\mathbf{1}\}) \approx \text{one of } X \text{ and } Y.$$

Then AC holds.

Again, this means that this principle is equivalent to AC.

Proof

We shall show that the dichotomy principle holds. Then by Theorem 9.10, AC holds.

Take any sets X, Y. If both sets are finite then there are natural numbers $\mathbf{m}, \mathbf{n}$ with $X \approx \mathbf{m}$ and $Y \approx \mathbf{n}$. As $\mathbf{m} \subseteq \mathbf{n}$ or $\mathbf{n} \subseteq \mathbf{m}$, it is easy to show that either $X \preceq Y$ or $Y \preceq X$. Otherwise at least one of X, Y is infinite, so that by supposition

$$(X \times \{\mathbf{0}\}) \cup (Y \times \{\mathbf{1}\}) \approx \text{one of } X \text{ and } Y.$$

In the case that $(X \times \{\mathbf{0}\}) \cup (Y \times \{\mathbf{1}\}) \approx X$, we have

$$\begin{aligned}Y &\approx Y \times \{\mathbf{1}\} \\ &\subseteq (X \times \{\mathbf{0}\}) \cup (Y \times \{\mathbf{1}\}) \\ &\approx X,\end{aligned}$$

so that $Y \preceq X$. Similarly if $(X \times \{\mathbf{0}\}) \cup (Y \times \{\mathbf{1}\}) \approx Y$, we have $X \preceq Y$. Thus the dichotomy principle holds, from which it follows that AC holds. ■

Exercise 9.15

Prove the corresponding result to the above theorem for 'multiplication', as follows. Suppose that for any sets X, Y, at least one of which is infinite,

$$X \times Y \approx \text{one of } X \text{ and } Y.$$

Show that AC holds.

We can go on in this way, showing that various principles of cardinal arithmetic are actually equivalent to AC. For instance, the principle that for any infinite set X,

$$X \times X \approx X,$$

which follows using AC from Theorem 8.21 of Section 8.5, itself implies AC.

The details can be found in Rubin and Rubin [24], a compendium of remarkably many equivalents to AC.

We have in this section essentially finished the recreation of Cantor's theory of cardinals and their arithmetic. And we have hit one of Cantor's major problems arising from this theory: assuming AC (which Cantor didn't realize was needed), which cardinal $\aleph_\gamma$ is the size of the continuum $\mathbb{R}$ (or equivalently of $\mathbf{2}^{\aleph_0}$)? We shall look further at this question in the next, and final, section of this book.

Further exercises

Exercise 9.16

Assuming AC, show that $\aleph_1 \preceq \mathbf{2}^{\aleph_0}$. [Your argument could now be very much shorter than that suggested in the hint to Exercise 8.22 of Section 8.2: this exercise deals with ω_1 and $\mathscr{P}(\mathbb{N})$, but these are essentially the same as or equivalent to $\aleph_1$ and $\mathbf{2}^{\mathbb{N}}$.]

Exercise 9.17

Let X be a set of cardinals. Show that $\bigcup X$ is a cardinal. [Hints: X is also a set of ordinals, so that $\bigcup X$ is an ordinal. Investigate two cases: first, when X contains a greatest cardinal; and second, when it doesn't.]

Exercise 9.18

Let X be any set and let $\text{Sym}(X)$ be the set of all bijections from X to itself.

We could equivalently regard $\text{Sym}(X)$ as the set of all *permutations* of X.

(a) A *transposition* in $\text{Sym}(X)$ is a permutation swapping two elements of X and leaving all the remaining elements of X fixed. Let X be infinite. Assuming AC, find $\text{Card}(\{f : f \text{ is a transposition in } \text{Sym}(X)\})$. [Hint: the answer is one of $\text{Card}(X)$ and $\mathbf{2}^{\text{Card}(X)}$.]

(b) Show that if $X \approx Y$ then $\text{Sym}(X) \approx \text{Sym}(Y)$.

(c) The result of part (b) means that, assuming AC, we can define a *factorial* operation on cardinals by

$$\kappa! = \text{Card}(\text{Sym}(\kappa)),$$

for all cardinals κ. Show, assuming AC, that $\kappa! = \mathbf{2}^\kappa$.

In everyday maths, the number of permutations of a finite set with n elements is $n!$.

Exercise 9.19

Let $\{A_i : i \in I\}$ and $\{B_i : i \in I\}$ both be indexed families of non-empty sets, with $A_i \prec B_i$ for each $i \in I$.

(a) Assuming AC, show that

$$\bigcup\{A_i : i \in I\} \preceq \prod_{i\in I} B_i,$$

by constructing a suitable one–one function.

(b) Show that $\bigcup\{A_i : i \in I\} \prec \prod_{i\in I} B_i$ as follows. Suppose that $\theta\colon \bigcup\{A_i : i \in I\} \longrightarrow \prod_{i\in I} B_i$ is any one–one function. Define a function θ_i with domain A_i for each $i \in I$ by

$$\theta_i\colon A_i \longrightarrow B_i$$
$$a \longmapsto \theta(a)(i).$$

(i) Explain why θ_i is not onto. You may assume AC.

(ii) Assuming AC, show that θ is not onto. [Hint: consider an element of $\prod_{i\in I}(B_i \setminus \text{Range}(\theta_i))$ and show that it cannot equal $\theta(a)$ for any $a \in \bigcup\{A_i : i \in I\}$.]

(c) Let X be a set. Put $I = X$. Put $A_i = \{i\}$ and $B_i = \mathbf{2}$ for each $i \in I$. Use the result above to deduce Cantor's theorem, $X \prec \mathbf{2}^X$.

This result is called König's theorem.

The definition of $\prod_{i\in I} B_i$ gives that $\theta(a)$ is a function with domain I such that $\theta(a)(i) \in B_i$ for each $i \in I$.

Thus there is no bijection from $\bigcup\{A_i : i \in I\}$ to $\prod_{i\in I} B_i$.

9.4 The continuum hypothesis

In this final section of the book we shall look at what is probably the major problem posed by Cantor but not resolved during his lifetime, the continuum problem.

As part of his analytic work on subsets of the real line, Cantor came to suspect that any infinite subset of the continuum is equinumerous with either the natural numbers or itself:

if $A \subseteq \mathbb{R}$ with A infinite, then $A \approx \mathbb{N}$ or $A \approx \mathbb{R}$.

All of the infinite subsets of $\mathbb{R}$ that he encountered fell into one of these two cases. One version of his *continuum hypothesis* is that the above is true.

For Cantor and his contemporaries, a 'continuum' was essentially any non-trivial segment of the real line and *the continuum* was the segment consisting of the real line itself. Of course Cantor's work on cardinals showed that any two such segments have the same size.

If $\mathbb{R}$ can be well-ordered (for instance, if we assume AC) then one can ask which of the $\aleph_\gamma$s equals Card($\mathbb{R}$), or equivalently equals the cardinal $\mathbf{2}^{\aleph_0}$. For there to be only the two sorts of infinite subset of $\mathbb{R}$, Cantor's continuum hypothesis, usually abbreviated as CH, then becomes as follows.

The continuum hypothesis

(CH) $\mathbf{2}^{\aleph_0} = \aleph_1$.

Once Cantor had hypothesized that the cardinal exponential $\mathbf{2}^{\aleph_0}$ equals the next biggest cardinal after $\aleph_0$, it was natural to suggest that the same pattern might hold for all cardinals, i.e. $2^{\aleph_\alpha}$ equals the next biggest cardinal after $\aleph_\alpha$. This is called the *generalized continuum hypothesis*, usually abbreviated as GCH:

The generalized continuum hypothesis

(GCH) $2^{\aleph_\alpha} = \aleph_{\alpha^+}$, for all ordinals α.

Proving CH was the first of Hilbert's list of 23 problems which he presented in 1900 as the major unsolved problems facing mathematicians for the 20th century. The issue of whether $\mathbb{R}$ can be well-ordered was not far behind in its interest for mathematicians – indeed the subsequent introduction of the axiom of choice and the discovery of all its ramifications made AC a more significant issue than CH. The resolution of these issues was perhaps even more momentous than Hilbert had anticipated!

Hilbert's problems, presented at the Paris International Congress of Mathematicians in 1900, both set and reflected much of the agenda for mathematical research in the 20th century. Doubtless there'll be some competition in the near future to attempt the same task for the next millennium!

Theorem 9.12 (Gödel, 1940)

If ZF is consistent, then so is ZF with AC and GCH.

This means that, assuming ZF is consistent (which, by Gödel's incompleteness theorems of 1930–31, essentially cannot be proved), neither AC nor GCH can be *disproved* from ZF. And then in 1963 the American mathematician Paul Cohen showed, again assuming that ZF is consistent, that AC and CH (and thus GCH) cannot be *proved* from ZF – indeed CH cannot be proved from ZF even if one adds AC as an extra axiom.

For Cohen's account of his work, see Cohen [25].

Theorem 9.13 (Cohen, 1963)

If ZF is consistent, then so are ZF with the negation of AC, and also ZF along with AC and the negation of CH.

This means that AC is *independent* of ZF, meaning that AC cannot be proved from ZF. Likewise CH is independent of ZF + AC.

Cohen's work exploited a method he called *forcing*, which we shall not attempt to describe in this book. This method has been very fruitful in showing the independence of various set-theoretic principles from axiom systems like ZF. We shall, however, give a brief idea of how Gödel proved the consistency of AC and CH with the ZF axioms. The key is to mimic the construction of the cumulative hierarchy of sets, the $\mathscr{V}_\alpha$s of Section 8.3, to produce what is called the *constructible hierarchy of sets*, with the stages of the hierarchy written as $\mathscr{L}_\alpha$ for each ordinal α.

A further account of forcing can be found in e.g. Kunen [26].

One way of regarding the cumulative hierarchy is that it shows how, given the ordinals, one can construct all possible sets by building up from the empty set, using power sets and unions. Gödel's constructible hierarchy does the same sort of thing, producing a more limited class of sets, with a much tighter idea of having constructed sets at any level from the sets at lower levels. Gödel's construction of the $\mathscr{L}_\alpha$s does the same as that of the $\mathscr{V}_\alpha$s at the $\mathbf{0}$ and limit ordinal stages. It's at the successor ordinal stage, where arguably the meaty business is done, that the constructions differ.

$$\mathscr{V}_0 = \varnothing,$$
$$\mathscr{V}_{\gamma^+} = \mathscr{P}(\mathscr{V}_\gamma),$$
$$\mathscr{V}_\lambda = \bigcup\{\mathscr{V}_\gamma : \gamma \in \lambda\} \text{ (limit } \lambda).$$

Taking power sets gives one really big sets; whereas taking unions seems a much more mundane way of getting infinite sets!

The construction of $\mathscr{V}_{\gamma^+}$ includes *all* the subsets of $\mathscr{V}_\gamma$. This is something of a bludgeon, as it doesn't of itself give much information about what these subsets look like. Gödel's $\mathscr{L}_{\gamma^+}$ is more like a rapier, in that it includes only those subsets of $\mathscr{L}_\gamma$ that are definable using formulas of the formal language in quite a strong sense, as follows. Let $\phi(z)$ be a formula with free variable z and possibly referring to named sets in $\mathscr{L}_\gamma$. This formula could be used, as in the axiom of separation, to define the subset of $\mathscr{L}_\gamma$ consisting of those zs in $\mathscr{L}_\gamma$ such that $\phi(z)$ holds. But Gödel's construction uses it in a subtler way, by insisting that in assessing the truth of $\phi(z)$, the quantified variables are taken as ranging over the elements of $\mathscr{L}_\gamma$ rather than over the universe $\mathscr{V}$ of all sets – we shall describe this by saying that $\phi(z)$ *holds in* $\mathscr{L}_\gamma$. Let's look at an example to try to explain this subtle distinction.

Normally we regard a statement about sets like $\forall x\theta(x)$ as true if $\theta(x)$ holds for all sets x: likewise $\exists x\theta(x)$ is true if $\theta(x)$ is true for some set x. In the restricted context of forming $\mathscr{L}_{\gamma^+}$ from $\mathscr{L}_\gamma$, we regard $\forall x\theta(x)$ (respectively $\exists x\theta(x)$) as true when $\theta(x)$ is true for all sets (respectively for some set) x in $\mathscr{L}_\gamma$.

Our definition will result in $\mathscr{L}_1$ being the set $\{\varnothing\}$, which is the same as $\mathscr{V}_1$. Let us investigate the subset of $\mathscr{L}_1$ defined by the formula $\phi(z)$ given by

$$\exists x(z \in x).$$

In the normal unrestricted sense, this subset would consist of those zs in $\mathscr{L}_1$ for which there is some set x such that $z \in x$ is true. As $\mathscr{L}_1 = \{\varnothing\}$, the only z to be considered is $z = \varnothing$; and there are plenty of candidates for a suitable set x such that $\varnothing \in x$, e.g. $\{\varnothing\}$ or the set $\mathbb{N}$ in *ZF*. Thus $\phi(z)$ defines the subset $\{\varnothing\}$ of $\mathscr{L}_1$. What happens in the subtle restricted sense that we have introduced? For $z = \varnothing$, is there a set x *in* $\mathscr{L}_1$ *itself* for which $\varnothing \in x$? The answer is 'No', as the only element x in $\mathscr{L}_1$ is $\varnothing$, and $z = \varnothing \notin \varnothing = x$. So the set of zs in $\mathscr{L}_1$ for which this $\phi(z)$ 'holds in $\mathscr{L}_1$' is just $\varnothing$.

The definition of the $\mathscr{L}_\alpha$s is as follows.

Definitions

The *constructible hierarchy of sets*, $\mathscr{L}_\alpha$, for each ordinal α is defined by:

$$\mathscr{L}_0 = \varnothing,$$
$$\mathscr{L}_{\gamma^+} = \{y \subseteq \mathscr{L}_\gamma : y = \{z \in \mathscr{L}_\gamma : \phi(z) \text{ holds in } \mathscr{L}_\gamma\} \text{ for some formula } \phi(z) \text{ with } z \text{ free which can refer to named sets in } \mathscr{L}_\gamma\}$$
$$\mathscr{L}_\lambda = \bigcup\{\mathscr{L}_\gamma : \gamma \in \lambda\} \quad \text{(for a limit ordinal } \lambda).$$

A set is *constructible* if it is an element of $\mathscr{L}_\alpha$ for some α. The class of all constructible sets is called the *constructible universe* and is written as $\mathscr{L}$.

Like the $\mathscr{V}_\alpha$s, the $\mathscr{L}_\alpha$s have many nice properties, for instance:

(i) if $\alpha < \beta$ then $\mathscr{L}_\alpha \subseteq \mathscr{L}_\beta$;

(ii) each $\mathscr{L}_\alpha$ is $\in$-transitive;

(iii) $\alpha \subseteq \mathscr{L}_\alpha$ and $\alpha \in \mathscr{L}_{\alpha^+}$.

These are similar to some of the results in Exercises 8.32–8.36 at the end of Section 8.3.

It is easy to show that $\mathscr{L}_\mathbf{n}$ is finite for each $\mathbf{n} \in \omega$ and that $\mathscr{L}_\omega$ is countably infinite, similar to the results for $\mathscr{V}_\mathbf{n}$ and $\mathscr{V}_\omega$. But whereas $\mathscr{V}_{\omega^+}$ has cardinality $\mathbf{2}^{\aleph_0}$ and includes all the subsets of $\mathbb{N}$ (as $\omega \subseteq \mathscr{V}_\omega$), $\mathscr{L}_{\omega^+}$ is only countably infinite. This is because there are only countably many formulas $\phi(z)$ of the right sort available at the $\mathscr{L}_{\omega^+}$ stage, so that at most countably many subsets of $\mathscr{L}_\omega$ are produced at this stage – $\mathscr{L}_{\omega^+}$ is the set of just these subsets.

So one has to ask whether there are more than countably many constructible subsets of $\mathbb{N}$. There are but it takes lots of stages for them to appear. In fact they have all appeared by the time one gets to $\mathscr{L}_{\omega_1}$, but the proof is non-trivial.

Exercise 9.20

Explain why there are only countably many formulas $\phi(z)$ at the stage of forming $\mathscr{L}_{\omega^+}$.

Solution

Such a formula is built up using finitely many symbols out of the formal language, where some of the free variables are replaced by named sets out of the countable set $\mathscr{L}_\omega$. The formal language uses symbols which are brackets '(' and ')' and the logical connectives ($\wedge$, $\vee$, $\neg$, $\rightarrow$, $\leftrightarrow$, $\forall$ and $\exists$), along with variables: no more than finitely many variables can appear in any one formula, so that a countable set of such variables, like $\{x_0, x_1, x_2, \ldots\}$, is all that's really needed by the language. Using a specific enumeration of the countably many symbols of the language along with the countably many sets in $\mathscr{L}_\omega$, we can specify an enumeration of all the finite strings made up of these symbols and sets. As the formulas $\phi(z)$ of the sort we want appear in this list (along with some non-formulas!), there are just countably many such formulas.

Using the enumeration of the symbols and the sets, the argument is essentially the same as in Exercise 6.39 in Section 6.4 showing that the set of all finite sequences of natural numbers is countable.

Our hand-waving solution above gives an idea of a key step in showing that each $\mathscr{L}_\alpha$ is well-ordered. We would of course use transfinite induction on α. For the successor stage we would assume that $\mathscr{L}_\alpha$ is well-ordered and try to show that $\mathscr{L}_{\alpha^+}$ is also well-ordered. If $\mathscr{L}_\alpha$ is well-ordered, we can then adjoin the symbols of the formal language in an organized way to $\mathscr{L}_\alpha$ so that the symbols and sets from which we construct the $\phi(z)$s for the next stage are well-ordered. As each $\phi(z)$ uses only finitely many of these symbols, it is relatively straightforward to well-order the set of the $\phi(z)$s. Each $\phi(z)$ defines an element of $\mathscr{L}_{\alpha^+}$, and the well-order of the $\phi(z)$s can thus be used to well-order $\mathscr{L}_{\alpha^+}$. Well-ordering $\mathscr{L}_\lambda$ for a limit ordinal λ, assuming that each $\mathscr{L}_\gamma$ for $\gamma < \lambda$ can be well-ordered, is also straightforward, as these $\mathscr{L}_\gamma$s form a $\subseteq$-chain and $\mathscr{L}_\lambda$ is the union of this chain – the well-order relation on $\mathscr{L}_\lambda$ (regarded as a set of pairs) can be taken as the union of the relations (suitably chosen to be compatible) of all the $\mathscr{L}_\gamma$s for $\gamma < \lambda$.

Once we have shown that each $\mathscr{L}_\alpha$ is well-ordered, it follows that each constructible set (which is a subset of some $\mathscr{L}_\alpha$) is also well-ordered. The relevance of this to the proof of Theorem 9.12 is as follows. Assuming that there is a universe $\mathscr{V}$ of sets in which all the axioms of ZF are true, it can be shown that these axioms are also true in the subclass $\mathscr{L}$ of $\mathscr{V}$ consisting of all constructible sets; and as every element of $\mathscr{L}$ can be well-ordered, the axiom of choice is also true in $\mathscr{L}$. Putting this in terms of consistency, if ZF is consistent, then so is ZF with AC added to it as an extra axiom.

The connection between the truth of formal axioms like those of ZF and consistency stems from yet another of Gödel's achievements, the completeness theorem for first-order predicate calculus, which says that a set of formulas in a formal language such as the one we have used for ZF is consistent if and only if it has a model, a 'something' in which all the formulas are true. A major catch, also due to Gödel through his incompleteness theorems, is that for a powerful theory like ZF, in which a formal theory of $\mathbb{N}$ and its arithmetic can be created, there are effectively insuperable barriers to proving by finite means that the theory is consistent. Hence the need for the condition 'if ZF is consistent' in the statements of Theorems 9.12 and 9.13.

Furthermore, as it transpires that all constructible subsets of $\mathbb{N}$ appear in an $\mathscr{L}_\alpha$ for some $\alpha < \omega_1$, the number of such subsets is at most

$$\mathrm{Card}\left(\bigcup\{\mathscr{L}_\alpha : \alpha < \omega_1\}\right),$$

which, as $\mathscr{L}_\alpha$ for $\alpha < \omega_1$ is countable (for reasons similar to those in Exercise 9.20 above), equals $\aleph_0 \cdot_{\mathscr{C}} \aleph_1$, which in turn equals $\aleph_1$. Thus within the constructible universe $\mathscr{L}$ the cardinality of $\mathscr{P}(\mathbb{N})$, i.e. $2^{\aleph_0}$, equals $\aleph_1$, so that the continuum hypothesis holds in $\mathscr{L}$. Similarly, it can be shown that the generalized continuum hypothesis holds in $\mathscr{L}$. Hence, in terms of consistency, if *ZF* is consistent, then so is *ZF* with GCH added to it as an extra axiom.

Because the construction of $\mathscr{L}$ has a sufficient claim to be 'natural' in the eyes of some mathematicians, $\mathscr{L}$ has a higher status than just that of a technical device used in the proof of Theorem 9.12. For instance, every set required by everyday pure mathematics is in $\mathscr{L}$. It is tempting to add the proposition that $\mathscr{L}$ is the whole universe of sets to the list of axioms for set theory: this proposition, called the *axiom of constructibility*, is written as

As the $\mathscr{L}_\alpha$s are definable within *ZF*, the proposition can be expressed within our formal language by $\forall x \exists \alpha(\alpha \text{ is an ordinal } \wedge x \in \mathscr{L}_\alpha)$, and can thus be used as an axiom.

$$\mathscr{V} = \mathscr{L},$$

and the content of Theorem 9.12 becomes rephrased as follows.

Theorem 9.12

If *ZF* is consistent, then $\mathscr{V} = \mathscr{L}$ entails both AC and GCH.

Part of the temptation to assume that $\mathscr{V} = \mathscr{L}$ is that it resolves several other set-theoretic and mathematical problems that cannot be resolved from *ZF* alone, besides AC and GCH. But it cannot resolve every problem of set theory, not least because of the limitations of Gödel's incompleteness theorems. And part of the richness of modern set theory, a fruitful vein of mathematical research, stems from investigating alternatives to $\mathscr{V} = \mathscr{L}$, and indeed alternatives to some of the axioms of *ZF* itself.

For an accessible introductory account of this see Devlin [4]. The detailed work is impressive and very difficult! A short, readable survey of the whole development of set theory to this point can be found in Kanamori [26].

One of the aims of this book was to give accounts, from a modern perspective, of Dedekind's and Cantor's theory of the nature of real numbers, crucial to underpinning the calculus, and of Cantor's theory of infinite sets. The questions which so quickly arose as part of these theories, of whether the real numbers can be well-ordered and of determining the possible sizes of infinite subsets of $\mathbb{R}$, turned out to be really good questions, with unexpectedly ambivalent answers. Every mathematician should know at least that much of what's called set theory! And we hope that you might be interested in finding out, and indeed developing, some more!

For further reading, you might like to try e.g. Devlin [4], Moschovakis [5] or Kunen [27].

BIBLIOGRAPHY

1. Herbert B. Enderton *Elements of Set Theory*, Academic Press, 1977.
2. Alan G. Hamilton *Numbers, Sets and Axioms*, Cambridge University Press, 1982.
3. Patrick Suppes *Axiomatic Set Theory*, Van Nostrand, 1960.
4. Keith Devlin *The Joy of Sets*, Springer-Verlag, 1993.
5. Yiannis N. Moschovakis *Notes on Set Theory*, Springer-Verlag, 1994.
6. Paul R. Halmos *Naive Set Theory*, Van Nostrand, 1960.
7. Joseph Warren Dauben *Georg Cantor: His Mathematics and Philosophy of the Infinite*, Princeton University Press, 1979.
8. Adrian W. Moore *The Infinite*, Routledge, 1990.
9. Shaughan Lavine *Understanding the Infinite*, Harvard University Press, 1994.
10. John Fauvel and Jeremy Gray *The History of Mathematics, a Reader*, Macmillan, 1987.
11. Jean van Heijenoort *From Frege to Gödel, a Source Book in Mathematical Logic 1879–1931*, Harvard University Press, 1967.
12. Georg Cantor (trans. Phillip E. B. Jourdain) *Contributions to the Founding of the Theory of Transfinite Numbers*, Dover, 1955.
13. Michel Spivak *Calculus* 2nd Edition, Publish or Perish, 1980.
14. Rod Haggarty *Fundamentals of Mathematical Analysis*, Addison-Wesley 1992.
15. Richard Dedekind (trans. Wooster W. Beman) *Essays on the Theory of Numbers*, Dover, 1963.
16. Elliott Mendelson *Introduction to Mathematical Logic*, Van Nostrand, 1964.
17. Penelope Maddy *Believing the Axioms I*, Journal of Symbolic Logic, Volume 53 (2), Pages 481–511, 1988.
18. Penelope Maddy *Realism in Mathematics*, Oxford University Press, 1990.
19. Abraham A. Fraenkel, Yehoshua Bar-Hillel, Azriel Levy *Foundations of Set Theory*, 2nd Edition, North-Holland, 1973.
20. G.T. Kneebone *Mathematical Logic and the Foundations of Mathematics*,Van Nostrand, 1963.
21. Bertrand Russell *My Philosophical Development*, George Allen & Unwin, 1959.
22. Gregory H. Moore *Zermelo's Axiom of Choice*, Springer-Verlag, 1982.
23. Michael Hallett *Cantorian Set Theory and Limitation of Size*, Oxford University Press, 1984.

24. H. Rubin and J. Rubin *Equivalents of the Axiom of Choice*, North-Holland, 1963.

25. Paul J. Cohen *Set Theory and the Continuum Hypothesis*, W. A. Benjamin, 1966.

26. Akihiro Kanamori *The Mathematical Development of Set Theory from Cantor to Cohen*, Bulletin of Symbolic Logic, Volume 2, 1996, pages 1 to 71.

27. Kenneth Kunen *Set Theory: An Introduction to Independence Proofs*, North-Holland, 1980.

INDEX

Printed in the United States
by Baker & Taylor Publisher Services